Operator Theory
Advances and Applications
Vol. 76

Editor
I. Gohberg

Asymptotics of Nonlinearities and Operator Equations

Alexander M. Krasnosel'skii

Translated from the Russian
by Mircea Martin

Birkhäuser Verlag
Basel · Boston · Berlin

Author's address:

Institute for Information Transmission Problems
Russian Academy of Sciences
19 Bol. Karetnyi per.
101447 Moscow
Russia
e-mail: amk@ippi.msk.su

Originally published in 1992 by Nauka, Moskva.

A CIP catalogue record for this book is available from the Library of Congress, Washington D.C., USA

Deutsche Bibliothek Cataloging-in-Publication Data
Krasnosel'skij, Aleksandr M.:
Asymptotics of nonlinearities and operator equations /
Alexander M. Krasnosel'skii. Transl. from the Russian by
Mircea Martin. – Basel ; Boston ; Berlin : Birkhäuser, 1995
  (Operator theory ; Vol. 76)
  ISBN-13:978-3-0348-9899-7   e-ISBN-13:978-3-0348-9082-3
  DOI: 10.1007/978-3-0348-9082-3

NE: GT

© 1995 Birkhäuser Verlag, P.O. Box 133, CH-4010 Basel, Switzerland
Softcover reprint of the hardcover 1st edition 1995
Printed on acid-free paper produced from chlorine-free pulp
Cover design: Heinz Hiltbrunner, Basel

ISBN-13:978-3-0348-9899-7

9 8 7 6 5 4 3 2 1

Table of Contents

Foreword

The subject matter of the present book sheds light on a new area in classical nonlinear analysis. The methods developed in its chapters are aimed at solving a wide range of problems connected with the asymptotic behavior of nonlinearities at infinity. These problems are related to various topics, such as solvability criteria for nonlinear operator equations, estimates of the solutions, conditions for the applicability of numerical procedures, variational methods, resonance phenomena, bifurcations and the occurrence of large norm solutions under the change of parameters, one-sided estimates of nonlinearities, weak nonlinearities, and so on. The main results, stated and proved in a general setting, are illustrated by applications to the study of forced oscillations in systems of various types (including single- and multi-loop control systems, and systems with delay) as well as to the study of nonlinear boundary value problems and integral equations.

The book is based on three essentially new observations, up to now only outlined in scientific articles in a rather concise form.

The primary observation consists of an unexpectedly strong connection between the admissible asymptotics of nonlinearities involved in various problems and the behavior near zero of the distributions of the normalized functions in some finite dimensional subspaces. To translate this observation into concrete and effective theorems requires a special analysis of the arrangements of some specific systems of functions, and the establishment of a priori estimates for the solutions of certain unusual (and apparently hitherto unnoticed) integral-functional inequalities.

The second observation refers to the possibility of using, in many problems, not only the absolute values of the leading eigenvalues of the linear operators involved in nonlinear problems, but also the arguments of those leading eigenvalues. This observation yields new fixed point principles. When applied to problems with delay, these principles lead to theorems in which the magnitude of the delay is essentially implicated.

Finally, we emphasize the role of some new classes of normal linear operators which are singled out and studied, such as positive operators, strictly positive operators, and potentially strictly positive operators. They are especially important in the study of forced periodic oscillations in control systems.

The results presented in this book lead to new problems; the way to solve many of them is quite clear, and thus the author refrains from including an unnecessary amount of examples. Too many might obscure the main ideas, preventing one from seeing the forest for the trees.

The author would like to express his gratitude to V. M. Alexeev†, N. A. Bobylev, and A. V. Pokrovskii whose interest in his work actively stimulated it. He is particularly thankful to his mentor in life as well as in mathematics — his father, Mark Alexandrovich Krasnoselskii.

Moscow   *A. M. Krasnoselskii*

TRANSLATOR'S NOTE: For the reader's convenience, I added to the list of references, when possible, the English translations of some papers and books initially published in other languages.

It would have been impossible to make this translation available to the English reader without the help of my wife, Larisa, who typeset the book in TEX.

Lawrence, Kansas   *Mircea Martin*

Chapter 1

Norm estimates for solutions
of integral-functional inequalities

§1. Distribution functions

1.1. DEFINITIONS. In the sequel, Ω will denote a set equipped with a finite measure μ. We will consider measurable functions $f(t)$ $(t \in \Omega)$ with values in the finite dimensional Euclidean space $\mathbb{R}^n$. The inner product and the norm on $\mathbb{R}^n$ are denoted by $(\cdot\,,\cdot)$ and $|\cdot|$, respectively.

We associate to each $\mathbb{R}^n$-valued measurable function $f(t)$ the real-valued function

$$\chi(\delta) = \mu\{t \in \Omega : |f(t)| \leqslant \delta\}, \quad 0 \leqslant \delta < \infty. \tag{1.1}$$

We will call it the *distribution* of the function $f(t)$ — often referred to as the "distribution function of $f(t)$" ([Feller, 1950]). Sometimes it is convenient to use the more complete notations $\chi(\delta; f)$ or $\chi(\delta; f, \mu)$ for the distribution (1.1).

According to the definition above, it turns out that the function (1.1) is nonnegative, nondecreasing and continuous to the right. If the inequality $|f_1(t)| \leqslant |f_2(t)|$ holds true for almost all $t \in \Omega$, then $\chi(\delta; f_1) \geqslant \chi(\delta; f_2)$ for any $\delta \geqslant 0$. The next identity is also true:

$$\chi(r\delta; f) \equiv \chi(\delta; r^{-1}f), \quad \delta \geqslant 0, \ r > 0. \tag{1.2}$$

Two nonnegative real-valued functions $\varphi(t)$ and $\psi(s)$ (with or without the same domains of definition) are called *equimeasurable* if their distributions $\chi(\delta; \varphi)$ and $\chi(\delta; \psi)$ coincide. For any given measurable function $f(t)$ $(t \in \Omega)$ there exists a unique nondecreasing and continuous to the right function $f^*(s)$ $(0 \leqslant s < \mu\Omega)$ such that $|f(t)|$ and $|f^*(s)|$ are equimeasurable. This function is called the *nondecreasing arrangement* of the function $|f(t)|$. We have

$$\chi(\delta; f) = \sup\{s : f^*(s) \leqslant \delta\}, \quad f^*(s) = \sup\{\delta : \chi(\delta; f) \leqslant s\}. \tag{1.3}$$

1.2. EXAMPLES. We proceed by pointing out a few examples of distributions $\chi(\delta; f)$ of some specific functions $f(t)$. We also indicate some nondecreasing lower and upper estimates of the form

$$\chi_{\mathrm{L}}(\delta; f) \leqslant \chi(\delta; f) \leqslant \chi_{\mathrm{U}}(\delta; f), \quad \delta \geqslant 0. \tag{1.4}$$

As a rule, we will use estimates of distributions like in (1.4) for sufficiently small values of δ, only.

A. We start our discussion with the case of scalar-valued functions defined on the interval $\Omega = [0, \pi]$. The measure μ is the usual Lebesgue measure. Table 1.1 contains examples of functions for which both their distributions and nondecreasing arrangements are explicitly described. Table 1.2 deals with estimates of distributions of some functions.

Table 1.1

$f(t)$	$\chi(\delta; f)$	$f^*(t)$
t^α	$\delta^{\frac{1}{\alpha}}, \quad 0 \leqslant \delta \leqslant \pi^\alpha$	t^α
$\mathrm{e}^{-\frac{1}{t}}$	$-\dfrac{1}{\ln \delta}, \quad 0 \leqslant \delta \leqslant \mathrm{e}^{-\frac{1}{\pi}}$	$\mathrm{e}^{-\frac{1}{t}}$
$\sin kt$	$2\arcsin \delta, \quad 0 \leqslant \delta \leqslant 1$	$\sin \frac{1}{2}t$
$\mathrm{e}^t - 1$	$\ln(1 + \delta), \quad 0 \leqslant \delta \leqslant \mathrm{e}^\pi - 1$	$\mathrm{e}^t - 1$

Table1.2

$f(t)$	$\chi_{\mathrm{L}}(\delta; f)$	$\chi_{\mathrm{U}}(\delta; f)$
$\sin kt$	$c_1\delta, \quad 0 \leqslant \delta \leqslant 1$	$c_2\delta, \quad 0 \leqslant \delta \leqslant 1$
$\mathrm{e}^t - 1$	$c_1\delta, \quad 0 \leqslant \delta \leqslant \mathrm{e}^\pi - 1$	$c_2\delta, \quad 0 \leqslant \delta \leqslant \mathrm{e}^\pi - 1$
$t^n + t^{n+1} + \cdots + t^{n+k}$	$c_1\delta^{-\frac{1}{n}}, \quad 0 \leqslant \delta \leqslant \delta_0$	$c_2\delta^{-\frac{1}{n}}, \quad 0 \leqslant \delta \leqslant \delta_0$

B. We next consider the function $f(x, y) = \sin k_1 x \cdot \sin k_2 y$ $(0 \leqslant x,\ y \leqslant \pi)$. Its distribution satisfies the estimates

$$c_1 \delta \ln \frac{1}{\delta} \leqslant \chi(\delta; f) \leqslant c_2 \delta \ln \frac{1}{\delta}, \quad 0 \leqslant \delta \leqslant \delta_0.$$

C. Let $\Omega \subset \mathbb{R}^n$ be a bounded domain with smooth boundary $\partial\Omega$. We consider a continuously differentiable function $f(t)$ defined on $\bar{\Omega}$ that has positive values everywhere inside Ω, equals zero identically on $\partial\Omega$ and is such that the derivative $\dfrac{\partial f}{\partial \bar{n}}$ with respect to the unit inward normal $\bar{n}$ satisfies the estimate $\dfrac{\partial f}{\partial \bar{n}} \geqslant c > 0$ $(t \in \partial\Omega)$. In this case we get

$$c_1 \delta \leqslant \chi(\delta; f) \leqslant c_2 \delta, \quad 0 \leqslant \delta \leqslant \delta_0,$$

that is, we can set

$$\chi_{\mathrm{L}}(\delta; f) = c_1 \delta, \quad \chi_{\mathrm{U}}(\delta; f) = c_2 \delta, \quad 0 \leqslant \delta \leqslant \delta_0.$$

D. We conclude our discussion with a couple of simple remarks. First, if $|f(t)| \geqslant \gamma > 0$ for almost all $t \in \Omega$, then $\chi(\delta; f) = 0$ for $0 \leqslant \delta \leqslant \gamma$. Second, $f(t)$ equals zero on a set with positive measure, if and only if $\chi(0; f) > 0$.

1.3. UNIFORM ESTIMATES OF DISTRIBUTIONS. We will next consider estimates for distributions of functions $f(t)$ that belong to a certain family. Let $\mathcal{F}$ denote the given family of functions $f(t) : \Omega \to \mathbb{R}^n$. We set

$$\chi_{\mathrm{L}}(\delta; \mathcal{F}) = \inf_{f \in \mathcal{F}} \chi(\delta; f), \tag{1.5}$$

$$\chi_{\mathrm{U}}(\delta; \mathcal{F}) = \inf_{f \in \mathcal{F}} \chi(\delta; f). \tag{1.6}$$

The functions (1.5) and (1.6) are clearly nondecreasing, and the former one is also continuous to the right.

An explicit form of these functions can be obtained, of course, only in a few exceptional cases. Therefore we use estimates of these functions.

Assume, for instance, that $\mathcal{F}$ is a compact set in the space $C = C(\Omega)$ of all continuous functions on a compact set $\Omega \subset \mathbb{R}^m$. If the distribution $\chi(\delta; f)$ of each function $f(t) \in \mathcal{F}$ is positive for $\delta > 0$, then the function (1.5) is positive for $\delta > 0$, too.

We give now an important nontrivial example.

Let $\Omega = [a, b]$ and let $\mathcal{F}_1$ be the set of all solutions $x(t)$ of the linear differential equation

$$x^{(N)} + q_1(t)x^{(N-1)} + \cdots + q_N(t)x = 0, \tag{1.7}$$

with measurable and bounded on $[a, b]$ coefficients. In addition, let

$$\mathcal{F} = \left\{ x(t) : x(t) \in \mathcal{F}_1, \ \sup_{t \in [a, b]} |x(t)| = 1 \right\}.$$

THEOREM 1.1. *For every $N > 1$, there exist some constants $c_1, c_2, \delta_0 > 0$, such that the estimates*

$$c_2 \delta^{\frac{1}{N-1}} \leqslant \chi_U(\delta; \mathcal{F}) \leqslant c_1 \delta^{\frac{1}{N-1}}, \quad 0 \leqslant \delta \leqslant \delta_0 \tag{1.8}$$

hold true.

The numbers c_1, c_2, and δ_0 depend, of course, on the coefficients $q_j(t)$ in equation (1.7).

We notice that the set $\mathcal{F}_1$ is a finite dimensional vector space. The equivalence of all norms on such a space implies that the conclusion of Theorem 1.1, that is, the estimates (1.8), is still true for the set

$$\mathcal{F} = \{x(t) : x(t) \in \mathcal{F}_1, \ \|x(t)\| = 1\},$$

where $\| \cdot \|$ is an arbitrary norm on $\mathcal{F}_1$.

1.4. PROOF OF THEOREM 1.1. We will first prove an auxiliary result. Let J be an arbitrary interval included in $\Omega = [a, b]$. We associate to each measurable on Ω function $x(t)$, the function

$$\chi_J(\delta; x) = \mu\{t \in \Omega : t \in J, \ |x(t)| \leqslant \delta\}, \quad 0 \leqslant \delta < \infty. \tag{1.9}$$

When $J = \Omega$, obviously the function (1.9) coincides with the function (1.1).

LEMMA 1.1 *Let $x(t)$ be a k-times continuously differentiable function, let $J \subset \Omega$ be an arbitrary interval, and let $\delta > 0$ be such that $\chi_J(\delta; x) > 0$. Then there exist some points $\tau_r = \tau_r(\delta, J, x) \in J$ $(r = 0, 1, \ldots, k)$ such that*

$$\left| x^{(r)}(\tau_r) \right| \leqslant 2^{r(r+1)} [\chi_J(\delta; x)]^{-r} \delta. \tag{1.10}$$

Proof. The existence of the point τ_0 follows from the condition $\chi_J(\delta; x) > 0$, since the inequality $\mu\{t \in \Omega : t \in J, \ |x(t)| \leqslant \delta\} > 0$ implies $\{t \in \Omega : t \in J, \ |x(t)| \leqslant \delta\} \neq \emptyset$.

Assume now that if $0 \leqslant r \leqslant s < k$, then the points τ_r can be chosen for any $\delta > 0$ and any interval J. By induction, in order to prove our lemma, it is enough to prove the existence of the point τ_r, when $r = s + 1$.

We split the interval J into three disjoint and consecutive subintervals J_1, J_2, and J_3, such that

$$\chi_{J_1}(\delta; x) = \chi_{J_3}(\delta; x) = \frac{1}{4}\chi_J(\delta; x), \quad \chi_{J_2}(\delta; x) = \frac{1}{2}\chi_J(\delta; x).$$

According to the induction assumption, we can choose some points

$$\tau_{s,1} = \tau_s(\delta, J_1, x) \in J_1 \quad \text{and} \quad \tau_{s,3} = \tau_s(\delta, J_3, x) \in J_3,$$

such that

$$\left|x^{(s)}(\tau_{s,1})\right|, \ \left|x^{(s)}(\tau_{s,3})\right| \leqslant 2^{s(s+1)}\left[\frac{1}{4}\chi_J(\delta; x)\right]^{-s}\delta.$$

We clearly have $|\tau_{s,1} - \tau_{s,3}| \geqslant \mu J_2 \geqslant \frac{1}{2}\chi_J(\delta; x)$, and therefore

$$\frac{x^{(s)}(\tau_{s,1}) - x^{(s)}(\tau_{s,3})}{\tau_{s,1} - \tau_{s,3}} \leqslant 2^{s^2+3s+1}[\chi_J(\delta; x)]^{-s-1}\delta.$$

In its turn, this inequality together with the Mean Value Theorem implies the existence of a point $\tau \in J$ such that

$$|x^{(s+1)}(\tau)| \leqslant 2^{(s+1)(s+2)}[\chi_j(\delta; x)]^{-s-1}\delta.$$

We conclude the proof by setting $\tau_{s+1}(\delta, J, x) = \tau$.

We now return to the proof of Theorem 1.1. Let us start by proving the right hand side estimate in (1.8). Set

$$\gamma(t, x) = \max\{|x(t)|, |x'(t)|, \ldots, |x^{(N-1)}(t)|\}. \tag{1.11}$$

The function (1.11) is jointly continuous in the variables $t \in \Omega$ and $x \in \mathcal{F}$ and has positive values. Since each of its variables belongs to a specific compact set we have $\gamma(x, t) > \gamma_0 > 0$ ($t \in \Omega$, $x \in \mathcal{F}$), for a suitable constant γ_0.

Assume, on the contrary, that the right hand side estimate in (1.8) fails for any positive constants c_1 and δ_0. Then there exist a sequence of functions $x_n(t) \in \mathcal{F}$ and a sequence of positive numbers δ_n, such that

$$\mu\{t \in \Omega : |x_n(t)| \leqslant \delta_n\} \geqslant n\delta_n^{\frac{1}{N-1}}. \tag{1.12}$$

This inequality leads to the following estimates

$$\delta_n \leqslant (b-a)^{N-1} n^{1-N}, \quad n = 1, 2, \ldots, \tag{1.13}$$

therefore $\delta_n \to 0$. In view of the compactness of the set $\mathcal{F}$ (relatively to any norm), we may assume that the sequence $x_n(t)$ converges uniformly to a function $x_*(t) \in \mathcal{F}$, together with all the derivatives up to the order $N - 1$.

Since $\gamma(t, x) > \gamma_0$, the sets

$$\Pi_j = \left\{ t : \left| x_*^{(j)}(t) \right| > \gamma_0 \right\}, \quad j = 0, 1, \ldots, N-1, \tag{1.14}$$

are open in the induced topology of Ω and provide a covering of the compact interval Ω. Therefore we can find a finite covering for $[a, b]$ consisting of some intervals $J_1, \ldots, J_M$, with the property that each of them is included in one of the sets (1.14). Thus, each i $(i = 1, 2, \ldots, M)$ corresponds to an integer $0 \leqslant m_i \leqslant N - 1$, such that $\left| x_*^{(m_i)}(t) \right| > \gamma_0$, $t \in J_i$. Hence, there exists an integer n_0 such that

$$\left| x_n^{(m_i)}(t) \right| > \frac{1}{2}\gamma_0, \quad t \in J_i; \ i = 1, \ldots, M; \ n \geqslant n_0. \tag{1.15}$$

According to (1.12), for each $n = 1, 2, \ldots$ we can find an interval $J(n)$ in the covering $J_1, \ldots, J_M$, such that

$$\mu\{t : t \in J(n), \ |x_n(t)| \leqslant \delta_n\} \geqslant \frac{n}{M} \delta_n^{\frac{1}{N-1}}.$$

Consequently, there exist an infinite sequence of indices $n_1 < n_2 < n_3 < \cdots$ and a fixed interval J_{i_0} in the covering $J_1, \ldots, J_M$, for which

$$\mu\left\{t : t \in J_{i_0}, \ |x_{n_k}(t)| \leqslant \delta_{n_k}\right\} \geqslant \frac{n_k}{M} \delta_{n_k}^{\frac{1}{N-1}}, \quad k = 1, 2, \ldots,$$

or, using (1.9),

$$\chi_{J_{i_0}}(\delta_{n_k}; x_{n_k}) \geqslant \frac{n_k}{M} \delta_{n_k}^{\frac{1}{N-1}}, \quad k = 1, 2, \ldots. \tag{1.16}$$

Based on Lemma 1.1, from (1.16) it follows that there are some points $\tau_r(k) = \tau_r(\delta_{n_k}, J_{i_0}, x_{n_k}) \in J_{i_0}$ $(r = 0, 1, \ldots, N-1; \ k = 1, 2, 3, \ldots)$, such that

$$\left| x_{n_k}^{(r)}[\tau_r(k)] \right| \leqslant c \delta_{n_k}^{1 - \frac{1}{N-1}} n_k^{-r}, \quad c = 2^{r(r+1)} M^r.$$

On the other hand, from (1.15) we know that

$$\left| x_{n_k}^{(m_{i_0})}(t) \right| > \frac{1}{2}\gamma_0, \quad t \in J_{i_0}; \ k = 1, 2, \ldots; \ n_k \geqslant n_0.$$

Therefore, for a sufficiently large k we get

$$\gamma_0 < 2c\delta_{n_k}^{1-\frac{m_{i_0}}{N-1}} n_k^{-m_{i_0}}.$$

But, according to (1.13), this estimate contradicts the positiveness of γ_0. Thus the right hand side estimate in (1.8) holds for some c_1 and δ_0.

In order to prove the existence of some constants c_2 and δ_0 for which the left hand side estimate in (1.8) is true, it is enough to establish the existence of a nonidentically zero solution $x_0(t)$ of equation (1.7), such that

$$\chi(\delta; x_0 \geqslant c_2 \delta^{\frac{1}{N-1}}, \quad 0 \leqslant \delta \leqslant \delta_0. \tag{1.17}$$

The function $x_0(t)$ can be defined (using the assumption that the coefficients $q_j(t)$ are bounded) as the unique solution of equation (1.7), subject to the initial conditions

$$x_0(a) = 0, \ x_0'(a) = 0, \dots, x_0^{(N-2)}(a) = 0, \ x_0^{(N-1)}(a) = 1.$$

For this particular solution, if $\Delta t > 0$ is sufficiently small, then we have the following estimate

$$0 \leqslant x_0(t) \leqslant \frac{2}{(N-1)!}(t-a)^{n-1}, \quad a \leqslant t \leqslant a + \Delta t.$$

Therefore

$$\chi(\delta; x_0) \geqslant \mu\{t : t \in [a, a+\Delta t], \ x_0(t) \leqslant \delta\} \geqslant c_2 \delta^{\frac{1}{N-1}},$$

where

$$c_2 = \left[\frac{(N-1)!}{2}\right]^{\frac{1}{N-1}}.$$

The proof of the theorem is now complete.

§2. Estimates for solutions of the basic integral-functional inequality

2.1. PERMISSIBLE NONLINEARITIES. In the sequel we will consider superpositionally measurable functions $\Phi(t, u)$, defined for $t \in \Omega$ and real numbers $u \geqslant 0$. By *superpositional measurability* (see [Krasnoselskii, M. A., et al., 1966]) we mean that the function $\Phi(t, x(t))$ of a single variable $t \in \Omega$ is measurable, for any measurable function $x(t) : \Omega \to \mathbb{R}^1$. A sufficient condition for superpositional measurability is the Caratheodory condition: $\Phi(t, u)$ is measurable in t for each fixed u, and continuous in u for each fixed t.

We let $\mathfrak{N}(u_0)$ $(u_0 > 0)$ denote the class of all bounded functions $\Phi(t, u)$, which for $u \geqslant u_0$ satisfy the next conditions:

    a) $\Phi(t, u)$ is nonincreasing in u;

    b) $\Phi(t, u)$ is nonnegative;

    c) $\Phi(t, u)$ is continuous in u;

    d) there exists a subset $\Omega_0 \subset \Omega$ $(\mu\Omega_0 > 0)$ such that $\Phi(t, u) > 0$ for all $t \in \Omega_0$, $u \geqslant u_0$.

If $\Phi(t, u) \in \mathfrak{N}(u_0)$ then, for any nonnegative measurable function $x(t)$, we get the strict inequality

$$\int_\Omega \Phi[t, u_0 + x(t)]\mathrm{d}\mu > 0.$$

If we are interested in such an inequality for any nonnegative $x(t) \in L_\infty$, then condition d) above can be replaced by

$$\mu\left[\bigcap_{u=u_0}^{u_0+N} \{t : \Phi(t, u) > 0\}\right] > 0.$$

2.2. AN IMPORTANT PARTICULAR CASE. Let $e(t) : \Omega \to \mathbb{R}^n$ be a given arbitrary integrable function. Consider next the inequality

$$\|h(t)\|_{L_1} \leqslant -\int_\Omega \Phi[t, |\xi e(t) + h(t)|]\mathrm{d}\mu, \tag{2.1}$$

where $h(t) : \Omega \to \mathbb{R}^n$ is an integrable function on Ω and ξ is a real number.

THEOREM. 2.1. *Let* $\Phi(t, u) \in \mathfrak{N}(u_0)$ *and assume that*

$$\lim_{\delta \to 0} \frac{\chi(\delta; e)}{\displaystyle\int_\Omega \Phi[t, u_0 + R\delta^{-1}|e(t)|]\mathrm{d}\mu} = 0, \tag{2.2}$$

for any $R > 0$. *Then there exists a number* $c > 0$ *such that inequality* (2.1) *has no integrable solutions whenever* $|\xi| > c$.

In other words, equality (2.2) yields the estimate $|\xi| \leqslant c$ for the first component of a solution $\{\xi, h(t)\}$ $(h(t) \in L_1)$ of inequality (2.1).

The sufficient condition (2.2) is quite difficult to handle and in many cases it can be simplified. We illustrate this remark by some examples of functions $\Phi(t, u)$

that do not depend upon t for $u \geqslant u_0$ ($\Phi(t, u) \equiv \Phi(u)$ for $u \geqslant u_0$). For this kind of functions condition (2.2) can be rewritten as

$$\lim_{\delta \to 0} \frac{\chi(\delta; e)}{\displaystyle\int_0^\infty \Phi\left(u_0 + R\delta^{-1}\xi\right) \mathrm{d}\chi(\xi; e)} = 0. \tag{2.3}$$

If the function $\chi(\delta; e)$ admits the estimates

$$c_2\delta^\alpha \leqslant \chi(\delta; e) \leqslant c_1\delta^\alpha, \quad \alpha > 0; \; 0 \leqslant \delta \leqslant \delta_0, \tag{2.4}$$

then condition (2.3) is equivalent to

$$\int_{u_0}^\infty u^{\alpha-1}\Phi(u)\mathrm{d}u = \infty. \tag{2.5}$$

If $\chi(\delta_0; e) = 0$ for some $\delta_0 > 0$, then condition (2.2) is fulfilled for any function $\Phi(t, u) \in \mathfrak{N}(u_0)$.

We notice then whenever condition (2.2) is satisfied, then $\chi(0; e) = 0$, i.e., $\mu\{t : e(t) = 0\} = 0$, an equality that provides a necessary condition for (2.2).

The proof of Theorem 2.1 is omitted. Actually it follows from a more general result stated in Theorem 2.2 below.

2.1. THE MAIN THEOREM. We next consider the spaces $L_p = L_p(\Omega, \mathbb{R}^n)$ $(1 \leqslant p < \infty)$ of p-integrable functions $x(t) : \Omega \to \mathbb{R}^n$, with the usual norms

$$\|x(t)\|_{L_p} = \left(\int_\Omega |x(t)|^p \mathrm{d}\mu\right)^{\frac{1}{p}}.$$

Let $\mathfrak{F} \subset L_p$ denote a compact family of functions $e(t) : \Omega \to \mathbb{R}^n$. In some subsequent applications $\mathfrak{F}$ will be a normalized set in a finite dimensional subspace of L_p.

The family $\mathfrak{F}$ is called compatible with the function $\Phi(t, u)$, if for any $\beta > 0$, there exist a positive nonincreasing function $\alpha(u)$ $(u \geqslant 0)$ and a number $c = c(\beta) > 0$, such that the inequality

$$\|h(t)\|_{L_p}^p \leqslant -\beta \int_\Omega \Phi[t, |\xi e(t) + h(t)|]\mathrm{d}\mu + \beta\alpha\left(\|\xi e(t) + h(t)\|_{L_p}\right) \tag{2.6}$$

has no solutions $h(t) \in L_p$ for $e(t) \in \mathfrak{F}$ and $|\xi| > c$.

In other words, the family $\mathfrak{F}$ is compatible with the function $\Phi(t, u)$, if for any $\beta > 0$ there exists a positive nonincreasing function $\alpha(u)$ such that an a priori estimate $|\xi| \leqslant c$ for the first component of all solutions $\{\xi, h(t)\}$ of inequality (2.6) is true for any $e(t) \in \mathfrak{F}$. For the second component $h(t)$ we have the obvious estimate $\|h(t)\|_{L_p} < \text{const}$.

THEOREM 2.2. *Let* $\Phi(t, u) \in \mathfrak{N}(u_0)$ $(u_0 > 0)$, *and assume that*

$$\lim_{\delta \to 0} \sup_{e(t) \in \mathfrak{F}} \frac{\chi(\delta; e)}{\displaystyle\int_\Omega \Phi\left[t, u_0 + R\delta^{-1}|e(t)|\right] \mathrm{d}\mu} = 0, \qquad (2.7)$$

for any $R > 0$. *Then the family* $\mathfrak{F}$ *is compatible with the function* $\Phi(t, u)$.

When the family $\mathfrak{F}$ consists of only one function $e(t)$ that satisfies condition (2.2), then $\mathfrak{F}$ is compatible with the function $\Phi(t, u) \in \mathfrak{N}(u_0)$; therefore, if $p = 1$, $\beta = 1$, then there exist a function $\alpha(u)$ and a number c, such that for $|\xi| > c$ equation (2.6) has no solutions $h(t) \in L_1$. Consequently, there are no solutions $h(t)$ of inequality (2.1). Thus, Theorem 2.1 follows from Theorem 2.2. The latter will be proved in the next section.

To check condition (2.7), it is convenient to use the functions (1.5) and (1.6). For instance, if $\Phi(t, u)$ does not depend on t for $u \geqslant u_0$, then equality

$$\lim_{\delta \to \infty} \frac{\chi_U(\delta, \mathfrak{F})}{\displaystyle\int_0^\infty \Phi\left(u_0 + R\delta^{-1}\xi\right) \mathrm{d}\chi_L(\xi, \mathfrak{F})} = 0 \qquad (2.8)$$

is a sufficient condition for (2.7).

We notice a rough version of condition (2.8).

Since the functions in $\mathfrak{F}$ are bounded in the norm of L_p, using Chebyshev inequality

$$\mu\{t : t \in \Omega, \ |e(t)| \geqslant \nu\} \leqslant \frac{\|e\|^p}{\nu^p},$$

we get the estimate

$$\int_\Omega \Phi\left[t, u_0 + R\delta^{-1}|e(t)|\right] \mathrm{d}\mu \geqslant \int_{\{t:|e(t)|<\nu\}} \Phi\left[t, u_0 + R\delta^{-1}|e(t)|\right] \mathrm{d}\mu \geqslant$$

$$\geqslant \Phi\left[t, u_0 + R\delta^{-1}\nu\right] \mu\{t : |e(t)| < \nu\} \geqslant \Phi\left[t, u_0 + R\delta^{-1}\nu\right] \left[\mu\Omega - \|e\|_{L_p}^p \nu^{-p}\right].$$

For

$$\nu = \left[\frac{1}{2}\mu\Omega\right]^{-\frac{1}{p}} \sup_{e(t)\in\mathfrak{F}} \|e(t)\|_{L_p}$$

the above inequality implies that

$$\int_\Omega \Phi\left[t, u_0 + R\delta^{-1}|e(t)|\right] d\mu \geqslant \frac{1}{2}\mu\Omega\,\Phi\left[t, u_0 + R\delta^{-1}\nu\right].$$

In other words, condition (2.8) is satisfied for any $R > 0$ provided that the equality

$$\lim_{\delta\to 0} \frac{\chi_U(\delta; \mathfrak{F})}{\Phi\left(u_0 + R\delta^{-1}\right)} = 0 \tag{2.9}$$

holds true for all $R > 0$.

As in the case of condition (2.3), condition (2.8), as well as condition (2.9), can be simplified, if estimates for functions (1.5) and (1.6) are known. For instance, if $\mathfrak{F}$ is the set of normalized solutions of an ordinary differential equation of order n $(n > 1)$ (see Theorem 1.1), then for its compatibility with the function $\Phi(t, u) \in \mathfrak{N}(u_0)$ $(u_0 > 0)$ it is enough to have the equality

$$\lim_{u\to\infty} u^{\frac{1}{n-1}}\Phi(u) = \infty.$$

Recall that $\Phi(t, u) \equiv \Phi(u)$ for $u \geqslant u_0$.

Conditions (2.8) and (2.7) are not equivalent even if $\Phi(t, u)$ does not depend on u. For instance, if $\mathfrak{F}$ is the set of normalized linear functions $e(t) = at + b$ on the interval $\Omega = [0, 1]$, then condition (2.7) is satisfied when

$$\int_{u_0}^\infty \Phi(u)du = \infty,$$

whereas a necessary condition for (2.8) relies on the stronger restriction

$$\lim_{u\to\infty} u\Phi(u) = \infty.$$

We also notice that, as in the case of (2.2), a necessary condition for (2.7) is that the equality $\mu\{t : e(t) = 0\} = 0$ holds for every $e(t) \in \mathfrak{F}$.

2.4. A CONVERSE OF THEOREM 2.1. In this subsection we will prove that condition (2.2) is "almost necessary" for the lack of solutions $\{\xi, h(t)\}$ for inequality (2.1) with large absolute values of the component ξ. Analogous assertions can be formulated when one inestigates the necessity of condition (2.7) for the compatibility of a family $\mathfrak{F}$ with a function $\Phi(t, u)$.

Assume that a function $\Phi_0(t, u) \in \mathfrak{N}(u_0)$ $(u_0 > 0)$ satisfies the relation

$$\gamma\Phi_0(t, u) \leqslant \Phi_0(t, u + u_0), \quad u \geqslant u_0 \tag{2.10}$$

for some $\gamma > 0$. Such a constant $\gamma > 0$ can be found, for instance, if $\Phi(t, u)$ has one of the "forms" $c(t)e^{-u}$, $c(t)u^{-a}$ $(a > 0)$, and so on, for sufficiently large values of u.

We denote by $\mathfrak{W}(\Phi_0)$ the class of all nonlinearities $\Phi(t, u) \in \mathfrak{N}(u_0)$ which coincide with $\Phi_0(t, u) \in \mathfrak{N}(u_0)$ for $u > u_0$.

THEOREM 2.3. *Let $\Phi_0(t, u)$ be a function such that inequality (2.1) has no solutions $\{\xi, h(t)\}$ $(h(t) \in L_1)$ with arbitrarily large values of $|\xi|$, for any arbitrary function $\Phi(t, u) \in \mathfrak{W}(\Phi_0)$. Then for $R = u_0$ the next analog of (2.2) is true:*

$$\lim_{\delta \to 0} \frac{\chi(\delta; e)}{\displaystyle\int_\Omega \Phi_0\left[t, u_0 + R\delta^{-1}|e(t)|\right] d\mu} = 0. \tag{2.11}$$

2.5. PROOF OF THEOREM 2.3. We define the function

$$\Phi(t, u) = \begin{cases} -k, & \text{if } u \leqslant u_0 \\ \Phi_0(t, u), & \text{if } u > u_0, \end{cases} \tag{2.12}$$

where $k > 0$. This function is oviously superpositionally measurable and $\Phi(t, u) \in \mathfrak{W}(\Phi_0)$ for each k. Under the assumptions in Theorem 2.3, if $\xi > 0$ is sufficiently large, then the function $h(t) \equiv 0$ is not a solution for inequality (2.1), that is,

$$\int_\Omega \Phi_0(t, |\xi e(t)|) d\mu > 0, \quad \xi \geqslant \xi_0(k). \tag{2.13}$$

Consider Ω as a union of the disjoint sets

$$\Omega_0 = \{t : |\xi e(t)| > u_0\}$$

and

$$\Omega_1 = \{t : |\xi e(t) \leqslant u_0\}.$$

Since, on one hand,

$$\int_{\Omega_1} \Phi_0(t, |\xi e(t)|) d\mu = -k\mu\Omega_1 = -k\chi\left(\frac{u_0}{\xi}\right)$$

and, on the other hand, according to (2.10),

$$\int_{\Omega_0} \Phi_0(t, |\xi e(t)|) \mathrm{d}\mu \leqslant \frac{1}{\gamma} \int_{\Omega_0} \Phi_0(t, u_0 + |\xi e(t)|) \mathrm{d}\mu \leqslant$$

$$\leqslant \frac{1}{\gamma} \int_{\Omega} \Phi_0(t, u_0 + \xi |e(t)|) \mathrm{d}\mu,$$

setting $\delta = u_0 \xi^{-1}$, from (2.13) we get the following inequality

$$\frac{\chi(\delta)}{\int_{\Omega} \Phi_0\left[t, u_0 + R\delta^{-1}|e(t)|\right] \mathrm{d}\mu} \leqslant \frac{1}{\gamma k}, \quad \delta \leqslant u_0 [\xi_0(k)]^{-1}.$$

Hence (since the left hand side is independent of the arbitrarily large number $k > 0$) (2.11) follows. Theorem 2.3 is now completely proved.

§3. Proof of Theorem 2.2

3.1. THE FIRST STEP OF THE PROOF. Throughout this section we will use the notations introduced in §1 and §2; the norm in L_p is denoted by $\| \cdot \|$.

Let

$$M = \sup_{t \in \Omega, \ u \geqslant 0} |\Phi(t, u)|. \tag{3.1}$$

We associate to any function $x(t) = \xi e(t) + h(t) \in L_p$ the set

$$G[x] = \{t : t \in \Omega, \ |\xi e(t)| < u_0 + |h(t)|\}$$

and we prove the estimate

$$\int_{\Omega} \Phi[t, |x(t)|] \mathrm{d}\mu \geqslant \int_{\Omega} \Phi[t, u_0 + 2|\xi e(t)|] \mathrm{d}\mu - 2M\mu G[x]. \tag{3.2}$$

Let $t \notin G[x]$, that is, $|\xi e(t)| \geqslant u_0 + |h(t)|$. Then

$$|x(t)| \geqslant |\xi e(t)| - |h(t)| \geqslant u_0$$

and

$$|x(t)| \leqslant |\xi e(t)| + |h(t)| \leqslant 2|\xi e(t)| - u_0 \leqslant 2|\xi e(t)| + u_0.$$

Since the function $\Phi(t, u)$ is monotonous in u (for $u \geq u_0$), we have

$$\Phi[t, |x(t)|] \geq \Phi[t, u_0 + 2|\xi e(t)|], \quad t \notin G[x]. \tag{3.3}$$

But

$$\int_{\Omega \setminus G[x]} \Phi[t, u_0 + 2|\xi e(t)|] \mathrm{d}\mu = \int_{\Omega} \Phi[t, u_0 + 2|\xi e(t)|] \mathrm{d}\mu - \int_{G[x]} \Phi[t, u_0 + 2|\xi e(t)|] \mathrm{d}\mu$$

and, in view of (3.3),

$$\int_{\Omega} \Phi[t, |x(t)|] \mathrm{d}\mu = \int_{\Omega \setminus G[x]} \Phi[t, |x(t)|] \mathrm{d}\mu + \int_{G[x]} \Phi[t, |x(t)|] \mathrm{d}\mu \geq$$

$$\geq \int_{\Omega} \Phi[t, u_0 + 2|\xi e(t)|] \mathrm{d}\mu + \int_{G[x]} (\Phi[t, |x(t)|] - \Phi[t, u_0 + 2|\xi e(t)|]) \mathrm{d}\mu.$$

Thus, in order to prove the estimate (3.2), it is enough to use the inequality

$$\int_{G[x]} (\Phi[t, |x(t)|] - \Phi[t, u_0 + 2|\xi e(t)|]) \mathrm{d}\mu \leq 2M \cdot \mu G[x].$$

3.2. THE SECOND STEP OF THE PROOF. Let $\alpha(u)$ $(u \geq 0)$ be an arbitrary positive nonincreasing function.

LEMMA 3.1. *Assume that the function* $x(t) = \xi e(t) + h(t) \in L_p$ *satisfies inequality (2.6) and let*

$$\|\xi e(t)\| \geq 2\{\beta[M \cdot \mu\Omega + \alpha(0)]\}^{\frac{1}{p}}, \tag{3.4}$$

where M is as in (3.1). Then

$$\int_{\Omega} \Phi[t, u_0 + 2|\xi e(t)|]) \mathrm{d}\mu \leq 2M\chi\left(\frac{u_0 + (2M\beta)^{\frac{1}{p}}}{|\xi|}; e\right) + \alpha\left(\frac{1}{2}\|\xi e(t)\|\right). \tag{3.5}$$

Proof. From inequality (2.6) we have

$$\|h(t)\|^p \leq \beta[M\mu\Omega + \alpha(0)]. \tag{3.6}$$

Therefore by (3.4) we get the estimate

$$\|x(t)\| \geq \|\xi e(t)\| - \{\beta[M\mu\Omega + \alpha(0)]\}^{\frac{1}{p}} \geq \frac{1}{2}\|\xi e(t)\|.$$

Since $\alpha(u)$ is a nonincreasing function, this estimate yields

$$\alpha(\|x\|) \leqslant \alpha\left(\frac{1}{2}\|\xi e(t)\|\right). \tag{3.7}$$

Let

$$\Omega_1[x] = \left\{t : t \in \Omega, \ |h(t)| > (2M\beta)^{\frac{1}{p}}\right\}.$$

Then,

$$G[x] \setminus \Omega_1[x] = \left\{t : t \in \Omega, \ |\xi e(t)| < u_0 + |h(t)|, \ |h(t)| \leqslant (2M\beta)^{\frac{1}{p}}\right\} \subset$$
$$\subset \left\{t : t \in \Omega, \ |\xi e(t)| < u_0 + (2M\beta)^{\frac{1}{p}}\right\}.$$

Consequently,

$$G[x] \subset \Omega_1[x] \cup \left\{t : t \in \Omega, \ |\xi e(t)| < u_0 + (2M\beta)^{\frac{1}{p}}\right\}.$$

Therefore,

$$\mu G[x] \leqslant \mu\left\{t : t \in \Omega, \ |h(t)| > (2M\beta)^{\frac{1}{p}}\right\} + \mu\left\{t : t \in \Omega, \ |\xi e(t)| < u_0 + (2M\beta)^{\frac{1}{p}}\right\}$$

and further (since $\xi \neq 0$),

$$\mu G[x] \leqslant \mu\left\{t : t \in \Omega, \ |h(t)| > (2M\beta)^{\frac{1}{p}}\right\} + \chi\left(\frac{u_0 + (2M\beta)^{\frac{1}{p}}}{|\xi|}; e\right) + \alpha\left(\frac{1}{2}\|\xi e(t)\|\right).$$

Hence, and by Chebyshev inequality

$$\mu G[x] \leqslant \mu\left\{t : t \in \Omega, \ |h(t)| > (2M\beta)^{\frac{1}{p}}\right\} \leqslant \frac{\|h\|^p}{2M\beta},$$

we get the estimate

$$\mu G[x] \leqslant \frac{\|h\|^p}{2M\beta} + \chi\left(\frac{u_0 + (2M\beta)^{\frac{1}{p}}}{|\xi|}; e\right). \tag{3.8}$$

From (3.1) and (3.2) we obtain the relation

$$\int_\Omega \Phi[t, |x(t)|]d\mu \geqslant \int_\Omega \Phi[t, u_0 + 2|\xi e(t)|]d\mu - \frac{\|h\|^p}{\beta} - 2M\chi\left(\frac{u_0 + (2M\beta)^{\frac{1}{p}}}{|\xi|}; e\right),$$

which, when combined with (2.6), leads to the inequality

$$\int_{\Omega} \Phi[t, u_0 + 2|\xi e(t)|] \mathrm{d}\mu \leqslant \alpha(\|x\|) + 2M\chi\left(\frac{u_0 + (2M\beta)^{\frac{1}{p}}}{|\xi|}; e\right).$$

The estimate (3.5) follows from (3.7). Lemma 3.1 is proved.

3.3. THE CONSTRUCTION OF $\alpha(u)$. We now consider the functional

$$\Gamma(u; x) = \frac{1}{u+1}\int_{\Omega} \Phi\left[t, u_0 + u^2|x(t)|\right]\mathrm{d}\mu, \quad u \geqslant 0, \ x(t) \in L_p.$$

Assume that u_n is a convergent sequence of nonnegative numbers and $x_n(t)$ is a convergent sequence of functions in L_p. Let u_* and $x_*(t)$ denote their limits, respectively. Since the function $\Phi(t, u)$ is continuous in u (for $u \geqslant u_0$) and measurable in t, the sequence of functions $\Phi\left[t, u_0 + u_n^2|x_n(t)|\right]$ converges in measure to the function $\Phi\left[t, u_0 + u_*^2|x_*(t)|\right]$. The boundedness of $\Phi(t, u)$ allows us to consider the limit in the equality

$$\Gamma(u_n; x_n) = \frac{1}{1+u_n}\int_{\Omega} \Phi\left[t, u_0 + u_n^2|x_n(t)|\right]\mathrm{d}\mu.$$

Consequently, the functional $\Gamma(u; x)$ is jointly continuous.

Since the family $\mathfrak{F}$ is compact in L_p, the function

$$\alpha(u) = \min_{x \in \mathfrak{F}} \Gamma(u; x) \tag{3.9}$$

is well-defined. This function is positive and nonincreasing since each functional $\Gamma(u; x)$ (with $x(t) \in L_p$ fixed) has the same properties. In order to prove Theorem 2.2 we will consider the inequality (2.6) where $\alpha(u)$ is the function defined in (3.9).

By the monotony of $\Phi(t, u)$ in u for $u \geqslant u_0$, the estimate

$$\Gamma(R_1 u; x) \geqslant \frac{1}{1+R_1 u}\int_{\Omega} \Phi[t, u_0 + Ru|x(t)|]\mathrm{d}\mu, \quad u \geqslant RR_1^{-2}$$

is true for every $x(t) \in L_p$ and every positive numbers R and R_1. Therefore,

$$\alpha(R_1 u) \leqslant \frac{1}{1+R_1 u}\int_{\Omega} \Phi[t, u_0 + Ru|e(t)|]\mathrm{d}\mu, \quad u \geqslant RR_1^{-2}; \ e \in \mathfrak{F}. \tag{3.10}$$

3.4. COMPLETION OF THE PROOF. Assume that the function $h(t)$ and the number ξ satisfy inequality (2.6) with $e(t) \in \mathfrak{F}$, where the function $\alpha(u)$ is defined by (3.9). From (2.6) it follows that $h(t)$ satisfies inequality (3.6). Without any loss of generality we may assume that condition (3.4) in Lemma 3.1 is fulfilled. Accordingly, inequality (3.5) is also true. Let

$$\delta_0 = \frac{u_0 + (2M\beta)^{\frac{1}{p}}}{|\xi|}, \quad R = 2\left[u_0 + (2M\beta)^{\frac{1}{p}}\right], \quad R_1 = \frac{1}{4}R \cdot \inf_{e \in \mathfrak{F}} \|e(t)\|.$$

Then the estimate (3.5) can be written as

$$\int_\Omega \Phi\left[t, u_0 + R\delta_0^{-1}|e(t)|\right] d\mu \leqslant 2M\chi(\delta; e) + \alpha\left(R_1\delta_0^{-1}\right). \tag{3.11}$$

Suppose, in addition, that $|\xi| \geqslant 8\left[\inf_{e \in \mathfrak{F}} \|e(t)\|\right]^{-2}$. Then the inequality $\delta_0^{-1} \geqslant RR_1^{-2}$ holds. Therefore, by (3.10) and (3.11) we obtain the relation

$$\left(1 - \frac{1}{1 + R_1\delta_0^{-1}}\right) \int_\Omega \Phi\left[t, u_0 + R\delta_0^{-1}|e(t)|\right] d\mu \leqslant 2M\chi(\delta_0; e),$$

which, for the sake of convenience, can be written as

$$\frac{\chi(\delta_0; e)}{\int_\Omega \Phi\left[t, u_0 + R\delta_0^{-1}|e(t)|\right] d\mu} \geqslant \frac{R_1\delta_0^{-1}}{2M\left(1 + R_1\delta_0^{-1}\right)}.$$

Hence, since

$$\frac{R_1\delta_0^{-1}}{2M\left(1 + R_1\delta_0^{-1}\right)} \geqslant \frac{2}{5M},$$

the next estimate follows

$$\frac{\chi(\delta_0; e)}{\int_\Omega \Phi\left[t, u_0 + R\delta_0^{-1}|e(t)|\right] d\mu} \geqslant \frac{2}{5M}. \tag{3.12}$$

We use now condition (2.7) in Theorem 2.2. By this condition, there exists a positive number $\delta_* = \delta_*(R, \beta)$ such that

$$\sup_{e(t) \in \mathfrak{F}} \frac{\chi(\delta_0; e)}{\int_\Omega \Phi\left[t, u_0 + R\delta^{-1}|e(t)|\right] d\mu} < \frac{2}{5M}, \quad \delta \leqslant \delta_*.$$

Therefore from (3.12) it follows that $\delta_0 > \delta_*$, i.e., the estimate

$$|\xi| \leqslant \delta_*^{-1}\left[u_0 + (2M\beta)^{\frac{1}{p}}\right]$$

is true.

Thus, the component ξ in any solution $\{\xi, h(t)\}$ $(h(t) \in L_p)$ of the inequality (2.6), where $\alpha(u)$ is given by (3.8), satisfies the estimate

$$|\xi| \leqslant \max\left\{2\{\beta[M\mu\Omega + \alpha(0)]\}^{\frac{1}{p}}, 8\left[\inf_{e \in \mathfrak{F}} \|e(t)\|\right]^{-2}, \delta_*^{-1}\left[u_0 + (2M\beta)^{\frac{1}{p}}\right]\right\}.$$

The proof of Theorem 2.2 is complete.

§4. A second integral-functional inequality

4.1. SIGN-COMPATIBLE FUNCTIONS. Two measurable scalar-valued functions

$$e(t), \quad g(t) \tag{4.1}$$

will be called *sign-compatible* provided their usual product $e(t)g(t)$ is positive almost everywhere, that is,

$$\mu\{t : t \in \Omega, \ e(t)g(t) \leqslant 0\} = 0. \tag{4.2}$$

If a function $e(t)$ is different from zero almost everywhere, then it is sign-compatible with itself. Any nonnegative and almost everywhere different from zero functions (4.1) are sign-compatible. Basically, these simple examples of sign-compatible functions are the most interesting ones in applications.

In the sequel we consider only integrable (on Ω) sign-compatible functions (4.1). The second one is used to define a new measure Mes on Ω by

$$\text{Mes}\,G = \int_\Omega |g(t)|\mathrm{d}\mu, \quad G \subset \Omega. \tag{4.3}$$

This measure is defined for all subsets G of Ω that are measurable with respect to the measure μ. Further, for the first of the functions (4.1) we define the nondecreasing function

$$\chi(\delta; e, g) \overset{\text{def}}{=} \text{Mes}\{t : t \in \Omega, \ |e(t)| \leqslant \delta\}, \quad \delta \geqslant 0, \tag{4.4}$$

or, equivalently,

$$\chi(\delta; e, g) = \int\limits_{\{t: t \in \Omega, \ |e(t)| \leqslant \delta\}} |g(t)|\mathrm{d}\mu. \tag{4.5}$$

When $g(t) = \operatorname{sign} e(t)$ $(t \in \Omega)$, the measure (4.3) coincides with μ, and the function (4.4) equals the function $\chi(\delta; e)$ (see 1.1).

In the theorems stated below we will deal with the functions (4.5) only for small values of $\delta > 0$.

In some specific cases it will be enough to use upper estimates $\chi_{\mathrm{U}}(\delta; e, g)$ for the function (4.5), i.e., nondecreasing functions which satisfy the inequality

$$\chi_{\mathrm{U}}(\delta; e, g) \geqslant \chi(\delta; e, g), \quad 0 \leqslant \delta \leqslant \delta_0,$$

for small values of $\delta > 0$. Let

$$|g(t)| \leqslant g_1(t), \quad |e(t)| \geqslant e_1(t) \geqslant 0, \quad t \in \Omega. \tag{4.6}$$

Then

$$\{t : t \in \Omega, \ |e(t)| \leqslant \delta\} \subset \{t : t \in \Omega, \ |e_1(t)| \leqslant \delta\}, \quad \delta \geqslant 0,$$

and, therefore,

$$\chi(\delta; e, g) \leqslant \chi(\delta; e_1, g_1).$$

Consequently, whenever the estimates (4.6) are true, the function $\chi_{\mathrm{U}}(\delta; e, g)$ can be defined by

$$\chi_{\mathrm{U}}(\delta; e, g) = \chi(\delta; e_1, g_1). \tag{4.7}$$

EXAMPLE 4.1. Let $\Omega = [0, 1]$ and assume that

$$|g(t)| \leqslant t^\alpha, \quad |e(t)| \geqslant t^\beta, \quad 0 \leqslant t \leqslant 1,$$

for some $\alpha, \beta > 0$. Then, in view of (4.7), we can choose

$$\chi_{\mathrm{U}}(\delta; e, g) = \frac{1}{\alpha + 1} \delta^{\frac{1+\alpha}{\beta}}, \quad \delta \geqslant 0.$$

EXAMPLE 4.2. Let $\Omega = [0, 1]$ and assume that

$$|g(t)| \leqslant t^\alpha, \quad |e(t)| \geqslant (1 - t)^\beta, \quad 0 \leqslant t \leqslant 1,$$

for some $\alpha, \beta > 0$. Then, in view of (4.7), we can choose

$$\chi_{\mathrm{U}}(\delta; e, g) = c\delta^{\frac{1}{\beta}}, \quad \delta \geqslant 0,$$

where c is a positive number.

In some specific cases it will be useful to consider effective lower estimates $\chi_{\mathrm{L}}(\delta; e, g)$ of the function (4.5), that is, nonnegative, nondecreasing functions satisfying the inequality

$$\chi_{\mathrm{L}}(\delta; e, g) \leqslant \chi(\delta; e, g), \quad 0 \leqslant \delta \leqslant \delta_0,$$

for sufficiently small values of $\delta > 0$. If the estimates

$$|g(t)| \geqslant g_2(t) > 0, \quad |e(t)| \leqslant e_2(t), \quad t \in \Omega,$$

are true, then as $\chi_{\mathrm{L}}(\delta; e, g)$ we can take the function

$$\chi_{\mathrm{L}}(\delta; e, g) = \chi(\delta; e_2, g_2), \quad 0 \leqslant \delta \leqslant \delta_0.$$

For instance, when $\Omega = [0, 1]$ and

$$|g(t)| \leqslant t^\alpha, \quad |e(t)| \geqslant (1 - t)^\beta, \quad 0 \leqslant t \leqslant 1,$$

for some $\alpha, \beta > 0$, we can set

$$\chi_{\mathrm{L}}(\delta; e, g) = \frac{1}{\alpha + 1} \delta^{\frac{1+\alpha}{\beta}}, \quad \delta \leqslant \delta_0.$$

If

$$|g(t)| \leqslant t^\alpha, \quad |e(t)| \geqslant (1 - t)^\beta, \quad 0 \leqslant t \leqslant 1,$$

for some $\alpha, \beta > 0$, then a possible choice is

$$\chi_{\mathrm{L}}(\delta; e, g) = c\delta^{\frac{1}{\beta}}, \quad \delta \leqslant \delta_0,$$

where c is a positive number.

4.2. THE PROBLEM OF A PRIORI ESTIMATES. In the sequel, $\mathfrak{V}(\Phi, \Psi)$ will denote the class of all superpositionally measurable functions $\varphi(t, x)$ ($t \in \Omega$, $x \in \mathbb{R}^1$) subject to the following conditions

$$\varphi(t, x) \operatorname{sign} x \geqslant \Phi(t, |x|), \quad t \in \Omega, \ x \in \mathbb{R}^1; \tag{4.8}$$

$$|\varphi(t, x)| \leqslant \Psi(|x|) \quad t \in \Omega, \ x \in \mathbb{R}^1. \tag{4.9}$$

If not otherwise mentioned, we will always assume that the function $\Phi(t, u)$ ($t \in \Omega$, $u \geqslant 0$) belongs to a certain class $\mathfrak{N}(u_0)$ ($u_0 > 0$) and the function $\Psi(u)$ ($u \geqslant 0$) is positive and nondecreasing.

In this subsection we will be mainly concerned with the study of the inequality

$$\xi \int_\Omega g(t)\varphi[t, \xi e(t) + h(t)]\mathrm{d}\mu \leqslant 0, \qquad (4.10)$$

where the functions $e(t)$ and $g(t)$ are sign-compatible. and the function $\varphi(t, x)$ lies in the class $\mathfrak{V}(\varPhi, \varPsi)$. By a solution for inequality (4.10) we mean, as in §2, either a function $h(t)$ in a fixed space $L_p = L_p(\Omega, \mathbb{R}^1)$ $(1 \leqslant p < \infty)$ for a fixed real number ξ, or a pair $\{\xi, h(t)\}$ consisting of a real number ξ and a function $h(t) \in L_p$.

The norm in the space L_p, as well as in other spaces of integrable functions, corresponds to the initial measure μ and not to the measure (4.3).

We add to inequality (4.10) the following condition

$$\|h(t)\|_{L_p} \leqslant H(|\xi|), \qquad (4.11)$$

where $H(u)$ $(u \geqslant 0)$ is a positive locally bounded function such that

$$\lim_{u \to \infty} u^{-1} H(u) = 0. \qquad (4.12)$$

Our main problem consists in finding relations between the functions $\varPhi(t, u)$, $\varPsi(u)$ and $H(u)$ which guarantee the lack of solutions $h(t) \in L_p$ for sufficiently large values of $|\xi|$, or, equivalently, which provide general a priori estimates for the norm $|\xi| + \|h(t)\|_{L_p}$ of all solutions of inequality (4.10) that satisfy (4.11). Essentially we are interested only in the existence of general a priori estimates for the scalar component ξ of a solution $\{\xi, h(t)\}$.

In all the theorems in this section we assume that the function $\varphi(t, x)$ belongs to one of the classes $\mathfrak{V}(\varPhi, \varPsi)$, i.e., it satisfies the estimates (4.8) and (4.9).

4.3. BOUNDED SOLUTIONS. In this subsection we consider the case $p = \infty$, i.e., the second component in a solution $\{\xi, h(t)\}$ of inequality (4.10) is bounded.

THEOREM 4.1. *Assume that $\chi(\delta; e, g) > 0$ for $\delta > 0$. Then, a sufficient condition for the existence of an a priori estimate*

$$|\xi|, \ \|h(t)\|_{L_\infty} \leqslant const < \infty \qquad (4.13)$$

of any solution $\{\xi, h(t)\}$ of inequality (4.10), with sign-compatible functions (4.1), that satisfies

$$\|h(t)\|_{L_\infty} \leqslant H(|\xi|), \qquad (4.14)$$

is given by the equality

$$\lim_{u \to \infty} \frac{\int\limits_{\Omega} |g(t)| \Phi[t, u_0 + 2u|e(t)|] \mathrm{d}\mu}{\chi \left[\dfrac{u_0 + H(u)}{u}; e, g \right] \Psi[u_0 + 2H(u)]} = \infty. \tag{4.15}$$

Condition (4.14) is merely (4.11) for $p = \infty$.

THEOREM 4.2. *Assume that $\chi(\delta; e, g) = 0$ for a small $\delta > 0$. Then any solution $\{\xi, h(t)\}$ of inequality (4.10) with sign-compatible functions (4.1), that satisfies (4.14), admits the a priori estimate (4.13).*

The condition "$\chi(\delta; e, g) = 0$ for a small $\delta > 0$" means that the estimate $|e(t)| \geqslant c > 0$ $(t \in \Omega)$ is true for a certain positive number c.

We now give some examples to illustrate the meaning of the quite inconvenient form of condition (4.15) in Theorem 4.1. Condition (4.15) is analogous to conditions (2.2) and (2.7) and the examples will be analogous to those considered in Subsection 2.3.

Let $\Omega = [0, 1]$ and assume that for any $t \in [0, 1]$ we have

$$c_1 t^\beta \leqslant |e(t)| \leqslant c_2 t^\beta, \quad c_3 t^\alpha \leqslant |g(t)| \leqslant c_4 t^\alpha, \quad \alpha, \beta > 0. \tag{4.16}$$

Then the function (4.5) admits the estimates

$$c_5 \delta^{\frac{\alpha+1}{\beta}} \leqslant \chi(\delta; e, g) \leqslant c_6 \delta^{\frac{\alpha+1}{\beta}}, \quad 0 \leqslant \delta \leqslant \delta_0. \tag{4.17}$$

Further, suppose that

$$\Phi(t, u) \equiv \Phi(u), \quad H(u), \Psi(u) = \text{const.} \tag{4.18}$$

Then, condition (4.15) can be rewritten as

$$\int\limits_0^\infty u^{\frac{\alpha+1}{\beta}-1} \Phi(u) \mathrm{d}u = \infty.$$

Assume that instead of (4.18) we have

$$H(u) = c_7 (1 + u)^{\gamma_1}, \quad \Psi(u) = c_8 (1 + u)^{\gamma_2} \quad (\gamma_1 < 1).$$

Then

$$\chi\left[\frac{u_0 + H(u)}{u}; e, g\right]\Psi[u_0 + 2H(u)] \approx u^{(\gamma_1 - 1)\frac{\alpha+1}{\beta} + \gamma_1\gamma_2}$$

and for $\gamma_1\gamma_2 \geqslant (1 - \gamma_1)\dfrac{\alpha + 1}{\beta}$ the denominator in (4.15) does not converge to zero as $u \to \infty$. In this case equality (4.15) fails for any permissible function $\Phi(t, u)$.

If some estimates $\chi_U(\delta; e, g)$ and $\chi_L(\delta; e, g)$ are known, then condition (4.15) is satisfied whenever

$$\lim_{u\to\infty} \frac{\displaystyle\int_0^\infty \Phi[u_0 + 2u\delta]\mathrm{d}\chi_L(\delta; e, g)}{\chi_U\left[\dfrac{u_0 + H(u)}{u}; e, g\right]\Psi[u_0 + 2H(u)]} = \infty.$$

Theorems 4.1 and 4.2 can be put together in a general statement. Specifically, a sufficient condition for the existence of the a priori estimate (4.13) of every solution $\{\xi, h(t)\}$ of inequality (4.10), with sign-compatible functions (4.1), that satisfies (4.14), is given by the equality

$$\lim_{u\to\infty} \frac{\chi\left[\dfrac{u_0 + H(u)}{u}; e, g\right]\Psi[u_0 + 2H(u)]}{\displaystyle\int_\Omega |g(t)|\Phi[t, u_0 + 2u|e(t)|]\mathrm{d}\mu} = 0.$$

4.4. ESTIMATES FOR SOLUTIONS IN THE SPACE L_p ($p < \infty$). In this subsection we point out conditions for the existence of the a priori estimate

$$|\xi|, \ \|h(t)\|_{L_p} \leqslant c = \mathrm{const} < \infty, \tag{4.19}$$

when $p < \infty$. Throughout this subsection we will assume that the function $g(t)$ is not only integrable but it also belongs to a space L_q where $q > 1$; q' will denote the conjugate of q defined by $q' = q(q - 1)^{-1}$.

We suppose that the function $\chi(\delta; e, g)$ is positive for $\delta > 0$. Then the function

$$z(\delta) = \delta[\chi(\delta; e, g)]^{q'/p}, \quad \delta \geqslant 0 \tag{4.20}$$

is strictly increasing, and $z(0) = 0$. Therefore, for all $z \geqslant 0$, we can define a nondecreasing continuous function $\theta(z)$ as the inverse of function (4.20). The function $\theta(z)$ is given by the equality

$$\theta(z) = \sup\{\delta : z(\delta) \leqslant z\}. \tag{4.21}$$

THEOREM 4.3. *Assume that* $\chi(\delta; e, g) > 0$ *for* $\delta > 0$. *Let* $1 \leqslant p < \infty$ *and* $q > 1$. *In addition, suppose that the function* $\Psi(u^{\gamma})$ *is concave, where*

$$\gamma = \max\left\{1, \frac{q'}{p}\right\}. \tag{4.22}$$

Then the equality

$$\lim_{u \to \infty} \frac{\displaystyle\int_{\Omega} |g(t)||\Phi[t, u_0 + 2u|e(t)|]\mathrm{d}\mu}{\chi\left\{\dfrac{u_0}{u} + \theta\left[\dfrac{H(u)}{u}\right]; e, g\right\}\Psi\left\{\dfrac{2H(u) + (\mu\Omega)^{\frac{1}{p}}u_0}{\left(\chi\left\{\dfrac{u_0}{u} + \theta\left[\dfrac{H(u)}{u}\right]; e, g\right\}\right)^{\frac{q'}{p}}}\right\}} = \infty \tag{4.23}$$

is a sufficient condition for the existence of the a priori estimate (4.19) of any solution $\{\xi, h(t)\}$ *of inequality (4.10) with sign-compatible functions (4.1), that satisfies (4.11).*

Condition (4.23) can be simplified in some specific cases. In particular, if $\Psi(u) \equiv \text{const}$, condition (4.23) becomes

$$\lim_{u \to \infty} \frac{\displaystyle\int_{\Omega} |g(t)||\Phi[t, u_0 + 2u|e(t)|]\mathrm{d}\mu}{\chi\left\{\dfrac{u_0}{u} + \theta\left[\dfrac{H(u)}{u}\right]; e, g\right\}} = \infty. \tag{4.24}$$

If, in addition, we know that

$$\lim_{u \to \infty} u\theta\left[\frac{H(u)}{u}\right] = \infty, \tag{4.25}$$

then (4.24) follows from the equality

$$\lim_{u \to \infty} \frac{\displaystyle\int_{\Omega} |g(t)||\Phi[t, u_0 + 2u|e(t)|]\mathrm{d}\mu}{\chi\left\{2\theta\left[\dfrac{H(u)}{u}\right]; e, g\right\}} = \infty.$$

If, on the contrary, we know that

$$u\theta\left[\frac{H(u)}{u}\right] \leqslant \text{const} < \infty, \tag{4.26}$$

then (4.24) is fulfilled whenever

$$\lim_{u \to \infty} \frac{\displaystyle\int_{\Omega} |g(t)|\Phi[t, u_0 + 2u|e(t)|]\mathrm{d}\mu}{\chi(Ru^{-1}; e, g)} = \infty,$$

for any sufficiently large $R > 0$. Equality (4.25) is always true if

$$\varlimsup_{u \to \infty} H(u) > 0.$$

In its turn, equality (4.26) occurs if $H(u) \to 0$ sufficiently fast as $u \to \infty$.

Expression $\chi[\theta(\delta); e, g]$ can be written in a different form. Since

$$\chi[\theta(\delta); e, g]^{\frac{q'}{p}} = \frac{1}{\theta(\delta)}\theta(\delta)\chi[\theta(\delta); e, g]^{\frac{q'}{p}} = \frac{z[\theta(\delta)]}{\theta(\delta)} = \frac{\delta}{\theta(\delta)},$$

then

$$\chi[\theta(\delta); e, g] = \left[\frac{\delta}{\theta(\delta)}\right]^{\frac{q'}{p}}.$$

For "power" functions $\chi(\delta; e, g)$ the estimates

$$a_1\chi[\theta(\delta); e, g] \leqslant \chi[2\theta(\delta); e, g] \leqslant a_2\chi[\theta(\delta); e, g]$$

are true. Therefore, if $\varlimsup_{u \to \infty} H(u) > 0$, then equality (4.24) follows from

$$\lim_{u \to \infty} \left\{u\theta\left[\frac{H(u)}{u}\right]\right\}^{\frac{p}{q'}} \int_{\Omega} |g(t)|\Phi[t, u_0 + 2u|e(t)|]\mathrm{d}\mu = \infty.$$

We introduce now an example to illustrate how condition (4.23) can be simplified when power estimates of the functions $e(t)$ and $g(t)$ are known.

Assume again that the estimates (4.16) are true and, consequently, inequalities (4.17) hold. Under these assumptions, we can prove that the function (4.21) admits the estimates

$$c_9 z^{\overline{\frac{p\beta}{p\beta + q'(1+\alpha)}}} \leqslant \theta(z) \leqslant c_{10} z^{\overline{\frac{p\beta}{p\beta + q'(1+\alpha)}}}, \quad z \geqslant 0.$$

Further, if we suppose that equality (4.18) is also true, then the basic condition (4.23) in Theorem 4.3 can be rewritten as

$$\lim_{u \to \infty} u^{-\frac{q'(1+\alpha)^2}{\beta[p\beta + q'(1+\alpha)]}} \int_{u_0}^{u} z^{\frac{1+\alpha}{\beta} - 1}\Phi(z)\mathrm{d}z = \infty. \tag{4.27}$$

Condition (4.27) is true, for instance, if

$$\lim_{u\to\infty} u^{\frac{p(1+\alpha)}{p\beta+q'(1+\alpha)}}\Phi(u) = \infty. \tag{4.28}$$

Condition (4.27) follows from (4.28) by the well-known l'Hôpital rule. We notice that, in this particular example, equality (4.25) is fulfilled.

4.5. A SECOND THEOREM ON ESTIMATES FOR SOLUTIONS IN THE SPACE L_p $(p < \infty)$. The theorem in this subsection refers to the case when $\chi(\delta; e, g) = 0$ for small values of $\delta > 0$, a case when Theorem 4.3 can not be applied.

THEOREM 4.4. *Assume that* $\chi(\delta; e, g) = 0$ *for* $0 \leqslant \delta \leqslant \delta_0$ $(\delta_0 > 0)$. *Let* $1 \leqslant p < \infty$ *and* $q > 1$. *In addition, suppose that the function* $\Psi(u^\gamma)$ *is concave, where* γ *is the same number as in (4.22). Then the equality*

$$\lim_{u\to\infty} \frac{u^{\frac{p}{q'}}\displaystyle\int_{\Omega} |g(t)|\Phi[t, u_0 + 2u|e(t)|]\mathrm{d}\mu}{[H(u)]^{\frac{p}{q'}}\Psi\left\{R_1 u\dfrac{1 + H(u)}{H(u)}\right\}} = \infty, \tag{4.29}$$

satisfied for all $R_1 > 0$, *is a sufficient condition for the existence of the a priori estimate (4.19) of any solution* $\{\xi, h(t)\}$ *of inequality (4.10) with sign-compatible functions (4.1) that satisfies (4.11).*

If condition (4.29) is fulfilled, the equality

$$\lim_{u\to\infty} \frac{u^{\frac{p}{q'}}\displaystyle\int_{\Omega} |g(t)|\Phi[t, u_0 + Ru]\mathrm{d}\mu}{[H(u)]^{\frac{p}{q'}}\Psi\left\{R_1 u\dfrac{1 + H(u)}{H(u)}\right\}} = \infty$$

is necessarily true for small values of $R > 0$.

Condition (4.29) has a very simple equivalent form in case $\Psi(u) \equiv \mathrm{const}$ and $\Phi(t, u) \equiv \Phi(u)$ $(u \geqslant u_0)$, namely,

$$\lim_{u\to\infty}\left[\frac{u}{H(u)}\right]^{\frac{p}{q'}}\Phi(Ru) = \infty.$$

Condition (4.29) can be simplified in other cases, too. Suppose again that $\Phi(t, u) \equiv \Phi(u)$ $(u \geqslant u_0)$. Thus, if

$$\lim_{u\to\infty} H(u) = 0,$$

then (4.29) can be replaced by the equality

$$\lim_{u \to \infty} \frac{u^{\frac{p}{q'}} \Phi(Ru)}{[H(u)]^{\frac{p}{q'}} \Psi \left[R_1 \dfrac{u}{H(u)} \right]} = \infty,$$

and if

$$\varliminf_{u \to \infty} H(u) > 0,$$

then (4.29) is equivalent to

$$\lim_{u \to \infty} \frac{u^{\frac{p}{q'}} \Phi(Ru)}{[H(u)]^{\frac{p}{q'}} \Psi(R_1 u)} = \infty.$$

REMARK. In 4.3 above we introduced an example of functions $\Psi(u)$ and $H(u)$ with enough large values, so that condition (4.15) in Theorem 4.1 fails for any bounded function $\Phi(t, u)$. Analogous examples can be given relatively to conditions (4.23) and (4.29) in Theorem 4.3 and 4.4, respectively. In all these situations we can obtain similar theorems on a priori estimates, using the classes $\mathfrak{V}(\Phi, \Psi)$ with functions $\Phi(t, u)$ that are increasing in u for large values of u.

§5. Proofs of Theorems 4.1–4.4

5.1. THE SET $G(\xi)$. To prove Theorems 4.1–4.4 we will use the set

$$G(\xi) = \{t : t \in \Omega, \ |\xi e(t)| \leqslant u_0 + |h(t)|\}, \quad \xi \in \mathbb{R}^1. \tag{5.1}$$

We will establish below estimates for the measure (4.3) of this set. We will distinguish between the cases $h(t) \in L_\infty$ and $h(t) \in L_p$ $(1 \leqslant p < \infty)$. For all $1 \leqslant p \leqslant \infty$ the norm in L_p is denoted by $\| \cdot \|_p$. Without explicitly mentioning them, we will use the definitions and notations introduced in the preceding section.

LEMMA 5.1. *Assume that $h(t) \in L_\infty$. Then the inequality*

$$\operatorname{Mes} G(\xi) \leqslant \chi \left[\frac{u_0 + \|h(t)\|_\infty}{|\xi|}; e, g \right] \tag{5.2}$$

is true, for all $\xi \neq 0$.

Proof. We split the set $G(\xi)$ into two subsets:

$$G_1(\xi) = \{t : t \in G(\xi), \ |h(t)| \leqslant \|h(t)\|_\infty\}, \quad G_2(\xi) = G(\xi) \setminus G_1(\xi).$$

Since $\mu G_2(\xi) = 0$ and, consequently, Mes $G_2(\xi) = 0$, the inclusion

$$G_1(\xi) \subset \{t : t \in \Omega, \ |\xi| \cdot |e(t)| \leqslant u_0 + \|h(t)\|_\infty\}$$

implies the inequality

$$\text{Mes}\, G(\xi) \leqslant \text{Mes}\{t : t \in \Omega, \ |\xi| \cdot |e(t)| \leqslant u_0 + \|h(t)\|_\infty\},$$

which obviously concludes the proof of Lemma 5.1.

LEMMA 5.2. *Assume that $h(t) \in L_p$ $(1 \leqslant p < \infty)$ and $g(t) \in L_q$ $(q > 1)$. Then the inequality*

$$\text{Mes}\, G(\xi) \leqslant \chi \left[\frac{u_0 + \rho}{|\xi|}; e, g \right] + \|g(t)\|_q \left[\rho^{-1} \|h(t)\|_p \right]^{\frac{p}{q'}} \tag{5.3}$$

is true, for all $\xi \neq 0$ and $\rho > 0$.

Proof. We split the set $G(\xi)$ into two subsets:

$$G_1(\xi) = \{t : t \in G(\xi), \ |h(t)| \leqslant \rho\}, \quad G_2(\xi) = \{t : t \in G(\xi), \ |h(t)| > \rho\}.$$

The inclusions

$$G_1(\xi) \subset \{t : t \in \Omega, \ |\xi| \cdot |e(t)| \leqslant u_0 + \rho\}$$

and

$$G_2(\xi) \subset \{t : t \in \Omega, \ |h(t)| > \rho\}$$

imply the inequality

$$\text{Mes}\, G(\xi) \leqslant \chi \left[\frac{u_0 + \rho}{|\xi|}; e, g \right] + \text{Mes}\{t : t \in \Omega, \ |h(t)| > \rho\}.$$

Thus, in order to prove Lemma 5.2, it remains to show that

$$\text{Mes}\{t : t \in \Omega, \ |h(t)| > \rho\} \leqslant \|g(t)\|_q \left[\rho^{-1} \|h(t)\|_p \right]^{\frac{p}{q'}}. \tag{5.4}$$

Since $\rho > 0$, then $[\rho^{-1}|h(t)|]^{p/q'} > 1$ for all $t \in \{t : t \in \Omega, \ |h(t)| > \rho\}$. Therefore

$$\text{Mes}\{t : t \in \Omega, \ |h(t)| > \rho\} = \int\limits_{\{t:|h(t)|>\rho\}} |g(t)| d\mu \leqslant$$

$$\leqslant \rho^{-\frac{p}{q'}} \int\limits_{\{t:|h(t)|>\rho\}} |g(t)| \cdot |h(t)|^{\frac{p}{q'}} d\mu \leqslant \rho^{-\frac{p}{q'}} \int\limits_{\Omega} |g(t)| \cdot |h(t)|^{\frac{p}{q'}} d\mu.$$

Consequently, by Hölder inequality, we get

$$\text{Mes}\{t : t \in \Omega,\ |h(t)| > \rho\} \leqslant \rho^{-\frac{p}{q'}} \|g(t)\|_p \left\| h(t)^{\frac{p}{q'}} \right\|_{q'}.$$

In view of the identity

$$\left\| h(t)^{\frac{p}{q'}} \right\|_{q'} \equiv \|h(t)\|_{q'}^{\frac{p}{q'}}, \quad h(t) \in L_p,$$

the last inequality coincides with (5.4).

We notice that for $g(t) \equiv 1$, $q = \infty$ and $p = 1$, inequality (5.4) coincides with Chebyshev inequality.

5.2. A BASIC LEMMA. In this subsection we will establish a lower estimate for the left hand side of inequality (4.10) with sign-compatible functions (4.1). Assume that the function $\varphi(t, x)$ lies in one of the classes $\mathfrak{V}(\Phi, \Psi)$ and denote

$$M = \sup_{t \in \Omega,\ u \geqslant 0} |\Phi(t, u)|. \tag{5.5}$$

LEMMA 5.3. *If $\xi \neq 0$, then*

$$J \overset{\text{def}}{=} \int_{\Omega} g(t)\varphi[t, \xi e(t) + h(t)]\mathrm{d}\mu \cdot \text{sign}\,\xi \geqslant \int_{\Omega} |g(t)|\Phi[t, u_0 + 2|\xi e(t)|]\mathrm{d}\mu - $$
$$-2M \cdot \text{Mes}\,G(\xi) - \int_{G(\xi)} |g(t)|\Psi[u_0 + |h(t)|]\mathrm{d}\mu. \tag{5.6}$$

Proof. We will prove the lemma in case $\xi > 0$. For $\xi < 0$ the proof of inequality (5.6) is analogous. We split the set Ω into the next six disjoint subsets:

$$\Omega_1 = \{t : \xi e(t) + h(t) \geqslant u_0,\ e(t) > 0,\ g(t) > 0\},$$
$$\Omega_2 = \{t : \xi e(t) + h(t) \leqslant -u_0,\ e(t) < 0,\ g(t) < 0\},$$
$$\Omega_3 = \{t : e(t) \cdot g(t) \leqslant 0\},$$
$$\Omega_4 = \{t : \xi e(t) + h(t) \geqslant u_0,\ e(t) < 0,\ g(t) < 0\},$$
$$\Omega_5 = \{t : \xi e(t) + h(t) \leqslant -u_0,\ e(t) > 0,\ g(t) > 0\},$$
$$\Omega_6 = \{t : \xi e(t) + h(t) < u_0,\ e(t) \cdot g(t) > 0\}.$$

Further, we estimate the quantities

$$J_i = \int_{\Omega_i} |g(t)|\varphi[t, |\xi e(t) + h(t)|]\mathrm{d}\mu, \tag{5.7}$$

for each $i = 1, \dots, 6$.

For $i = 1, 2$, from the relations

$$|\xi e(t) + h(t)| \geqslant u_0 > 0, \quad \operatorname{sign}[g(t)] = \operatorname{sign}[\xi e(t) + g(t)] \quad (t \in \Omega_1 \cup \Omega_2)$$

and by (4.8), we get the inequalities

$$J_i \geqslant \int\limits_{\Omega_i} |g(t)| \Phi[t, |\xi e(t) + h(t)|] \mathrm{d}\mu, \quad i = 1, 2.$$

From these inequalities (since the function $\Phi(t, u)$ is nonincreasing in u, for $u \geqslant u_0$) we obtain the estimates

$$J_i \geqslant \int\limits_{\Omega_i} |g(t)| \Phi[t, u_0 + \xi|e(t)| + |h(t)|] \mathrm{d}\mu, \quad i = 1, 2. \tag{5.8}$$

For $i = 3$, we have $\mu\Omega_3 = 0$ (since the functions (4.1) are sign-compatible), hence

$$J_3 = 0. \tag{5.9}$$

For $i = 4, 5, 6$ we notice that $\Omega_i \subset G(\xi)$, therefore

$$J_4 + J_5 + J_6 \geqslant - \int\limits_{G(\xi)} |g(t)| \cdot |\varphi[t, \xi e(t) + h(t)]| \mathrm{d}\mu. \tag{5.10}$$

From (5.8), (5.9) and (5.10), and by the obvious equality

$$J = \sum_{i=1}^{6} J_i,$$

we get the estimate

$$J \geqslant \int\limits_{\Omega_1 \cup \Omega_2} |g(t)| \Phi[t, u_0 + \xi|e(t)| + |h(t)|] \mathrm{d}\mu - \int\limits_{G(\xi)} |g(t)| \cdot |\varphi[t, \xi e(t) + h(t)]| \mathrm{d}\mu$$

and, in view of (4.9) and the definition of the set (5.1), we conclude that

$$J \geqslant \int\limits_{\Omega_1 \cup \Omega_2} |g(t)| \Phi[t, u_0 + \xi|e(t)| + |h(t)|] \mathrm{d}\mu - \int\limits_{G(\xi)} |g(t)| \Psi[u_0 \dot{+} 2|h(t)|] \mathrm{d}\mu. \tag{5.11}$$

Since

$$\Omega \setminus (\Omega_1 \cup \Omega_2) \subset \Omega_3 \cup \Omega_4 \cup \Omega_5 \cup \Omega_6 \subset \Omega_3 \cup G(\xi)$$

and because $\mu\Omega_3 = 0$, it follows that

$$\int_{\Omega_1 \cup \Omega_2} |g(t)|\Phi[t, u_0 + \xi|e(t)| + |h(t)|]\mathrm{d}\mu = \int_{\Omega} |g(t)|\Phi[t, u_0 + \xi|e(t)| + |h(t)|]\mathrm{d}\mu -$$

$$- \int_{\Omega \setminus (\Omega_1 \cup \Omega_2)} |g(t)|\Phi[t, u_0 + \xi|e(t)| + |h(t)|]\mathrm{d}\mu \geqslant$$

$$\geqslant \int_{\Omega} |g(t)|\Phi[t, u_0 + \xi|e(t)| + |h(t)|]\mathrm{d}\mu - \int_{G(\xi)} |g(t)|\Phi[t, u_0 + \xi|e(t)| + |h(t)|]\mathrm{d}\mu,$$

and by (5.5) we obtain

$$\int_{\Omega_1 \cup \Omega_2} |g(t)|\Phi[t, u_0 + \xi|e(t)| + |h(t)|]\mathrm{d}\mu \geqslant$$

$$\geqslant \int_{\Omega} |g(t)|\Phi[t, u_0 + \xi|e(t)| + |h(t)|]\mathrm{d}\mu - M \cdot \mathrm{Mes}\, G(\xi).$$

This last inequality together with (5.11) yields the estimate

$$J \geqslant \int_{\Omega} |g(t)|\Phi[t, u_0 + \xi|e(t)| + |h(t)|]\mathrm{d}\mu -$$

$$- M \cdot \mathrm{Mes}\, G(\xi) - \int_{G(\xi)} |g(t)|\Psi[u_0 + 2|h(t)|]\mathrm{d}\mu. \tag{5.12}$$

We next split Ω into two subsets:

$$\Omega_7 = \{t : t \in \Omega,\ \xi|e(t)| \leqslant |h(t)|\}, \quad \Omega_8 = \{t : t \in \Omega,\ \xi|e(t)| > |h(t)|\}.$$

Since

$$\int_{\Omega_8} |g(t)|\Phi[t, u_0 + \xi|e(t)| + |h(t)|]\mathrm{d}\mu \geqslant \int_{\Omega_8} |g(t)|\Phi[t, u_0 + 2\xi|e(t)|]\mathrm{d}\mu$$

we obviously get that

$$\int_{\Omega} |g(t)|\Phi[t, u_0 + \xi|e(t)| + |h(t)|]\mathrm{d}\mu \geqslant \int_{\Omega_8} |g(t)|\Phi[t, u_0 + 2\xi|e(t)|]\mathrm{d}\mu.$$

Because $\Omega_7 \subset G(\xi)$ we obtain the estimate

$$\int_\Omega |g(t)|\Phi[t, u_0 + \xi|e(t)| + |h(t)|]d\mu \geqslant$$

$$\geqslant \int_\Omega |g(t)|\Phi[t, u_0 + 2\xi|e(t)|]d\mu - \int_{G(\xi)} |g(t)|\Phi[t, u_0 + 2\xi|e(t)|]d\mu.$$

Inequality (5.6) follows now from this last relation and (5.12). Lemma 5.3 is completely proved.

5.3. PROOF OF THEOREM 4.1. Let $\{\xi, h(t)\}$ be a solution for inequality (4.10) with sign-compatible functions (4.1), subject to condition (4.14). Without any loss of generality we can assume that $\xi \neq 0$.

By Lemma 5.3, for the solution $\{\xi, h(t)\}$ we have the following estimate

$$\int_\Omega |g(t)|\Phi[t, u_0 + 2\xi|e(t)|]d\mu \leqslant 2M \cdot \operatorname{Mes} G(\xi) + \int_{G(\xi)} |g(t)|\Psi[u_0 + 2|h(t)|]d\mu.$$

Hence (since the function $\Psi(u)$ is nondecreasing) we get

$$\int_\Omega |g(t)|\Phi[t, u_0 + 2\xi|e(t)|]d\mu \leqslant \operatorname{Mes} G(\xi)\{2M + \Psi[u_0 + 2\|h(t)\|_\infty]\},$$

which, based on (4.14), leads to the inequality

$$\int_\Omega |g(t)|\Phi[t, u_0 + 2\xi|e(t)|]d\mu \leqslant \operatorname{Mes} G(\xi)\{2M + \Psi[u_0 + 2H(|\xi|)]\}. \tag{5.13}$$

We use now Lemma 5.1. By (5.2) and (5.13) the following relation

$$\int_\Omega |g(t)|\Phi[t, u_0 + 2|\xi| \cdot |e(t)|]d\mu \leqslant$$

$$\leqslant \{2M + \Psi[u_0 + 2H(|\xi|)]\} \cdot \chi\left[\frac{u_0 + H(|\xi|)}{|\xi|}; e, g\right] \tag{5.14}$$

holds true. The assumption on the positiveness of the function $\chi(\delta; e, g)$ for $\delta > 0$, together with (5.14), provides the estimate

$$\frac{\displaystyle\int_\Omega |g(t)|\Phi[t, u_0 + 2|\xi| \cdot |e(t)|]d\mu}{\chi\left[\dfrac{u_0 + H(|\xi|)}{|\xi|}; e, g\right] \cdot \Psi[u_0 + 2H(|\xi|)]} = 1 + \frac{2M}{\Psi(0)} < \infty. \tag{5.15}$$

According to assumption (4.15) in Theorem 4.1 we get the existence of a number $\xi_0 < \infty$, such that inequality (5.15) implies the estimate $|\xi| < \xi_0$. Therefore,

$$|\xi|, \ \|h(t)\|_\infty \leqslant \max \left\{ \xi_0, \ \sup_{0 \leqslant u \leqslant \xi_0} H(u) \right\}.$$

The proof of Theorem 4.1 is complete.

5.4. THE PROOF OF THEOREM 4.2. Assume again that $\{\xi, h(t)\}$ is a solution for inequality (4.10) with sign-compatible functions (4.1), subject to condition (4.14), where $\xi \neq 0$. Based on Lemmas 5.1 and 5.3, under the assumptions in Theorem 4.2 (see the previous section), we easily observe that relation (5.14) is satisfied. But, under the very assumptions in Theorem 4.2 (by condition (4.12)) there exists $\xi_0 < \infty$ such that, whenever $|\xi| > \xi_0$, we have

$$\chi \left[\frac{u_0 + H(|\xi|)}{|\xi|} ; e, g \right] = 0.$$

Therefore, if $|\xi| > \xi_0$, then from (5.14) we get the inequality

$$\int_\Omega |g(t)| \Phi[t, u_0 + 2|\xi| \cdot |e(t)|] \mathrm{d}\mu \leqslant 0,$$

which contradicts the condition $\Phi(t, u) \in \mathfrak{N}(u_0)$ (see §2). Consequently, $|\xi| \leqslant \xi_0$. By (4.14) we obtain

$$\|h(t)\|_\infty \leqslant \sup_{0 \leqslant u \leqslant \xi_0} H(u).$$

Theorem 4.2 is now proved.

5.5. A SPECIAL LEMMA. In this subsection we will state and prove a special result that will be used to prove Theorems 4.3 and 4.4.

LEMMA 5.4. *Let $q > 1$ and $p < \infty$. Assume that the function $\Psi(u)$ $(u \geqslant 0)$ is positive, nondecreasing, and such that $\Psi(u^\gamma)$ is concave, where γ is the number defined by (4.22). Then we have*

$$\int_G |g(t)| \Psi[|h_1(t)|] \mathrm{d}\mu \leqslant \mathrm{Mes}\, G \cdot \Psi \left[\frac{\|h_1(t)\|_p \|g(t)\|_p^{\frac{q'}{p}}}{(\mathrm{Mes}\, G)^{\frac{q'}{p}}} \right], \tag{5.16}$$

for any functions $h_1(t) \in L_p$ and $g(t) \in L_q$ and for every measurable set $G \subset \Omega$ $(\mu G \neq 0)$.

Proof. We first prove our lemma in case $\gamma = 1$, that is, when $p \geqslant q'$ and the function $\Psi(u)$ is concave. To start the proof, we will use Jensen integral inequality (see [Zygmund, 1935])

$$\int_G \Psi[|h_1(t)|]\mathrm{dMes} \leqslant \mathrm{Mes}\,G \cdot \Psi\left[\frac{\displaystyle\int_G |h_1(t)|\mathrm{dMes}}{\mathrm{Mes}\,G}\right],$$

which in our case has the form

$$\int_G |g(t)|\Psi[|h_1(t)|]\mathrm{d}\mu \leqslant \mathrm{Mes}\,G \cdot \Psi\left[\frac{\displaystyle\int_G |g(t)| \cdot |h_1(t)|\mathrm{d}\mu}{\mathrm{Mes}\,G}\right]. \qquad (5.17)$$

Next we estimate the quantity $\int_G |g(t)| \cdot |h_1(t)|\mathrm{d}\mu$. From Hölder inequality we get the relations

$$\int_G 1 \cdot |h_1(t)|\mathrm{dMes} \leqslant \left\{\int_G [h_1(t)]^{\frac{p}{q'}}\mathrm{dMes}\right\}^{\frac{q'}{p}} \cdot (\mathrm{Mes}\,G)^{1-\frac{q'}{p}}$$

and

$$\int_G [h_1(t)]^{\frac{p}{q'}}|g(t)|\mathrm{d}\mu \leqslant \|g(t)\|_q \left\{\int_G [h_1(t)]^p\mathrm{d}\mu\right\}^{\frac{1}{q'}};$$

therefore,

$$\int_G |g(t)| \cdot |h_1(t)|\mathrm{d}\mu \leqslant (\mathrm{Mes}\,G)^{1-\frac{q'}{p}}\|g(t)\|_q^{\frac{q'}{p}}\|h_1(t)\|_p.$$

Inequality (5.16) follows from the last relation and (5.17).

Thus, Lemma 5.4 is completely proved in case $\gamma = 1$. Assume now that $\gamma > 1$, that is, $p > q'$ and $\gamma = \dfrac{q'}{p}$. In this case we can rewrite the estimate (5.16) as

$$\int_G |g(t)|\Psi_\gamma\left[\left|h_1^{\frac{1}{\gamma}}(t)\right|\right]\mathrm{d}\mu \leqslant \mathrm{Mes}\,G \cdot \Psi_\gamma\left[\frac{\left\|h_1^{\frac{1}{\gamma}}(t)\right\|_{p\gamma}\|g(t)\|_p^{\frac{q'}{p\gamma}}}{(\mathrm{Mes}\,G)^{\frac{q'}{p\gamma}}}\right], \qquad (5.18)$$

where $\Psi_\gamma(u)$ is the concave function given by $\Psi_\gamma(u) = \Psi(u^\gamma)$. In view of the equality $p\gamma = q'$, (5.18) can be rewritten in the form

$$\int_G |g(t)| \Psi_\gamma \left[\left| h_1^{\frac{1}{\gamma}}(t) \right| \right] d\mu \leqslant \operatorname{Mes} G \cdot \Psi_\gamma \left[\frac{\left\| h_1^{\frac{1}{\gamma}}(t) \right\|_{q'} \|g(t)\|_q}{\operatorname{Mes} G} \right]. \qquad (5.19)$$

The etimate (5.19) follows from Jensen inequality applied to the function $\Psi_\gamma(u)$, namely

$$\int_G \Psi_\gamma \left[\left| h_1^{\frac{1}{\gamma}}(t) \right| \right] d\operatorname{Mes} \leqslant \operatorname{Mes} G \cdot \Psi_\gamma \left[\frac{\int_G \left| h_1^{\frac{1}{\gamma}}(t) \right| d\operatorname{Mes}}{\operatorname{Mes} G} \right]$$

and from Hölder inequality

$$\int_G |g(t)| \Psi_\gamma \left[\left| h_1^{\frac{1}{\gamma}}(t) \right| \right] d\mu \leqslant \left\| h_1^{\frac{1}{\gamma}}(t) \right\|_{q'} \|g(t)\|_q.$$

The proof of Lemma 5.4 is complete.

We include an additional remark. Under the assumptions in Lemma 5.4 above, the function

$$d(u) = u\Psi(cu^{-\gamma}), \quad u > 0, \qquad (5.20)$$

is nondecreasing in u, for any $c > 0$. This fact follows from the general observation that the function $u\Phi_0(u^{-1})$ is nondecreasing whenever the function $\Phi_0(u)$ is concave, positive and nondecreasing for $u \geqslant 0$.

5.6. PROOF OF THEOREM 4.3. We will first give a more accurate estimate for $\operatorname{Mes} G(\xi)$ than that in (5.3). To this end, let

$$\rho = |\xi| \theta \left[\frac{H(|\xi|)}{|\xi|} \right],$$

where $\theta(z)$ is the function defined by (4.21). Then

$$\frac{\rho}{|\xi|} = \theta \left[\frac{H(|\xi|)}{|\xi|} \right]$$

and, since $\theta(z)$ was defined as the inverse of the function (4.20), it follows that

$$\frac{H(|\xi|)}{|\xi|} \leqslant \frac{\rho}{|\xi|} \left[\chi \left(\frac{\rho}{|\xi|}; e, g \right) \right]^{\frac{q'}{p}}.$$

Hence, we get the inequality

$$\left[\frac{H(|\xi|)}{|\xi|} \right]^{\frac{q'}{p}} \leqslant \chi \left(\frac{\rho}{|\xi|}; e, g \right).$$

Therefore, if ξ and $h(t)$ satisfy (4.11), due to Lemma 5.2 we obtain the estimate

$$\text{Mes}\, G(\xi) \leqslant (1 + \|g(t)\|_q) \chi \left(\frac{\rho + u_0}{|\xi|}; e, g \right),$$

that is,

$$\text{Mes}\, G(\xi) \leqslant (1 + \|g(t)\|_q) \chi \left\{ \frac{u_0}{|\xi|} + \theta \left[\frac{H(|\xi|)}{|\xi|} \right]; e, g \right\}. \tag{5.21}$$

We now return to the proof of Theorem 4.3. Let $\{\xi, h(t)\}$ $(\xi \neq 0)$ be a solution for inequality (4.10), with sign-compatible functions (4.1), subject to condition (4.11)

By Lemma 5.3, for $\{\xi, h(t)\}$ we have the following inequality

$$\int\limits_{\Omega} |g(t)| \Phi[t, u_0 + 2|\xi e(t)|] \mathrm{d}\mu \leqslant 2\text{Mes}\, G(\xi) + \int\limits_{G(\xi)} |g(t)| \Psi[u_0 + 2|h(t)|] \mathrm{d}\mu. \tag{5.22}$$

Therefore, $\text{Mes}\, G(\xi) > 0$, and from (5.22) and based on Lemma 5.4, we find the estimate

$$\int\limits_{\Omega} |g(t)| \Phi[t, u_0 + 2|\xi e(t)|] \mathrm{d}\mu \leqslant \text{Mes}\, G(\xi) \left\{ 2M + \Psi \left[\frac{\| u_0 + 2|h(t)| \|_p \|g(t)\|_q^{\frac{q'}{p}}}{[\text{Mes}\, G(\xi)]^{\frac{q'}{p}}} \right] \right\}.$$

But (see the remark at the end of the previous section) the right hand side in the last inequality is nondecreasing as a function of $\text{Mes}\, G(\xi)$. Consequently, inequality (5.21) implies the relation

$$\int\limits_{\Omega} |g(t)| \Phi[t, u_0 + 2|\xi e(t)|] \mathrm{d}\mu \leqslant (1 + \|g(t)\|_q) \chi \left\{ \frac{u_0}{|\xi|} + \theta \left[\frac{H(|\xi|)}{|\xi|} \right]; e, g \right\} \times$$

$$\times \left[2M + \Psi \left\{ \frac{u_0 \cdot (\mu\Omega)^{\frac{1}{p}} + 2\|h(t)\|_p}{\left(\chi \left\{ \frac{u_0}{|\xi|} + \theta \left[\frac{H(|\xi|)}{|\xi|} \right]; e, g \right\} \right)^{\frac{q'}{p}}} \cdot \left[\frac{\|g(t)\|_q}{1 + \|g(t)\|_q} \right]^{\frac{q'}{p}} \right\} \right].$$

This relation and (4.11) (since $\Psi(u)$ is nondecreasing) lead to the inequality

$$\int_\Omega |g(t)| \Phi[t, u_0 + 2|\xi e(t)|] \mathrm{d}\mu \leqslant (1 + \|g(t)\|_q) \chi \left\{ \frac{u_0}{|\xi|} + \theta \left[\frac{H(|\xi|)}{|\xi|} \right] ; e, g \right\} \times$$

$$\times \left[2M + \Psi \left\{ \frac{u_0 \cdot (\mu\Omega)^{\frac{1}{p}} + 2\|h(t)\|_p}{\left(\chi \left\{ \frac{u_0}{|\xi|} + \theta \left[\frac{H(|\xi|)}{|\xi|} \right] ; e, g \right\} \right)^{\frac{q'}{p}}} \right\} \right],$$

and thus, to the estimate

$$\frac{\displaystyle\int_\Omega |g(t)| \Phi[t, u_0 + 2|\xi| \cdot |e(t)|] \mathrm{d}\mu}{\chi \left\{ \dfrac{u_0}{|\xi|} + \theta \left[\dfrac{H(|\xi|)}{|\xi|} \right] ; e, g \right\} \Psi \left\{ \dfrac{u_0 \cdot (\mu\Omega)^{\frac{1}{p}} + 2\|h(t)\|_p}{\left(\chi \left\{ \dfrac{u_0}{|\xi|} + \theta \left[\dfrac{H(|\xi|)}{|\xi|} \right] ; e, g \right\} \right)^{\frac{q'}{p}}} \right\}} \leqslant$$

$$\leqslant (1 + \|g(t)\|_q) \left[1 + \frac{2M}{\Psi(0)} \right]. \tag{5.23}$$

By the assumption (4.23) in Theorem 4.3 it follows that there exists $\xi_0 < \infty$ such that, from inequality (5.23), we get the estimate $|\xi| \leqslant \xi_0$. Thus, since

$$\|h(t)\|_p \leqslant \sup_{0 \leqslant u \leqslant \xi_0} H(u),$$

the estimate (4.19) follows.

Theorem 4.3 is proved.

5.7. PROOF OF THEOREM 4.4. As in the proof of Theorem 4.3, we start by estimating appropriately $\mathrm{Mes}\, G(\xi)$. Accordingly, let

$$\rho = \frac{1}{2} \delta_0 |\xi|,$$

where $\delta_0 > 0$ is such that $\chi(\delta_0; e, g) = 0$. Then, for $|\xi| > 2u_0 \delta_0^{-1}$, we have

$$\chi \left(\frac{u_0 + \rho}{|\xi|}; e, g \right) = 0.$$

Therefore, if ξ and $h(t)$ satisfy (4.11), then Lemma 5.2 yields the inequality

$$\text{Mes}\, G(\xi) \leqslant 2\delta_0^{-1} \|g(t)\|_q \left[\frac{H(|\xi|)}{|\xi|}\right]^{\frac{p}{q'}}, \quad |\xi| > 2u_0\delta_0^{-1}. \tag{5.24}$$

Let $\{\xi, h(t)\}$, $|\xi| > 2u_0\delta_0^{-1}$, be a solution for inequality (4.10) with sign-compatible functions (4.1), satisfying (4.11).

By Lemma 5.3, for $\{\xi, h(t)\}$ we obtain the next estimate

$$\int\limits_{\Omega} |g(t)| \Phi[t, u_0 + 1|\xi e(t)|] \mathrm{d}\mu \leqslant \text{Mes}\, G(\xi) + \int\limits_{G(\xi)} |g(t)| \Phi[u_0 + 2|h(t)|] \mathrm{d}\mu.$$

Hence, $\text{Mes}\, G(\xi) \neq 0$ and therefore, Lemma 5.4 leads to the inequality

$$\int\limits_{\Omega} |g(t)| \Phi[t, u_0 + 2|\xi| \cdot |e(t)|] \mathrm{d}\mu \leqslant \text{Mes}\, G(\xi) + \left(2M + \Psi\left\{\frac{\|u_0 + 2h(t)\|_p \|g(t)\|_q^{\frac{q'}{p}}}{[\text{Mes}\, G(\xi)]^{\frac{q'}{p}}}\right\}\right).$$

This inequality and (5.24) imply

$$\int\limits_{\Omega} |g(t)| \Phi[t, u_0 + 2|\xi| \cdot |e(t)|] \mathrm{d}\mu \leqslant$$

$$\leqslant 2\delta_0^{-1} \|g(t)\|_q \left[\frac{H(|\xi|)}{|\xi|}\right]^{\frac{p}{q'}} \left[2M + \Psi\left\{|\xi| \cdot \frac{u_0(\mu\Omega)^{\frac{1}{p}} + 2H(|\xi|)}{2^{\frac{q'}{p}}\delta_0^{-\frac{q'}{p}} H(|\xi|)}\right\}\right].$$

The last relation can be rewritten as

$$\int\limits_{\Omega} |g(t)| \Phi[t, u_0 + 2|\xi| \cdot |e(t)|] \mathrm{d}\mu \leqslant$$

$$\leqslant 2\delta_0^{-1} \|g(t)\|_q \left[\frac{H(|\xi|)}{|\xi|}\right]^{\frac{p}{q'}} \left[2M + \Psi\left\{\left(\frac{\delta_0}{2}\right)^{\frac{q'}{p}} \cdot |\xi| \cdot \frac{u_0(\mu\Omega)^{\frac{1}{p}}}{H(|\xi|)}\right\}\right]. \tag{5.25}$$

Let us define R_1 by

$$R_1^p = \delta_0^q \cdot 2^{-q} \cdot \max\left\{2^p, u_0^p(\mu\Omega)\right\}.$$

Then, the following inequality

$$\left(\frac{\delta_0}{2}\right)^{\frac{q'}{p}} \cdot |\xi| \cdot \frac{u_0(\mu\Omega)^{\frac{1}{p}}}{H(|\xi|)} \leqslant R_1 \frac{1 + H(u)}{H(u)}$$

holds, and from (5.25) we get the estimate

$$\int_\Omega |g(t)||\Phi[t, u_0 + 2|\xi e(t)|] \mathrm{d}\mu \leqslant$$

$$\leqslant 2\delta_0^{-1} \|g(t)\|_q \left[\frac{H(|\xi|)}{|\xi|}\right]^{\frac{p}{q'}} \cdot \left[2M + \Psi\left\{\frac{1 + H(|\xi|)}{H(|\xi|)} R_1 |\xi|\right\}\right],$$

which, in its turn, leads to the relation

$$\frac{|\xi|^{\frac{p}{q'}} \int_\Omega |g(t)||\Phi[t, u_0 + 2|\xi e(t)|] \mathrm{d}\mu}{[H(|\xi|)]^{\frac{p}{q'}} \Psi\left[R_1|\xi| \dfrac{1 + H(|\xi|)}{H(|\xi|)}\right]} \leqslant 2\delta_0^{-1} \|g(t)\|_q \left[1 + \frac{2M}{\Psi(0)}\right]. \qquad (5.26)$$

By assumption (4.29) in Theorem 4.4, it follows that there exists $\xi_0 > 0$, such that from (5.23) we get the estimate $|\xi| \leqslant \xi_0$. The proof of Theorem 4.4 is complete.

§6. Additional remarks

6.1. SIGN-COMPATIBLE FAMILIES OF FUNCTIONS. We consider two compact (with respect to a certain norm) families $\mathfrak{F}_0$ and $\mathfrak{F}_1$ of integrable functions. The family $\mathfrak{F}_0$ is called *sign-compatible with the family* $\mathfrak{F}_1$ if there exists a mapping $\Pi : \mathfrak{F}_0 \to \mathfrak{F}_1$ such that

$$\mu\{t : t \in \Omega, \ e(t)\Pi e(t) \leqslant 0\} = 0, \qquad (6.1)$$

for every $e(t) \in \mathfrak{F}_0$. In other words, the family $\mathfrak{F}_0$ is sign-compatible with $\mathfrak{F}_1$ provided that for any function in $\mathfrak{F}_0$ there exists a sign-compatible with it function in $\mathfrak{F}_1$.

We will give a few illustrative examples.

Suppose that each family consists of but one function $e(t) \in \mathfrak{F}_0$ and $g(t) \in \mathfrak{F}_1$, and the functions $e(t)$ and $g(t)$ are sign-compatible. Then the family $\mathfrak{F}_0$ is sign-compatible with $\mathfrak{F}_1$.

Assume now that $\mathfrak{F}_0$ is a given family which does not contain functions that are identically zero on sets of positive measure. Then $\mathfrak{F}_0$ is sign-compatible with itself (as the mapping Π we can choose the identity on $\mathfrak{F}_0$).

Let $\Omega = [0, \pi]$, and let $\mathfrak{E}_0$ and $\mathfrak{E}_1$ be the linear spans of the functions $\{1, \sin t, \cos t\}$ and $\{1, t, t^2\}$, respectively. Denote by $\mathfrak{F}_i$ the unit ball in the three dimensional space $\mathfrak{E}_i$ ($i = 0, 1$). Then the family $\mathfrak{F}_0$ is sign-compatible with $\mathfrak{F}_1$, but the family $\mathfrak{F}_1$ is not sign-compatible with $\mathfrak{F}_0$ (for the function $a(t-1) \in \mathfrak{F}_1$ there are no sign-compatible functions in $\mathfrak{F}_0$).

Suppose now that the family $\mathfrak{F}_0$ is sign-compatible with a family $\mathfrak{F}_1$. We introduce the following notation

$$\chi(\delta; e) = \chi(\delta; e, \mathfrak{F}_0, \mathfrak{F}_1) = \int\limits_{\{t: t \in \Omega,\, |e(t)| \leqslant \delta\}} \Pi e(t) \mathrm{d}\mu \quad (e \in \mathfrak{F}_0,\ \delta \geqslant 0). \tag{6.2}$$

The functions (6.2) and (1.1) do not coincide, although they are denoted by the same symbol.

As a generalization of inequality (4.10), we consider the inequality

$$\xi \int\limits_{\Omega} \Pi e(t) \varphi[t, \xi e(t) + h(t)] \mathrm{d}\mu \leqslant 0, \tag{6.3}$$

where again the function $\varphi(t, x)$ lies in one of the classes $\mathfrak{V}(\Phi, \Psi)$, and $e(t) \in \mathfrak{F}_0$. We assume that $\mathfrak{F}_0 \subset L_p$ and $\mathfrak{F}_1 \subset L_q$.

By a solution for inequality (6.3) we mean a pair $\{\xi, h(t)\}$ ($\xi \in \mathbb{R}^1$, $h(t) \in L_p$) that satisfies this inequality. We add to inequality (6.3) an analog of condition (4.11), namely,

$$\|h(t)\|_p \leqslant H(|\xi|). \tag{6.4}$$

For solutions $\{\xi, h(t)\}$ as above, we can find a priori norm estimates.

6.2. A PRIORI ESTIMATES. In this subsection we state a result on the existence of a priori norm estimates for solutions $\{\xi, h(t)\}$ of inequality (6.3). We assume that $p = \infty$, and in this respect, the next result is a generalization of Theorem 4.1. Analogous generalizations can be found for all the other theorems in §4.

THEOREM 6.1. *Assume that the family $\mathfrak{F}_0$ is sign-compatible with the family $\mathfrak{F}_1$, and the function $\chi(\delta; e)$ defined by (6.2) is positive when $\delta > 0$ and $e(t) \in \mathfrak{F}_0 \subset L_\infty$. In addition, let us suppose that condition (6.4) has the form (4.10), i.e., $p = \infty$. Further, assume that the inequality*

$$\lim_{u \to \infty} \sup_{f(t) \in \mathfrak{F}_0} \frac{\displaystyle\int\limits_{\Omega} |\Pi e(t)| \Phi[t, u_0 + 2u|e(t)|] \mathrm{d}\mu}{\chi\left[\dfrac{u_0 + H(u)}{u}; e\right] \cdot \Psi[u_0 + 2H(u)]} = \infty \tag{6.5}$$

holds true. Then all the solutions of inequality (6.3) that satisfy (6.4) admit the general a priori estimate

$$|\xi|,\ \|h(t)\|_\infty \leqslant \mathrm{const} < \infty. \tag{6.6}$$

Obviously, if $\mathfrak{F}_0$ consists of but one function $e(t)$, then condition (6.5) coincides with (4.15), where $g(t) = \Pi e(t)$.

6.3. ANOTHER INTEGRAL-FUNCTIONAL INEQUALITY. Let us consider the inequality

$$\int_\Omega \Phi[t, |\xi e(t) + h(t)|]\mathrm{d}\mu \leqslant 0, \tag{6.7}$$

where $\Phi(t, u) \in \mathfrak{N}(u_0)$, together with the inequality

$$\|h(t)\|_{L_p} \leqslant H(|\xi|). \tag{6.8}$$

Assume that the function $H(u)$ satisfies condition (4.12).

Let the pair $\{\xi, h(t)\}$, where $\xi > 0$ and $h(t) \in L_p$, be a solution for inequalities (6.7) and (6.8). We consider the function

$$s(t) = \begin{cases} 1, & \text{if } e(t) \geqslant 0, \\ -1, & \text{if } e(t) < 0, \end{cases} \tag{6.9}$$

which is measurable on the set Ω. Let us introduce now the following notations

$$\varphi(t, u) = \begin{cases} \Phi(t, u) & \text{for } u \geqslant 0, \\ -\Phi(t, -u) & \text{for } u < 0, \end{cases} \tag{6.10}$$

$g(t) \equiv 1$, $\tilde{e}(t) = |e(t)|$, $\tilde{h}(t) = s(t)h(t)$. The pair $\{\xi, \tilde{h}(t)\}$ is a solution for an inequality of the same type as (4.10), namely,

$$\xi \int_\Omega g(t)\varphi[t, \xi\tilde{e}(t) + \tilde{h}(t)]\mathrm{d}\mu \leqslant 0 \tag{6.11}$$

with sign-compatible functions $g(t)$ and $\tilde{e}(t)$, and where $\varphi(t, u) \in \mathfrak{V}(\Phi, \mathrm{const})$. If the corresponding assumptions in Theorems 4.1–4.4 are satisfied, then there exist a priori estimates for the norms $|\xi|$, $\|\tilde{h}(t)\|_{L_p}$.

We indicate some of the possible results.

THEOREM 6.2. *Let* $p = \infty$ *and assume that*

$$\lim_{u \to \infty} \frac{\chi\left[\dfrac{u_0 + H(u)}{u}; e\right]}{\displaystyle\int_\Omega \Phi[t, u_0 + 2u|e(t)|]\mathrm{d}\mu} = 0, \tag{6.12}$$

where $\chi(\delta; e)$ *is the function* (1.1). *Then, for the solution* $\{\xi, h(t)\}$ *of inequalities* (6.7) *and* (6.8) *we have the a priori estimate*

$$|\xi|, \ \|h(t)\|_{L_p} \leqslant \text{const} < \infty.$$

THEOREM 6.3. *Suppose that the function* $\chi(\delta; e)$ *is positive for* $\delta > 0$. *Let* $p < \infty$ *and assume that*

$$\lim_{u \to \infty} \frac{\displaystyle\int_\Omega \Phi[t, u_0 + 2u|e(t)|]\mathrm{d}\mu}{\chi\left\{\dfrac{u_0}{u} + \theta\left[\dfrac{H(u)}{u}\right]; e\right\}} = \infty, \tag{6.13}$$

where θ *is the function defined by* (4.20) *and* (4.21) *using the the function* $\chi(\delta; e, 1) = \chi(\delta; e)$ *and* $q' = 1$. *Then, for the solution* $\{\xi, h(t)\}$ *of inequalities* (6.7) *and* (6.8) *we have the a priori estimate*

$$|\xi|, \ \|h(t)\|_{L_p} \leqslant \text{const} < \infty.$$

6.4. This chapter was based mainly on results from [Krasnoselskii, A. M., 1986; Krasnoselskii, A. M., 1991].

Chapter 2

Two-sided estimates for nonlinearities

§7. Equations with self-adjoint and normal operators

7.1. SELF-ADJOINT OPERATORS ON L_2. We consider the space $L_2 = L_2(\Omega, \mathbb{R}^n)$ of all square-integrable functions $x(t) : \Omega \to \mathbb{R}^n$. Recall that the norm and the inner product on $\mathbb{R}^n$ are denoted by $|\cdot|$ and $(\cdot, \cdot)$, respectively. In their turn, the norm and the inner product on L_2 will be denoted by $\|\cdot\|$ and $[\cdot, \cdot]$. The norm on L_2 is defined by the usual equality

$$\|x(t)\|^2 = \int_\Omega |x(t)|^2 \mathrm{d}\mu, \quad x(t) \in L_2.$$

We next consider a self-adjoint completely continuous linear operator A acting on L_2. Let us first recall some properties of such an operator.

For any self-adjoint completely continuous operator A there exists an orthonormal basis $\{e_n(t)\}$ for L_2 consisting of eigenvectors of the operator A. Each $e_n(t)$ corresponds to a real eigenvalue λ_n. The spectrum of the operator A is given by all these numbers λ_n, together with the number 0 (0 can also occur among the numbers λ_n). Only the eigenvalue 0 may have an infinite multiplicity. The equality $\lim \lambda_n = 0$ holds true. The operator A can be represented as

$$Ax(t) = \sum_{n=0}^{\infty} \lambda_n [e_n, x] e_n(t). \tag{7.1}$$

A self-adjoint completely continuous operator A is called positive definite (resp., positive semidefinite) if all its eigenvalues λ_n are positive (resp., nonnegative). For a positive semidefinite operator A we define its powers A^τ, $\tau > 0$, by

$$A^\tau x(t) = \sum_{n=0}^{\infty} \lambda_m^\tau [e_n, x] e_n(t). \tag{7.2}$$

For each $\tau > 0$, the operator A^τ is a self-adjoint, completely continuous, and positive semidefinite operator on L_2. The norm of A^τ as an operator from L_2 into L_2 is given by

$$\|A^\tau\|_{L_2 \to L_2} = \max_n \lambda_n^\tau = \left(\max_n \lambda_n\right)^\tau = \|A\|_{L_2 \to L_2}^\tau. \tag{7.3}$$

The operators (7.2) are used mostly when $\tau = \frac{1}{2}$.

Analogously, we introduce the concept of a self-adjoint, completely continuous, negative definite (resp., negative semidefinite) operator on L_2.

An important class of self-adjoint completely continuous operators on L_2 consists of the integral operators given by

$$Ax(t) = \int_\Omega K(t, s) x(s) \mathrm{d}\mu(s), \tag{7.4}$$

with kernels of the form

$$K(t, s) = \sum_{n=0}^{\infty} \lambda_n e_n(s) e_n(t).$$

The series in the right hand side of the last equality converges in mean to a square-integrable on $\Omega \times \Omega$ function $K(t, s)$, if and only if

$$\sum_{n=0}^{\infty} \lambda_n^2 < \infty.$$

7.2. OPERATORS OF TWO-POINT BOUNDARY VALUE PROBLEMS. We introduce a first simple example of a completely continuous, self-adjoint operator acting on the space $L_2 = L_2([0, T], \mathbb{R}^1)$.

Let us consider the equation

$$Lx(t) \stackrel{\text{def}}{=} x'' + p(t)x' + q(t)x = u(t), \tag{7.5}$$

where the coefficients $p(t)$ and $q(t)$ are continuous on the interval $0 \leqslant t \leqslant T$. We assume that the equation $Lx(t) = 0$ has no nontrivial solutions satisfying the conditions

$$x(0) = x(T) = 0. \tag{7.6}$$

Then, any function $u(t)$ integrable on $[0, T]$ corresponds to an absolutely continuous differentiable solution $x(t) = Au(t)$ of equation (7.5), subject to condition (7.6). The solution $x(t)$ can be defined by the formula

$$x(t) = Au(t) = \int_\Omega G(t, s)u(s)\mathrm{d}s, \tag{7.7}$$

where the kernel $G(t, s)$ is given by the equality

$$G(t, s) = \omega(s)x_2^{-1}(0)\begin{cases} x_1(s)x_2(t) & \text{for } 0 \leqslant s \leqslant t, \\ x_1(t)x_2(s) & \text{for } t < s \leqslant T. \end{cases} \tag{7.8}$$

In (7.8) above, $\omega(s)$ stands for the function

$$\omega(s) = \exp\left\{\int_0^s p(\tau)\mathrm{d}\tau\right\},$$

and $x_1(t)$ and $x_2(t)$ denote the solutions of equation $Lx(t) = 0$ corresponding to the Cauchy conditions

$$x_1(0) = 0, \quad x_1'(0) = 1, \quad x_2(T) = 0, \quad x_2'(T) = -1.$$

Since the kernel $G(t, s)$ is continuous, the operator (7.7) is a completely continuous operator on L_2. Moreover, the operator (7.7) sends any integrable function into an absolutely continuous function whose second derivative is integrable on $[0, T]$, and any continuous function into a twice continuously differentiable function.

The operator (7.7) is called *the operator of the two-point boundary value problem*, or briefly, *the operator of the two-point problem*.

Together with the Lebesgue measure, we consider on $[0, T]$ the measure μ defined by

$$\mu G = \int_G \omega(s)\mathrm{d}s. \tag{7.9}$$

The measure (7.9) defines the space $\mathcal{L}_2$ consisting of functions $x(t) \in L_2$, the norm $\|\cdot\|_\mu$ and the the inner product $[\cdot, \cdot]_\mu$ on this space being defined by

$$\|x(t)\|_\mu^2 = \int_0^T x^2(t)\mathrm{d}\mu(t) \equiv \int_0^T \omega(t)x^2(t)\mathrm{d}t$$

and

$$[x(t), y(t)]_\mu = \int_0^T x(t)y(t)\mathrm{d}\mu(t) \equiv \int_0^T \omega(t)x(t)y(t)\mathrm{d}t,$$

respectively. Obviously, the norms $\| \cdot \|$ and $\| \cdot \|_\mu$ are equivalent. The operator (7.7) on $\mathcal{L}_2$ is self-adjoint. The spectrum of A consists of 0 and the sequence of its simple eigenvalues $\lambda_n \neq 0$, which converges to zero. Each eigenvalue λ_n corresponds to a eigenfunction $e_n(t)$. The functions $e_n(t)$ are orthogonal in $\mathcal{L}_2$. For any pair $\{\lambda_n, e_n(t)\}$ the equality

$$\lambda_n[e_n'' + p(t)e_n' + q(t)e_n] = e_n. \tag{7.10}$$

holds. If the interval $[0, T]$ is a *nonoscillation interval* for the differential operator $Lx(t)$ (i.e., any nontrivial solution $x(t)$ of the equation $Lx(t) = 0$ vanishes no more than once on the interval $(0, T)$), then all the numbers λ_n are negative. In this case the operator of the two-point problem is negative definite. In the general case, the operator (7.7) may have a finite number of positive eigenvalues.

For the distributions $\chi(\delta; e_n)$ of each eigenfunction $e_n(t)$ of the operator of the two-point problem the estimates

$$c_1\delta \leqslant \chi(\delta; e_n, \text{mes}), \chi(\delta; e_n, \mu) \leqslant c_2\delta, \quad 0 \leqslant \delta \leqslant \delta_0, \tag{7.11}$$

are true. By c_j in (7.11) and in the sequel we denote some positive numbers, whose specific values are irrelevant, their existence being the only thing that matters.

Let us consider two different eigenvalues λ_0 and λ_1 of the operator (7.4), and let $e_0(t)$ and $e_1(t)$ be their corresponding eigenfunctions. We next denote by $\mathfrak{F} = \mathfrak{F}(\lambda_0, \lambda_1; \| \cdot \|)$ the set of all the functions of the form $\xi_0 e_0(t) + \xi_1 e_1(t)$ $(\xi_0, \xi_1 \in \mathbb{R}^1)$ normalized with respect to a certain norm $\| \cdot \|$.

LEMMA 7.1. *Assume that the coefficients $p(t)$ and $q(t)$ in the differential operator $Lx(t)$ are twice continuously differentiable on the interval $[0, T]$. Then, for every $\delta_0 > 0$ there are some constants c_j such that the functions (1.5) and (1.6) admit the estimates*

$$c_3\delta \leqslant \chi_\mathrm{L}(\delta; \mathfrak{F}) \leqslant c_4\delta, \quad 0 \leqslant \delta \leqslant \delta_0, \tag{7.12}$$

and

$$c_5\delta^{\frac{1}{3}} \leqslant \chi_\mathrm{U}(\delta; \mathfrak{F}) \leqslant c_6\delta^{\frac{1}{3}}, \quad 0 \leqslant \delta \leqslant \delta_0. \tag{7.13}$$

Proof. The inequality in the right rand side of (7.12) follows straightforwardly from (7.11).

For the proof of the inequality in the left hand side of (7.12) it is enough to notice that all functions $e(t) \in \mathfrak{F}$ are continuously differentiable and

$$\sup_{e(t) \in \mathfrak{F}} \max_{0 \leqslant t \leqslant T} |e'(t)| = c < \infty.$$

Therefore (since, under our assumptions, any function $e(t)$ of the form $\xi_0 e_0 + \xi_1 e_1$ satisfies the condition $e(0) = 0$), the following estimate

$$|e(t)| \leqslant ct, \quad 0 \leqslant t \leqslant T; \ e(t) \in \mathfrak{F},$$

holds true.

From equalities (7.10) it follows that all functions $e(t) \in \mathfrak{F}$ are solutions of the next fourth order linear differential equation

$$\left[\frac{\mathrm{d}^2}{\mathrm{d}t^2} + p(t)\frac{\mathrm{d}}{\mathrm{d}t} + q(t) - \lambda_1^{-1} \right] \circ \left[\frac{\mathrm{d}^2}{\mathrm{d}t^2} + p(t)\frac{\mathrm{d}}{\mathrm{d}t} + q(t) - \lambda_2^{-1} \right] e(t) = 0.$$

Thus the inequality in the right hand side of (7.13) follows from Theorem 1.1.

To prove the remaining part of inequality (7.13) we must find a function $e_*(t) \in \mathfrak{F}$ that satisfies the estimate

$$\chi(\delta; e_*) \geqslant c_5 \delta^{\frac{1}{3}}, \quad 0 \leqslant \delta \leqslant \delta_0. \tag{7.14}$$

Such a function can be defined by the equality

$$e_*(t) = \frac{e_2'(0)e_1(t) - e_1'(0)e_2(t)}{\|e_2'(0)e_1(t) - e_1'(0)e_2(t)\|}.$$

This function satisfies the conditions $e_*(0) = e_*'(0) = e_*''(0) = 0$. Consequently, if t is sufficiently small, then we get the estimate $|e_*(t)| \leqslant ct^3$, which implies (7.14). Lemma 7.1 is now proved.

We notice that the functions (1.5) and (1.6) (as well as the function $\chi(\delta; e)$) depend on the choice of the measure used in their construction. Inequalities (7.12) and (7.13) were proved for the Lebesgue measure, but they are still true (with other constants c_j, of course) for the measure (7.9).

The simplest form of the operator of the two-point problem occurs when $T = \pi$ and $p(t) \equiv q(t) \equiv 0$. In this case the kernel $G(t,s)$ of the operator (7.7) is given by

$$G_0(t,s) = -\frac{1}{\pi} \begin{cases} (\pi - t)s & \text{if } 0 \leqslant s \leqslant t, \\ (\pi - s)t & \text{if } t < s \leqslant T. \end{cases} \tag{7.15}$$

The operator (7.7) is self-adjoint and negative definite. Its eigenvalues are $\lambda_n = -n^{-2}$, and the corresponding eigenfunctions are

$$e_n(t) = \sin nt \quad (0 \leqslant t \leqslant \pi; \ n = 1, 2, \ldots).$$

Obviously, if the functions $p(t)$ and $q(t)$ are not constant, then it is quite impossible, generally speaking, to describe explicitly the eigenvalues λ_n and the corresponding eigenfunctions.

7.3. NORMAL OPERATORS ON L_2. A linear operator A acting on the space $L_2 = L_2(\Omega; \mathbb{R}^n)$ is called *normal* if it commutes with its adjoint A^*. In particular, every self-adjoint operator is normal.

In the sequel we will describe some spectral properties of completely continuous normal operators on real Hilbert spaces. These properties are stated for operators acting on L_2, but they are true in general.

Any such operator A determines a sequence $\Pi_n \subset L_2$ ($n = 0, 1, 2, \ldots$) of mutually orthogonal invariant subspaces for A, whose dimensions are 1 or 2. Each one-dimensional subspace Π_n consists of eigenfunctions e_n of the operator A corresponding to real eigenvalues λ_n, that is, $A e_n = \lambda_n e_n$ ($e_n \in \Pi_n$, $\dim \Pi_n = 1$). On each two-dimensional subspace Π_n the operator A is defined by the equality

$$A e_n(t) = |\lambda_n| U_n e_n(t) \quad (e_n(t) \in \Pi_n, \ \dim \Pi_n = 2), \tag{7.16}$$

where U_n denotes the rotation in the plane Π_n determined by an angle φ_n. Any Π_n with $\dim \Pi_n = 2$ corresponds to a pair of complex conjugate eigenvalues $\lambda_n = |\lambda_n|(\cos \varphi_n + i \sin \varphi_n)$ and $\bar{\lambda}_n = |\lambda_n|(\cos \varphi_n - i \sin \varphi_n)$. The subspaces Π_n ($n = 0, 1, 2, \ldots$) are mutually orthogonal in L_2. The family of all subspaces Π_n is complete, in the sense that any function $x(t) \in L_2$ can be represented as a series

$$x(t) = \sum_{n=0}^{\infty} x_n(t)$$

that converges in L_2, where $x_n(t) \in \Pi_n$. We let P_n denote the orthogonal projection onto the subspace Π_n. The operator A has a spectral representation analogous to (7.1), namely,

$$A x(t) = \sum_{n=0}^{\infty} |\lambda_n| U_n P_n x(t), \quad x(t) \in L_2, \tag{7.17}$$

where, for $\dim \Pi_n = 1$, U_n denotes the operator given by $U_n e_n(t) = (\operatorname{sign} \lambda_n) e_n(t)$. The eigenvalue 0 is the only one that may have infinite multiplicity. The equality

$\lim |\lambda_n| = 0$ holds true. The operator A^* — the adjoint of A — has the following representation

$$A^* x(t) = \sum_{n=0}^{\infty} |\lambda_n| U_n^{-1} P_n x(t), \quad x(t) \in L_2. \tag{7.18}$$

The spectrum $\sigma(A)$ of the completely continuous normal operator A is given by the union of all its eigenvalues together with 0.

If the real eigenvalues λ_n of the operator A (corresponding to the one-dimensional invariant subspaces Π_n) are nonnegative, then the fractional powers A^τ ($\tau > 0$) of the completely continuous normal operator A can be defined on L_2.

We denote by U_n^τ the operator U_n if $\dim \Pi_n = 1$, or the operator corresponding to a rotation of angle $\tau\varphi_n$, if $\dim \Pi_n = 2$. The fractional powers A^τ ($\tau > 0$) of the operator A are defined by

$$A^\tau x(t) = \sum_{n=0}^{\infty} |\lambda_n|^\tau U_n^\tau P_n x(t), \quad x(t) \in L_2. \tag{7.19}$$

The operator A^τ is also normal and completely continuous, for each $\tau > 0$. The spectrum $\sigma(A^\tau)$ of the operator A^τ consists of the numbers λ_n^τ (if λ_n are real), together with $|\lambda_n^\tau|(\cos\tau\varphi_n \pm \mathrm{i}\sin\tau\varphi_n)$ and 0.

The next simple formulas for the norms of the operators A and A^τ follow from the representations (7.17) and (7.19):

$$\|A\|_{L_2 \to L_2} = \max_{\lambda \in \sigma(A)} |\lambda|,$$
$$\|A^\tau\|_{L_2 \to L_2} = \max_{\lambda \in \sigma(A)} |\lambda^\tau| = \|A\|_{L_2 \to L_2}^\tau.$$

For the description of completely continuous normal operators on a real Hilbert space H we could also use the complexification of H, i.e., the complex space $H^{\mathbb{C}}$ consisting of all the elements $z = x + iy$ ($x, y \in H$).

7.4. OPERATORS OF PERIODIC PROBLEMS. Let

$$L(p) = p^l + a_1 p^{l-1} + \cdots + a_l, \tag{7.20}$$

$$M(p) = b_0 p^m + b_1 p^{m-1} + \cdots + b_m \tag{7.21}$$

be two given polynomials with constant real coefficients. We assume in the sequel that $b_0 \neq 0$, that the polynomials (7.20) and (7.21) are coprime, and that their degrees satisfy the condition

$$l > m. \tag{7.22}$$

We will investigate the equation

$$L\left(\frac{\mathrm{d}}{\mathrm{d}t}\right)x(t) = M\left(\frac{\mathrm{d}}{\mathrm{d}t}\right)u(t). \tag{7.23}$$

For sufficiently smooth functions $u(t)$, the solution of equation (7.23) is meant in the usual sense. We will consider in the sequel the equation (7.23) corresponding to functions u(t) which are supposed to be measurable and locally integrable only. In this case there are a few different specific ways to define the notion of a solution for equation (7.23).

We introduce the notations

$$W(p) = \frac{M(p)}{L(p)}, \tag{7.24}$$

and

$$\omega_k = 2k\pi T^{-1}, \tag{7.25}$$

where $T > 0$ is a fixed number. In control theory the function (7.24) is called *the transfer function of the linear link*.

We consider in $L_2 = L_2([0, T], \mathbb{R}^1)$ the orthonormal basis provided by the functions

$$e_0(t) = \sqrt{\frac{1}{T}}, \ g_k(t) = \sqrt{\frac{2}{T}}\cos\omega_k t, \ e_k(t) = \sqrt{\frac{2}{T}}\sin\omega_k(t), \quad k = 1, 2, \dots, \tag{7.26}$$

where ω_k are given by (7.25). We let $\varPi_0$ denote the one-dimensional subspace of all constant functions and $\varPi_k$ (for $k = 1, 2, \dots$) the two-dimensional subspace generated by the functions e_k and g_k. For each k, let P_k denote the orthogonal projection from L_2 onto $\varPi_k$, i.e.,

$$P_0 u = [u, e_0]e_0, \quad P_k u = [u, e_k]e_k + [u, g_k]g_k, \quad u \in L_2; \ k = 1, 2, \dots. \tag{7.27}$$

We assume that $L(\omega_k\mathrm{i}) \neq 0$, that is, $W(\omega_k\mathrm{i})$ is well-defined (here and further on $W(p)$ is the function defined by (7.24)).

If $W(\omega_k\mathrm{i}) \neq 0$, then by U_k we denote the isometry on the space $\varPi_k$ given by

$$\begin{aligned} U_k(\xi e_k + \eta g_k) = \frac{1}{|W(\omega_k\mathrm{i})|} &\times \{[\mathrm{Re}\,W(\omega_k\mathrm{i})\xi + \mathrm{Im}\,W(\omega_k\mathrm{i})\eta]e_k + \\ &+ [-\mathrm{Im}\,W(\omega_k\mathrm{i})\xi + \mathrm{Re}\,W(\omega_k\mathrm{i})\eta]g_k\}. \end{aligned} \tag{7.28}$$

When $k = 0$, the subspace $\varPi_0$ is one-dimensional and U_0 is defined as the multiplication by sign $W(0)$. If $W(\omega_k\mathrm{i}) = 0$, then the corresponding U_k stands for the identity operator on $\varPi_k$.

We now set

$$Au(t) = \sum_{k=0}^{\infty} |W(\omega_k \mathrm{i})| U_k P_k u(t), \quad u(t) \in L_2. \tag{7.29}$$

Since (7.22) implies that $W(\omega_k \mathrm{i}) \to 0$ as $k \to \infty$, the operator given by (7.29) is defined on the whole space L_2 and is completely continuous. From (7.29) we get

$$\|A\|_{L_2 \to L_2} = w(T), \tag{7.30}$$

where

$$w(T) = \max_{k=0,\pm 1,\ldots} |W(\omega_k \mathrm{i})|. \tag{7.31}$$

The quantity (7.31) may depend on T.

Let us recall that under our assumptions the polynomial (7.20) has no roots of the form $\omega_k \mathrm{i}$. In this case equation (7.23) with $u(t) = \xi e_k(t) + \eta g_k(t)$ has, for each $k = 0, 1, 2, \ldots$, a unique periodic solution with period T. A straightforward computation shows that the solution is given by the formula

$$x(t) = [\operatorname{Re} W(\omega_k \mathrm{i})\xi + \operatorname{Im} W(\omega_k \mathrm{i})\eta]\, e_k(t) +$$
$$+ [-\operatorname{Im} W(\omega_k \mathrm{i})\xi + \operatorname{Re} W(\omega_k \mathrm{i})\eta]\, g_k(t) = Au(t).$$

Thus we conclude that the values of the continuous operator (7.29) on the dense subset in L_2 of all trigonometric polynomials of the form

$$u_N(t) = \xi_0 e_0(t) + \sum_{k=1}^{N} [\xi_k e_k(t) + \eta_k g_k(t)], \tag{7.32}$$

provide solutions of equation (7.23) with $u(t) = u_N(t)$. It can be proved that the function $x(t) = Au(t)$, where $u(t)$ is the restriction on $[0, T]$ of an m-times continuously differentiable T-periodic function, is a solution of equation (7.23) for periodic boundary conditions. If $u(t)$ is a locally integrable T-periodic function, then the function $Au(t)$ is called the solution of equation (7.23).

The operator (7.29) is called *the periodic problem operator*, or more explicitly, *the operator of the periodic problem for the linear link with the fractional-rational transfer function* (7.24).

The operator of the periodic problem admits the integral representation

$$Au(t) = \int_0^T G(t - s; T) u(s) \mathrm{d}s, \tag{7.33}$$

where the function $G(\tau; T)$ is called *the unit impulse respose of the linear link with the transfer function* $W(p)$.

We notice some properties of the operator of the periodic problem.

The operator (7.29) is completely continuous as an operator from any L_p $(p > 1)$ into C^{l-m-1}, as well as an operator from L_1 into C^k $(k < l - m - 1)$. The operator (7.29) acts continuously from L_1 into C^{l-m-1}, but it is no longer completely continuous. The operator (7.29) is completely continuous and normal on L_2. Its spectrum $\sigma(A)$ consists of the numbers $W(\omega_k \mathrm{i})$ $(k = 0, \pm 1, \dots)$ and zero.

In the following sections we will consider equations containing the operator of the periodic problem. The estimates that are used for the distributions of functions from invariant subspaces of the operator (7.29) will be established in Section 11.

7.5. NONLINEAR EQUATIONS. The main object we are concerned with is the class of nonlinear operator equations of the form

$$x(t) = A\mathfrak{f}x(t), \tag{7.34}$$

where $\mathfrak{f}x(t)$ is (generally speaking) a nonlinear operator acting from a space B_1 of functions $x(t) : \Omega \to \mathbb{R}^n$ into a space B_2 of functions of the same kind, and A is a linear operator acting from B_2 into B_1. In this case, the operator $A\mathfrak{f}$ acts on the space B_1. If this operator has "nice" properties, it can be proved that there exist solutions $x(t) \in B_1$ of equation (7.34). Moreover, we will indicate different procedures to construct approximate solutions, and conditions for the uniqueness of the solutions.

We will use spaces B_1 and B_2 (often $B_1 = B_2$) of integrable functions.

The nonlinear operator $\mathfrak{f}$ will be usually the superposition operator

$$\mathfrak{f}x(t) = f[t, x(t)]. \tag{7.35}$$

The nonlinear function $f(t, x) : \Omega \times \mathbb{R}^n \to \mathbb{R}^n$ in (7.35) above is supposed to satisfy the Caratheodory condition (it is continuous in x for any fixed t and measurable in t for each fixed x). Sometimes different additional conditions on continuity and smoothness are required.

In this chapter we assume that the function $f(t, x)$ satisfies the two-sided constraint

$$|f(t, x)| \leqslant c_1 |x| + c_2(t), \quad t \in \Omega, \ x \in \mathbb{R}^n, \tag{7.36}$$

where $c_1 > 0$ and the function $c_2(t)$ belongs to one of the spaces $L_p = L_p(\Omega, \mathbb{R}^1)$.

In this case the superposition operator (7.35) acts continuously on L_p. If the function $c_2(t)$ is essentially bounded ($c_2(t) \in L_\infty$), then the operator (7.33) acts on each L_p $(1 \leqslant p \leqslant \infty)$.

We next give a simple result on the solvability of the equation (7.34).

THEOREM 7.1. *Let the nonlinear operator $\mathfrak{f}$ be the superposition operator (7.35), where $f(t, x)$ satisfies the Caratheodory condition and condition (7.36), with $c_2(t) \in L_p$ ($1 \leqslant p \leqslant \infty$). Assume that the linear operator A acts on L_p and is completely continuous. If*

$$c_1 \|A\|_{L_p \to L_p} < 1, \tag{7.37}$$

then equation (7.34) has at least one solution $x(t) \in L_p$.

Proof. Consider the ball

$$\mathfrak{B} = \left\{ x(t) \in L_p, \ \|x(t)\|_{L_p} \leqslant \frac{\|c_2(t)\|_{L_p} \cdot \|A\|_{L_p \to L_p}}{1 - c_1 \|A\|_{L_p \to L_p}} \right\}.$$

Let $x(t) \in \mathfrak{B}$. Then

$$\|A\mathfrak{f}x(t)\|_{L_p} \leqslant \|A\|_{L_p \to L_p} \cdot \|\mathfrak{f}x(t)\|_{L_p} \leqslant \|A\|_{L_p \to L_p} \cdot \left[c_1 \|x(t)\|_{L_p} + \|c_2(t)\|_{L_p} \right] \leqslant$$

$$\leqslant \|A\|_{L_p \to L_p} \cdot \left[c_1 \frac{\|c_2(t)\|_{L_p} \cdot \|A\|_{L_p \to L_p}}{1 - c_1 \|A\|_{L_p \to L_p}} + \|c_2(t)\|_{L_p} \right] = \frac{\|c_2(t)\|_{L_p} \cdot \|A\|_{L_p \to L_p}}{1 - c_1 \|A\|_{L_p \to L_p}},$$

that is, the operator $A\mathfrak{f}$ sends $\mathfrak{B}$ into itself. Since the operator $A\mathfrak{f}$ is completely continuous on the space L_p, in view of the Schauder Principle it has a fixed point in L_p, and this is a solution of equation (7.32).

Theorem 7.1, or its versions, can be found in many monographs and textbooks on nonlinear analysis. Even this simple theorem provides important criterions for the existence of solutions of boundary value problems for quasilinear ordinary or partial differential equations.

Theorem 7.1 does not take into account any specific properties of the operators A and $\mathfrak{f}$. If we take into consideration the structure of the eigenvalues of the operator A, the fact that A may be self-adjoint or normal, the size of the arguments of the leading eigenvalues of the normal operator A, the positivity or strict positivity of the operator A, as well as, the weak nonlinearities that may occur in the superposition operator, then more subtle results than Theorem 7.1 can be found.

A natural analog of Theorem 7.1 can be established in case the continuous operator $\mathfrak{f}x(t)$ satisfies the estimate

$$\|\mathfrak{f}x(t)\|_{B_2} \leqslant c_1 \|x(t)\|_{B_1} + c_2,$$

but it is not necessarily a superposition operator, and the norm of the operator A is such that

$$c_1 \|A\|_{B_2 \to B_1} < 1.$$

In this chapter we consider the equation (7.34) in the space L_p for $p = 2$. The study of this equation in the Hilbert space L_2 enables us to take full advantage of the fact that the linear operator A is either self-adjoint or normal and to use the specific properties of the space itself.

§8. Solvability of equations in case the solutions do not admit a priori norm estimates

8.1. PRELIMINARIES. In this section we will use the space $L_2 = L_2(\Omega, \mathbb{R}^n)$ of all square integrable vector-functions $x(t) : \Omega \to \mathbb{R}^n$. We keep the notations $| \cdot |$ and $(\cdot, \cdot)$ for the norm and the inner product on $\mathbb{R}^n$, respectively. The norm and the inner product on L_2 are denoted by $\| \cdot \|$ and $[\cdot, \cdot]$.

We continue to study the equation

$$x(t) = A\mathfrak{f}x(t), \tag{8.1}$$

where A is a completely continuous linear operator acting on L_2, and $\mathfrak{f}$ is the superposition operator (7.35) corresponding to a function $f(t, x)$ from $\Omega \times \mathbb{R}^n$ into $\mathbb{R}^n$ that is continuous in $x \in \mathbb{R}^n$ for any $t \in \Omega$ and measurable in $t \in \Omega$ for each $x \in \mathbb{R}^n$. We assume that the operator $\mathfrak{f}$ acts (and that means that it is also continuous) on the space L_2, i.e., it satisfies the estimates

$$|f(t, x)| \leqslant c_1|x| + c_2(t), \quad t \in \Omega, \ x \in \mathbb{R}^n, \tag{8.2}$$

where $c_2(t) \in L_2(\Omega, \mathbb{R}^1)$. Consequently, the nonlinear operator $A\mathfrak{f}$ acts and is completely continuous on L_2.

Relatively to the operator A, we assume that it has two invariant orthogonal subspaces E_0 and E_1 ($E_0 \oplus E_1 = L_2$), with E_0 finite dimensional. The orthogonal projections onto the subspaces E_0 and E_1 are denoted by P and Q, respectively. Clearly, $P + Q = I$, where I is the identity operator. The operators P and Q commute with the operator A. We suppose that the estimates

$$\|APx\| \leqslant \lambda_0\|Px\|, \quad \|AQx\| \leqslant \lambda_1\|Qx\|, \qquad x \in L_2, \tag{8.3}$$

hold true, where $0 < \lambda_1 < \lambda_0$. From (8.3) we get the inequality $\|A\| \leqslant \lambda_0$.

The estimates (8.3) are fulfilled, for instance, if the operator A is normal and the subspace E_0 is defined by the equality

$$E_0 = \{x(t) : x(t) \in L_2, \ \|Ax(t)\| = \|A\| \cdot \|x(t)\|\}.$$

In this case $\lambda_0 = \|A\|$ and

$$\lambda_1 = \max_{\lambda \in \sigma(A); \, |\lambda| \neq \lambda_0} |\lambda|,$$

where $\sigma(A)$ is the spectrum of A.

We give yet another example. Let $n = 1$, i.e., the space L_2 consists of scalar-valued functions. Assume that the linear operator A has a simple eigenvalue $\lambda_0 > 0$ and let $e(t)$ be a corresponding eigenfunction. Since the operator A^* acts on L_2, too, λ_0 is a simple eigenvalue of A^* also. Let $g(t)$ be one of its corresponding eigenfunctions. We introduce the notations

$$E_0 = \left\{ x(t) : x(t) \in L_2, \; x(t) = \xi e(t), \; \xi \in \mathbb{R}^1 \right\},$$

and

$$E_1 = \{ x(t) : x(t) \in L_2, \; [x(t), g(t)] = 0 \}.$$

Both E_0 and E_1 are invariant subspaces for the operator A. If $g(t) \in E_0$, then $E_0 \perp E_1$ and if λ_0 is sufficiently large, then the estimates (8.3) are true.

Assume next that the functions $e(t)$ and $g(t)$ are sign-compatible (see Subsection 1.1), and

$$\alpha_1 g(t) \leqslant e(t) \leqslant \alpha_2 g(t) \quad (t \in \Omega), \tag{8.4}$$

for some $\alpha_1, \alpha_2 > 0$. Then we can choose a new measure μ_1 on the set Ω, such that the subspaces E_0 and E_1 are orthogonal in $L_2(\mu_1)$. Conditions (8.3) are fulfilled for a sufficiently large λ_0. The measure μ_1 can be defined for any set G that is measurable with respect to the measure μ, by the equality

$$\mu_1 G = \int_{\Omega} \frac{g(t)}{e(t)} \mathrm{d}\mu.$$

If in the previously introduced setting we also have that $\lambda_0 c_1 < 1$, then, in view of Theorem 7.1, equation (8.1) has at least one solution $x(t) \in L_2$ that lies in the ball

$$\mathfrak{B} = \left\{ x(t) : x(t) \in L_2, \; \|x\| \leqslant \rho_0 = \|c_2\| \cdot \lambda_0 \cdot (1 - \lambda_0 c_1)^{-1} \right\}.$$

8.2. THE MAIN THEOREM. Recall that by $\mathfrak{N}(u_0)$ $(u_0 > 0)$ we denote the class of bounded superpositionally measurable functions $\Phi(t, u)$, subject to conditions a)–d) in Subsection 2.1. Suppose that the operator A satisfies the assumptions mentioned in the previous subsection.

THEOREM 8.1. *Assume that the estimate*

$$|f(t,x)| \leqslant k|x| - \Psi(t,|x|), \quad t \in \Omega, \ x \in \mathbb{R}^n, \tag{8.5}$$

holds true, where the function $\Psi(t,u)$ *is bounded, and the function*

$$\Phi_0(t,u) = u\Psi(t,u), \quad t \in \Omega, \ u \geqslant 0, \tag{8.6}$$

belongs to a class $\mathfrak{N}(u_0)$ $(u_0 > 0)$. *In addition, suppose that*

$$\lim_{\delta \to 0} \sup_{e(t) \in E_0; \, \|e\|=1} \frac{\chi(\delta; e)}{\int\limits_\Omega \Phi_0\left[t, u_* + R\delta^{-1}|e(t)|\right] d\mu} = 0, \tag{8.7}$$

for every $R > 0$ *and any* $u_* \geqslant u_0$.

     Then there exists a number

$$k_0 = k_0(E_0, \Psi_0, \lambda_0, \lambda_1) > \lambda_0^{-1},$$

such that the inequality $k \leqslant k_0$ *together with condition (8.5) imply the existence of at least one solution of equation (8.1) in a certain ball* $\{\|x\| \leqslant \rho\} \subset L_2$. *The number* ρ *depends only on the subspace* E_0, *the function* $\Psi(t,u)$ *and on* λ_0, λ_1 *and* k_0. *If* $k \leqslant \lambda_0^{-1}$ *then the set of all solutions of equation (8.1) admits a general a priori estimate in the norm of* L_2.

     The proof of Theorem 8.1 is given in Section 9.

     Condition (8.7) reseambles condition (2.7) in Theorem 2.2. As for (2.7), in some specific cases condition (8.7) has more simpler analogs.

8.3. LACK OF A PRIORI NORM ESTIMATES FOR THE SOLUTIONS OF EQUATION (8.1) UNDER THE ASSUMPTIONS OF THEOREM 8.1. In this subsection we give an example of a family of equations (8.1), each of them satisfying the conditions stated in Theorem 8.1 and such that there are no a priori norm estimates of their solutions.

     We will consider a particular form of equation (8.1) in the space $L_2 = L_2(\Omega, \mathbb{R}^1)$ of scalar-valued functions. Let E_0 denote a finite dimensional subspace of L_2 whose dimension is odd. We assume that $Ae = \lambda_0 e$ $(\lambda_0 > 0)$ for any $e \in E_0$, that the equality

$$\lim_{\delta \to 0} \max_{e(t) \in E_0; \, \|e\|=1} \chi(\delta; e) = 0 \tag{8.8}$$

holds true, that the subspace $E_1 = E_0^\perp$ is invariant for the operator A, and that the right hand side estimate in (8.3) is satisfied for some $\lambda_1 \in (0, \lambda_0)$.

Condition (8.8) is fulfilled if any nonzero function $e(t) \in E_0$ vanishes only on a set of measure zero.

Let

$$f(t,x) = kx - h(t,x), \quad t \in \Omega, \ x \in \mathbb{R}^1, \tag{8.9}$$

where $h(t,x)$ is a bounded function that is continuous in x and measurable in t and satisfies the estimate

$$xh(t,x) \geqslant \gamma > 0, \quad t \in \Omega, \ x \in \mathbb{R}^1; \ |x| \geqslant u_0. \tag{8.10}$$

Then the estimate (8.5) is true for the function $f(t,x)$, with

$$\Psi(t,u) = \gamma u^{-1}, \quad u \geqslant u_0.$$

Therefore, equality (8.8) can be looked upon as condition (8.7) in Theorem 8.1.

In view of Theorem 8.1, there exists $k_0 > \lambda_0^{-1}$ such that for any k satisfying the conditions $0 < \varepsilon_0 \leqslant k \leqslant k_0$, the equation

$$x(t) = A[kx(t) - h(t,x)] \tag{8.11}$$

has at least one solution in L_2. We will next show that among the solutions of this equation there exist functions whose norms are as large as we wish them to be.

Set

$$B_k x(t) = kAx(t) - Ah[t, x(t)], \quad x(t) \in L_2.$$

The operator B_k is jointly completely continuous with respect to the variables $x(t) \in L_2$ and $k \in [\varepsilon_0, k_0]$. Since $h(t,x)$ is bounded, we get the equality

$$\lim_{\rho \to \infty} \max_{\varepsilon_0 \leqslant k \leqslant k_0; \, \|x\| = \rho} \frac{B_k x - kAx}{\|x\|} = 0.$$

Thus, the operator B_k has an asymptotic derivative at infinity equal to kA (see [Krasnoselskii, Zabreiko, 1975]). On the other hand 1 is an eigenvalue of odd multiplicity of the linear operator $\lambda_0^{-1}A$. We can now use M. A. Krasnoselskii's theorem on asymptotic bifurcation points (see again [Krasnoselskii, Zabreiko, 1975]). According to this theorem, on every sphere

$$S_\rho = \{x(t) : x(t) \in L_2, \ \|x(t)\| = \rho\}$$

with a sufficiently large radius ρ, there exists at least one solution $x_\rho(t)$ of equation (8.11) for a certain $k = k(\rho)$, such that $k(\rho) \to \lambda_0^{-1}$ as $\rho \to \infty$. Consequently, $k(\rho) \in [0, k_0]$ for any sufficiently large ρ, and thus our claim is proved.

By Theorem 8.1, and from the previously introduced constructions, it follows that for some values of k sufficiently close to λ_0^{-1}, equation (8.11) has at least two solutions.

8.4. NECESSITY OF CONDITION (8.7). In this subsection we will state a result that shows the "naturalness" of condition (8.7).

We assume that the operator A satisfies all the conditions in Theorem 8.1, where $\dim E_0 = 1$, and let $e_0(t) \in E_0$ with $\|e_0(t)\| = 1$ be fixed. Then condition (8.7) can be rewritten as

$$\lim_{\delta \to 0} \frac{\chi(\delta; e_0)}{\displaystyle\int_\Omega \Phi_0\left[t, u_* + R\delta^{-1}|e_0(t)|\right] d\mu} = 0. \tag{8.12}$$

We will suppose that the function $e_0(t)$ is bounded, $\chi(0; e_0) = 0$, $\chi(\delta; e_0) > 0$ for $\delta > 0$, and the estimate

$$\chi(2\delta; e_0) \geqslant \beta\chi(\delta; e_0), \quad 0 \leqslant \delta \leqslant \delta_0, \tag{8.13}$$

holds true for some $\delta_0 > 0$ and $\beta > 1$. For example, if the function $\chi(\delta; e_0)$ equals δ^a, or $\delta^a|\ln \delta|^r$, where $a > 0$ and r is an arbitrary real number, then the estimate (8.13) is true for any $\beta \in (1, 2^a)$, with a proper choice of the corresponding $\delta_0 = \delta_0(\beta)$. To prove this statement, it is enough to check the equality

$$\lim_{\delta \to 0} \frac{\chi(2\delta; e_0)}{\chi(\delta; e_0)} = 2^a.$$

However, the estimate (8.13) fails for any $\delta_0 > 0$ and $\beta > 1$, if $\chi(\delta; e_0) = |\ln \delta|^{-1}$.

Let $\Psi_0(t, u)$, $t \in \Omega$, $u \geqslant u_0 > 0$, be a given bounded function that is continuous in u and measurable in t, and is such that the function $\Phi_0(t, u) = u\Psi_0(t, u)$ belongs to the class $\mathfrak{N}(u_0)$. Moreover, assume that

$$\gamma\Psi_0(t, u) \leqslant \Psi_0(t, u_* + u), \quad t \in \Omega, \ u \geqslant u_0, \tag{8.14}$$

for some $\gamma > 0$ and $u_* > u_0$. For instance, the estimate (8.14) can be easily verified (with an appropriate value of γ) for $\Psi_0(t, u)$ given by e^{-u}, u^{-a}, or $u^{-q}\ln^r u$ $(a > 0)$. For the function $\Psi_0(t, u) = \exp(-u^2)$ this estimate fails for every $\gamma > 0$.

We denote by $\mathfrak{L}(\Psi_0, E_0)$ the class of all equations (8.1) with completely continuous linear operators A satisfying the estimate (8.3), and with nonlinearities $f(t, x)$

satisfying the estimate (8.5), where $k = \lambda_0^{-1}$, and the function $\Psi(t, u)$ is bounded and coincides with $\Psi_0(t, u)$ for $u \geqslant u_0$.

THEOREM 8.2. *Suppose that any equation in the class* $\mathfrak{L}(\Psi_0, E_0)$ *has solutions in* L_2. *Then*

$$\lim_{\delta \to 0} \frac{\chi(\delta; e_0)}{\displaystyle\int_\Omega \Phi_0 \left[t, u_* + R\delta^{-1}|e_0(t)|\right] d\mu} = 0, \tag{8.15}$$

for a certain $R > 0$.

The proof of Theorem 8.2 will be given in Section 9.

Equality (8.15) differs from (8.12) only by using the lower limit instead of the usual one, and by the fact that the numbers $u_* = u_0$ and $R > 0$ in (8.15) are no longer arbitrary. Thus, the solvability of all equations in the class $\mathfrak{L}(\Psi_0, E_0)$ implies an equality that is "almost the same" as condition (8.7) in Theorem 8.1.

The assumption that E_0 is an one-dimensional subspace is irrelevant, and it was introduced merely to simplify the subsequent proof.

§9. Proofs of Theorems 8.1 and 8.2

9.1. LEMMAS. We assume that all the conditions in Theorem 8.1 are fulfilled. For a further use, we introduce the notations

$$M = \sup_{t \in \Omega, \, u \geqslant 0} |\Psi(t, u)|, \tag{9.1}$$

where $\Psi(t, u)$ is the same function as in condition (8.5), and

$$u_* = \max\{u_0, M \cdot \lambda_0\}, \tag{9.2}$$

where u_0 is the positive number defining the class $\mathfrak{N}(u_0)$ which contains the function $\Psi_0(t, u) \equiv u\Psi(t, u)$ given as in (8.6). We also set

$$\Phi(t, u) = \begin{cases} -M^2 - 2Mu_* \left(\dfrac{1}{\lambda_0} + 1\right) & \text{if } 0 \leqslant u < u_*, \\[2mm] \dfrac{1}{\lambda_0}\Phi_0(t, u) & \text{if } u \geqslant u_*. \end{cases} \tag{9.3}$$

LEMMA 9.1. *Assume that inequality (8.5) is fulfilled, where the coefficient* k *satisfies the estimates*

$$\frac{1}{\lambda_0} \leqslant k \leqslant \frac{1}{\lambda_0} + 1. \tag{9.4}$$

Then

$$|f(t,x)|^2 \leqslant k^2|x|^2 - \Phi(t,u), \quad t \in \Omega, \ x \in \mathbb{R}^n. \tag{9.5}$$

Proof. By taking the squares of both sides in inequality (8.5) we get the inequality

$$|f(t,x)|^2 \leqslant k^2|x|^2 - 2k|x|\Psi(t,|x|) + |\Psi(t,|x|)|^2. \tag{9.6}$$

Thus, inequality (9.5) for $|x| < u_*$ follows from the right hand side estimate in (9.4) and from the definition of the number (9.1).

Suppose next that $|x| \geqslant u_*$. Then (9.6) can be written as

$$|f(t,x)|^2 \leqslant k^2|x|^2 - \Phi(t,u) - 2\left(k - \frac{1}{\lambda_0}\right)|x|\Psi(t,|x|) - \frac{\Psi(t,|x|)}{|x|}\left[\frac{1}{\lambda_0}|x|^2 - \Phi(t,|x|)\right].$$

But for $|x| \geqslant u_*$ the function $\Psi(t,|x|)$ has nonnegative values, and based on (9.4) we have

$$2\left(k - \frac{1}{\lambda_0}\right)|x|\Psi(t,|x|) \geqslant 0, \quad |x| \geqslant u_*.$$

In addition, if $|x| = u_*$, we have the inequality $\Phi(t,|x|) \leqslant \lambda_0^{-1}u_*^2$. Since for $u \geqslant u_*$ the function u^2 is increasing and $\Phi(t,u)$ is nonincreasing, then

$$\frac{\Psi(t,|x|)}{|x|}\left[\frac{1}{\lambda_0}|x|^2 - \Phi(t,|x|)\right] \geqslant 0, \quad |x| \geqslant u_0.$$

Thus, the estimate (9.6) holds for $|x| \geqslant u_*$, too. Lemma 9.1 is proved.

The function (9.3) lies in the class $\mathfrak{N}(u_*)$. According to the conditions in Theorem 8.1, it follows that condition (2.7) in Theorem 2.2 is satisfied for this function, where

$$\mathfrak{F} = \{e(t) : e(t) \in E_0, \ \|e\| = 1\}. \tag{9.7}$$

In view of Theorem 2.2, the family $\mathfrak{F}$ is compatible with the function (9.3). Let

$$\beta = \lambda_0^2\lambda_1^2\left(\lambda_0^2 - \lambda_1^2\right)^{-1}. \tag{9.8}$$

By the definition of compatibility, there exist a positive nonincreasing function $\alpha(u)$ $(u \geqslant 0)$ and a number $c = c(\beta)$, such that inequality (2.6) with $p = 2$ has no solutions $h(t) \in L_p$, for $e(t) \in \mathfrak{F}$ and $|\xi| > c$. From now on, both the function $\alpha(u)$ and the number c will be fixed until the end of the proof of Theorem 8.1.

Let us denote by X the set of all functions $x(t) \in L_2$ such that, for some $\theta = \theta(x) \in [0, 1]$ and some $Fx \in L_2$, the equality

$$x = \theta AFx \tag{9.9}$$

and the estimate

$$\|Fx\|^2 \leqslant \lambda_0^{-2}\|x\|^2 - \int_\Omega \Phi[t, |x(t)|]\mathrm{d}\mu + \alpha(\|x\|) \tag{9.10}$$

are true.

LEMMA 9.2. *The estimate*

$$\|x(t)\| \leqslant \rho \overset{\text{def}}{=} \sqrt{c^2 + \beta\left[\sup_{t\in\Omega,\, u\geqslant 0} |\Phi(t,u)|\mu\Omega + \alpha(0)\right]}, \quad x(t) \in X, \tag{9.11}$$

holds true, for β as in (9.8).

Proof. Let $x \in X$, i.e., $x = \theta AFx$, where $\theta \in [0,1]$ and (9.10) holds. From (9.9) we get

$$\|x\|^2 = \theta^2\|APFx\|^2 + \theta^2\|AQFx\|^2 =$$
$$= \theta^2\|APFX\|^2 + \frac{\theta^2\lambda_0^2}{\lambda_1^2}\|AQFx\|^2 + \theta^2\frac{\lambda_1^2 - \lambda_0^2}{\lambda_1^2}\|AQFx\|^2,$$

and, since $Qx = \theta AQFx$, we have

$$\|x\|^2 = \theta^2\|APFx\|^2 + \frac{\theta^2\lambda_0^2}{\lambda_1^2}\|AQFx\|^2 + \left(1 - \frac{\lambda_0^2}{\lambda_1^2}\right)\|Qx\|^2 \leqslant$$
$$\leqslant \|APFx\|^2 + \frac{\lambda_0^2}{\lambda_1^2}\|AQFx\|^2 + \left(1 - \frac{\lambda_0^2}{\lambda_1^2}\right)\|Qx\|^2.$$

Hence, and from estimate (8.3), it follows that

$$\|x\|^2 \leqslant \lambda_0^2\|PFx\|^2 + \lambda_0^2\|QFx\|^2 + \left(1 - \frac{\lambda_0^2}{\lambda_1^2}\right)\|Qx\|^2,$$

or, equivalently,

$$\|x\|^2 + \left(\frac{\lambda_0^2}{\lambda_1^2} - 1\right)\|Qx\|^2 \leqslant \lambda_0^2\|Fx\|^2.$$

Now, using the estimate (9.10) we get

$$\|x\|^2 + \left(\frac{\lambda_0^2}{\lambda_1^2} - 1\right)\|Qx\|^2 \leqslant \|x\|^2 - \lambda_0^2\int_\Omega \Phi[t, |x(t)|]\mathrm{d}\mu + \lambda_0^2\alpha(\|x\|),$$

that is, inequality (2.6) is satisfied for β given by (9.8), and with $h(t) = Qx(t)$, $\xi e(t) + h(t) = x(t)$. The compatibility of the family (9.7) with the function (9.3) implies the estimate $|\xi| \leqslant c$, which, in its turn, leads to the estimate (9.11). Lemma 9.2 is proved.

9.2. CONCLUSION OF THE PROOF OF THEOREM 8.1. Let ρ be the number defined by formula (9.11). We define on the space L_2 the continuous linear operator

$$Fx(t) = \begin{cases} f[t, x(t)] & \text{if } \|x(t)\| \leqslant \rho, \\ (1 + \rho - \|x\|)f[t, x(t)] & \text{if } \rho < \|x(t)\| < \rho + 1, \\ 0 & \text{if } \rho + 1 \leqslant \|x(t)\|. \end{cases} \qquad (9.12)$$

For the completely continuous operator AF we have the obvious estimate

$$\|AFx(t)\| \leqslant \lambda_0 \sup_{\|x\| \leqslant \rho+1} \|f[t, x(t)]\| = d < \infty, \quad x \in L_2.$$

Therefore, according to the Schauder Principle, the equation

$$x = AFx \qquad (9.13)$$

is solvable in the space L_2, for any superposition operator f acting on L_2 that defines the operator (9.12). We set

$$k_0 = \min \left\{ 1 + \lambda_0^{-1}, \sqrt{\frac{1}{\lambda_0^2} + \frac{\alpha(\rho+1)}{(\rho+1)^2}} \right\}.$$

Then, the number k_0 satisfies, besides condition (9.4), the additional inequality

$$\left(k_0^2 - \lambda_0^{-2} \right)(\rho+1)^2 \leqslant \alpha(\rho+1). \qquad (9.14)$$

We show that under the conditions in Theorem 8.1, equation (8.1) is solvable in the space L_2, for $k \leqslant k_0$. In case $k < \lambda_0^{-1}$, this assertion follows from Theorem 7.1. We next assume that $\lambda_0^{-1} \leqslant k \leqslant k_0$. To prove the solvability of equation (8.1), it will be enough to show that all the solutions of the corresponding equation (9.13) lie in the ball $\|x\| \leqslant \rho$. Indeed, in view of (9.12), these solutions satisfy equation (8.1) as well.

We assume that the solution $x_*(t)$ of equation (9.13) lies in the domain $\rho < \|x\| < \rho + 1$. Then, $x_*(t)$ is a solution of equation (9.9), where $\theta = 1 + \rho - \|x_*(t)\|$ and $Fx = fx(t)$. Therefore, based on Lemma 9.1, we get

$$\|f[t, x_*(t)]\|^2 \leqslant \lambda_0^{-2}\|x_*\|^2 - \int_{\Omega} \Phi[t, |x_*(t)|] d\mu + \left(k_0^2 - \lambda_0^{-2} \right)\|x_*\|^2.$$

Hence (by (9.14))

$$\|f[t, x_*(t)]\|^2 \leqslant \lambda_0^{-2}\|x_*\|^2 - \int_\Omega \Phi[t, |x_*(t)|]\mathrm{d}\mu + \alpha(\rho + 1).$$

But the monotony of the function $\alpha(u)$ implies that $\alpha(\rho + 1) \leqslant \alpha(\|x_*\|)$. Therefore,

$$\|f[t, x_*(t)]\|^2 \leqslant \lambda_0^{-2}\|x_*\|^2 - \int_\Omega \Phi[t, |x_*(t)|]\mathrm{d}\mu + \alpha(\|x_*\|).$$

Hence $x_*(t) \in X$, and by Lemma 9.2 we conclude that $\|x_*\| \leqslant \rho$ — a contradiction.

Equation (9.13) clearly has no solutions in the set $\|x\| \geqslant \rho + 1$ since on this set the operator AF equals zero.

The solvability of equation (8.1) is proved. To conclude the proof of our theorem, it remains to show that for $k \leqslant \lambda_0^{-1}$ the solutions of equation (8.1) admit a general a priori estimate. If $k < \lambda_0^{-1}$, the existence of such an estimate follows from Theorem 7.1. Suppose that $k = \lambda_0^{-1}$. Then, based on Lemma 9.1, we have the estimate (9.5), where $k = \lambda_0^{-1}$. Therefore, the estimate

$$\|fx_*(t)\|^2 \leqslant \lambda_0^{-2}\|x_*\|^2 - \int_\Omega \Phi[t, |x_*(t)|]\mathrm{d}\mu$$

is true, for any solution $x_*(t)$ of equation (8.1), i.e., $x_* \in X$. In view of Lemma 9.2, $x_*(t)$ satisfies the condition $\|x_*(t)\| \leqslant \rho$.

Theorem 8.1 is completely proved.

9.3. REMARKS. In the statement of Theorem 8.1 we assumed that equality (8.7) holds for every $u_* \geqslant u_0$. From the proof of this theorem, it turns out that we used equality (8.7) only for the value of u_* defined by formula (9.2).

The proof of Theorem 8.1 also shows that under its assumptions the rotation of the compact vector field $x - Afx$ on some sphere $\|x\| = \rho$ in L_2 equals 1.

9.4. PROOF OF THEOREM 8.2. For the sake of simplicity we will prove this theorem for $\lambda_0 = 1$ and assuming that

$$\Psi_0(t, u_0) < \frac{1}{2}u_0, \quad t \in \Omega. \tag{9.15}$$

We define, for each $k = 1, 2, \ldots$, the function

$$g_k(t, u) = \begin{cases} -k & \text{if } 0 \leqslant u \leqslant u_0, \\ -\dfrac{k(u - u_0)}{u_* - u_0} + \dfrac{u - u_0}{u_* - u_0}\Psi(t, u_*) & \text{if } u_0 < u < u_*, \\ \Psi_0(t, u) & \text{if } u_* \leqslant u. \end{cases}$$

The function $g_k(t, u)$ is continuous in u for each $t \in \Omega$, on the intervals $[0, u_0]$ and $[u_*, \infty)$ it coincides with $-k$ and $\Psi_0(t, u)$, respectively, and on the interval (u_0, u_*) is linear. By (9.15) we get the estimate

$$u - g_k(t, u) \geqslant \min \left\{ 1, \frac{1}{2} u_0 \right\} = c_1 > 0, \quad u \geqslant 0. \tag{9.16}$$

We next consider, for each $k = 1, 2, \ldots$, the equation

$$x(t) = P f_k[t, x(t)], \tag{9.17}$$

where P denotes the orthogonal projection onto the subspace E_0, that is, $Px(t) = e_0(t)[x, e_0]$, and

$$f_k(t, x) = \{|x| - g_k(t, |x|)\}\mathrm{sign}[e_0(t)].$$

Each of the equations (9.17) lies in the class $\mathfrak{L}(\Psi_0, E_0)$, and, according to our assumptions, has a solution $x_k(t)$ in the space L_2. In view of (9.17), this solution is given by $x_k = \xi_k e_0(t)$, where

$$\xi_k = \int\limits_{\Omega} \{|x_k(t)| - g_k[t, |x_k(t)|]\}|e_0(t)|d\mu. \tag{9.18}$$

From (9.16) and (9.18) we find the estimate

$$\xi_k \geqslant c_1 \int\limits_{\Omega} |e_0(t)|d\mu = c_2 > 0. \tag{9.19}$$

On the other hand,

$$\int\limits_{\Omega} |x_k(t)| \, |e_0(t)|d\mu = \xi_k,$$

i.e., $\|e_0\| = 1$, and therefore, equality (9.18) can be written as

$$\int\limits_{\Omega} g_k[t, |x_k(t)|]|e_0(t)|d\mu = 0, \quad k = 1, 2, \ldots . \tag{9.20}$$

Let

$$\Omega_k = \{t : t \in \Omega, \ |e_0(t)|\xi_k \geqslant u_0\}, \quad k = 1, 2, \ldots .$$

Since

$$\int\limits_{\Omega} \Phi_0[t, u_* + \xi_k|e_0(t)|]d\mu \geqslant \int\limits_{\Omega_k} \Phi_0[t, u_* + \xi_k|e_0(t)|]d\mu \geqslant$$

$$\geqslant \xi_k \int\limits_{\Omega_k} \Phi_0[t, u_* + \xi_k|e_0(t)|]|e_0(t)|d\mu,$$

based on (8.14) we have

$$\int_{\Omega} \Phi_0[t, u_* + \xi_k |e_0(t)|] \mathrm{d}\mu \geqslant \gamma \xi_k \int_{\Omega_k} \Phi_0[t, \xi_k |e_0(t)|] |e_0(t)| \mathrm{d}\mu,$$

and, according to the definition of $g_k(t, u)$ ($g_k(t, u) \leqslant \Psi_0(t, u)$ for $u \geqslant u_0$), we get

$$\int_{\Omega} \Phi_0[t, u_* + \xi_k |e_0(t)|] \mathrm{d}\mu \geqslant \gamma \xi_k \int_{\Omega_k} g_k[t, \xi_k |e_0(t)|] |e_0(t)| \mathrm{d}\mu.$$

Consequently, from (9.20) we obtain the estimate

$$\int_{\Omega} \Phi_0[t, u_* + \xi_k |e_0(t)|] \mathrm{d}\mu \geqslant - \gamma \xi_k \int_{\Omega \setminus \Omega_k} g_k[t, \xi_k |e_0(t)|] |e_0(t)| \mathrm{d}\mu,$$

i.e., the estimate

$$\int_{\Omega} \Phi_0[t, u_* + \xi_k |e_0(t)|] \mathrm{d}\mu \geqslant k \gamma \xi_k \int_{\Omega \setminus \Omega_k} |e_0(t)| \mathrm{d}\mu. \tag{9.21}$$

If the sequence ξ_k would contain a bounded subsequence $\xi_{k(j)}$ ($j = 1, 2, \ldots$) then, from $\xi_{k(j)} \leqslant \eta < \infty$ and from inequalities (9.19) and (9.21), we would conclude that

$$\int_{\Omega} \Phi_0[t, u_* + c_2 |e_0(t)|] \mathrm{d}\mu \geqslant k(j) \gamma c_2 \int_{\Omega_*} |e_0(t)| \mathrm{d}\mu,$$

where $\Omega_* = \{t : t \in \Omega, \ \eta |e_0(t)| \leqslant u_0\}$. Actually, the last estimate can not be reached for all j, since the left hand side is a finite number, whereas the right hand side increases unboundedly together with $k(j)$ because the integral in the right hand side is positive according to our hypothesis: $\chi(0; e_0) = 0$ and $\chi(\delta; e_0) > 0$, for $\delta > 0$. Thus,

$$\lim_{k \to \infty} \xi_k = \infty. \tag{9.22}$$

By (9.21) we find that

$$\int_{\Omega} \Phi_0[t, u_* + \xi_k |e_0(t)|] \mathrm{d}\mu \geqslant k \gamma \xi_k \int_{\{t : \frac{1}{2} u_0 < \xi_k |e_0(t)| \leqslant u_0\}} |e_0(t)| \mathrm{d}\mu \geqslant$$

$$\geqslant \frac{1}{2} u_0 k \gamma \cdot \mu \left\{ t : t \in \Omega, \ \frac{1}{2} u_0 \xi_k^{-1} < |e_0(t)| \leqslant \xi_k^{-1} u_0 \right\} \geqslant$$

$$\geqslant \frac{1}{2} u_0 k \gamma \left[\chi \left(u_0 \xi_k^{-1}; e_0 \right) - \chi \left(\frac{1}{2} u_0 \xi_k^{-1}; e_0 \right) \right].$$

But (in view of (8.13) and (9.22)), for sufficiently large values of k, the inequality

$$\chi\left(u_0\xi_k^{-1}; e_0\right) - \chi\left(\frac{1}{2}u_0\xi_k^{-1}; e_0\right) \geqslant (\beta - 1)\chi\left(\frac{1}{2}u_0\xi_k^{-1}; e_0\right)$$

holds true. Therefore, for large values of k, we get the estimate

$$\int_\Omega \Phi_0[t, u_* + \xi_k|e_0(t)|]\mathrm{d}\mu \geqslant \frac{1}{2}u_0k\gamma(\beta - 1)\chi\left(\frac{1}{2}u_0\xi_k^{-1}; e_0\right),$$

that is,

$$\frac{\chi(\delta_k; e)}{\displaystyle\int_\Omega \Phi_0\left[t, u_* + R\delta_k^{-1}|e_0(t)|\right]\mathrm{d}\mu} \leqslant \frac{2}{u_0k\gamma(\beta - 1)}, \tag{9.23}$$

where

$$\delta_k = \frac{1}{2}u_0\xi_k^{-1}, \quad R = \frac{1}{2}u_0.$$

From (9.23) it follows that

$$\lim_{k\to\infty} \frac{\chi(\delta_k; e)}{\displaystyle\int_\Omega \Phi_0\left[t, u_* + R\delta_k^{-1}|e_0(t)|\right]\mathrm{d}\mu} = 0,$$

whence relation (8.15) follows, since (9.22) implies that $\delta_k \to 0$. Theorem 8.2 is proved.

§10. Two-point boundary value problems

10.1. QUASILINEAR EQUATIONS. In Section 7 we studied the operator A of the two-point boundary value problem, and we established estimates for the distributions of the eigenfunctions of this operator. This section is concerned with the study the quasilinear equation containing the operator A.

We consider the quasilinear equation

$$Lx \overset{\text{def}}{=} x'' + p(t)x' + q(t)x = f(t, x), \tag{10.1}$$

where the coefficients $p(t)$ and $q(t)$ are continuous on $[0, T]$, and, at the same time, the function $f(t, x)$ is jointly continuous (in the variables $t \in [0, T]$ and $x \in \mathbb{R}^1$).

We will be interested in finding conditions under which equation (10.1) has solutions satisfying the boundary conditions

$$x(0) = x(T) = 0. \tag{10.2}$$

This problem has been considered by a lot of authors (C. N. Bernstein, Birkhoff, Tonelli, M. A. Krasnoselskii, and many others).

We assume that the equation $Lx = 0$ has no nontrivial solutions satisfying (10.2). Then, we may define the operator A of the two-point boundary value problem. We let Λ denote the spectral radius of the operator A, i.e.,

$$\Lambda = \sup_{\lambda \in \sigma(A)} |\lambda|$$

(by $\sigma(A)$ we denote the spectrum of the operator A). The subspace

$$E_0 = \{e(t) : e(t) \in L_2,\ \|Ae(t)\|_\mu = \Lambda \|e(t)\|_\mu\} \tag{10.3}$$

is invariant for A and has dimension 1 or 2. Obviously, $\dim E_0 = 1$ if and only if either $\Lambda \in \sigma(A)$ and $-\Lambda \notin \sigma(A)$, or $-\Lambda \in \sigma(A)$ and $\Lambda \notin \sigma(A)$, whereas $\dim E_0 = 2$ whenever $\Lambda \in \sigma(A)$ and $-\Lambda \in \sigma(A)$.

For example, if $T = \pi$ and $p(t) \equiv q(t) \equiv 0$, then $\Lambda = 1$ and $\dim E_0 = 1$; if $T = \pi$, $p(t) \equiv 0$, and $q(t) \equiv -2.5$, then $\Lambda = 2/3$ and $\dim E_0 = 2$.

The orthogonal complement of E_0 in L_2 is also invariant for the operator A. The estimates (8.3) are satisfied for a certain $\lambda_1 < \lambda_0 = \Lambda$.

We suppose that $\Psi(u)$ ($u \geqslant u_0 > 0$) is a continuous, positive and nonincreasing function. Let $[a, b] \subset [0, T]$ be a subinterval; one or both of its endpoints a and b may coincide with 0 or T. Let

$$\Psi(t, u) = \begin{cases} 0 & \text{if } t \notin [a, b], \\ \Psi(u) & \text{if } t \in [a, b], \end{cases} \quad u \geqslant u_0, \tag{10.4}$$

and assume that the function $\Psi(t, u)$ is bounded for $0 \leqslant u \leqslant u_0$. Then the function $u\Psi(t, u)$ belongs to the class $\mathfrak{N}(u_0)$.

THEOREM 10.1. *Assume that the function $\Psi(u)$ satisfies one of the following three conditions:*

$$\int_{u_0}^{\infty} u\Psi(u)\,du = \infty, \tag{10.5}$$

if $\dim E_0 = 1$ *and the function* $e_0(t) \in E_0$ *vanishes on* $[a, b]$;

$$\lim_{u \to \infty} u^2 \Psi(u) = \infty, \qquad (10.6)$$

if $\dim E_0 = 1$ *and* $e_0(t) \neq 0$ $(e_0(t) \in E_0, t \in [a, b])$;

$$\lim_{u \to \infty} u^{\frac{4}{3}} \Psi(u) = \infty, \qquad (10.7)$$

if $\dim E_0 = 2$.

Then there exists a number $\varepsilon_0 > 0$ such that, whenever

$$|f(t, x)| \leqslant \left(\frac{1}{\Lambda} + \varepsilon_0 \right) |x| - \Psi(t, |x|), \quad 0 \leqslant t \leqslant T; \ x \in \mathbb{R}^1, \qquad (10.8)$$

where $\Psi(t, u)$ is defined by equality (10.4), the existence of at least one solution $x(t) \in C^2$ of the problem (10.1)–(10.2) follows.

Proof. Since any solution $x(t) \in L_2$ of the equation $x(t) = Af[t, x(t)]$, where A is the operator of the of the two-point boundary value problem, is a solution $x(t) \in C^2$ of the problem (10.1)–(10.2), the conclusion of Theorem 10.1 follows easily from Theorem 8.1. All we need to prove is that condition (8.7) follows from conditions (10.5)–(10.7), for any $R > 0$ and $u_* \geqslant u_0$, where the function $\Psi(t, u)$, for $u \geqslant u_0$, is given by equality (10.4), and the subspace E_0 is defined by (10.3). Below, the function $u\Psi(u)$ will be denoted by $\Phi(u)$ $(u \geqslant u_0)$.

Assume first that $\dim E_0 = 1$. In this case, condition (8.7) can be rewritten as

$$\lim_{\delta \to 0} \frac{\chi(\delta; e)}{\displaystyle\int_\Omega \Phi_0[t, u_* + R\delta^{-1}|e_0(t)|]d\mu} = 0. \qquad (10.9)$$

Suppose, in addition, that $e_0(t_0) = 0$, for some $t_0 \in [a, b]$. Without any loss of generality, we may assume that $t_0 \neq T$. Since $e_0'(t_0) \neq 0$, then, for a sufficiently small $\varepsilon > 0$, we have the estimate

$$|e_0(t)| \leqslant c(t - t_0), \quad t_0 \leqslant t \leqslant t_0 + \varepsilon,$$

and, therefore,

$$\int_a^b \Phi[u_0 + R\delta^{-1}|e_0(t)|]dt \geqslant \int_{t_0}^{t_0+\varepsilon} \Phi[u_0 + R\delta^{-1}c(t - t_0)]dt. \qquad (10.10)$$

Equality (10.9) follows from (10.5), in view of (7.11) and (10.10).

Assume next that $e_0(t) > 0$, for all $t \in [a, b]$. Since $e_0(t)$ is bounded, then

$$\int\limits_a^b \Phi[u_0 + R\delta^{-1}|e_0(t)|]dt \geqslant \Phi\left[u_0 + R\delta^{-1} \cdot \sup_t |e_0(t)|\right],$$

and equality (10.9) follows from (10.6), in view of (7.11).

If dim $E_0 = 2$, then condition (8.7) follows from the uniform boundedness on $[a, b]$ of the functions $e(t) \in E_0$, $|e(t)| = 1$, and from Lemma 7.1 (the right hand side inequality in (7.13)), and condition (10.7). Theorem 10.1 is proved.

The strongest of the conditions (10.5)–(10.7) is condition (10.7). Therefore, whenever condition (10.7) is satisfied, the conclusion of Theorem 10.1 is true, regardless the dimension of the subspace E_0 and the behavior of the functions from E_0 on the interval $[a, b]$.

10.2. NONLINEARITIES DEPENDING ON DERIVATIVES. In this subsection we consider a more general equation than (10.1) above, namely,

$$Lx \overset{\text{def}}{=} x'' + p(t)x' + q(t)x = f(t, x, x'). \tag{10.11}$$

We suppose that the function $f(t, x, y)$ is jointly continuous in all its variables $t \in [0, T]$ and $x, y \in \mathbb{R}^1$, and that the equation $Lx = 0$ has no nontrivial solutions satisfying the boundary condition (10.2).

THEOREM 10.2. *Assume that the function $\Psi(u)$ satisfies the conditions in Theorem 10.1. Then there exists a number $\varepsilon_0 > 0$ such that, whenever*

$$|f(t, x, y)| \leqslant \left(\frac{1}{\Lambda} + \varepsilon_0\right)|x| - \Psi(t, |x|), \quad 0 \leqslant t \leqslant T; \ -\infty < x, y < \infty, \tag{10.12}$$

where $\Psi(t, u)$ is defined by equality (10.4), the existence of at least one solution $x(t) \in C^2([0, T])$ of the problem (10.11), (10.2) follows.

We consider the operator A of the two-point boundary value problem constructed out of the differential operator in the left hand side of (10.11). We define a new operator A_1 by the equality

$$A_1 x(t) = \frac{\mathrm{d}}{\mathrm{d}t} A x(t). \tag{10.13}$$

The operator (10.13) acts on L_2, is completely continuous, and sends any function from L_2 into a continuous one.

The operator

$$Bz(t) = f[t, Az(t), A_1 z(t)] \tag{10.14}$$

also acts on the space L_2, is completely continuous, and any solution $z(t) \in L_2$ of the equation $z = Bz$ is continuous on $[0, T]$ and defines a twice differentiable solution $x(t) = Az(t)$ of the problem (10.11), (10.2).

To prove Theorem 10.2, we have to show that under its assumptions there exists an $\varepsilon_0 > 0$, such that (10.12) implies the existence of a solution $z(t) \in L_2$ of the equation $z = Bz$. The proof is analogous to the proof of Theorem 8.1.

10.3. DIRICHLET PROBLEM. Theorem 8.1 provides new criteria for the existence of solutions of boundary value problems for partial differential equations. We present two simple examples. To keep the previously introduced notations, we let the variable t denote a point in the plane $\mathbb{R}^2$.

For our first example, let

$$\Omega = \{t = \{t_1, t_2\},\ 0 \leqslant t_1, t_2 \leqslant \pi\}$$

be a square in the plane. We consider the Dirichlet problem

$$\Delta x + \alpha x = f(t, x) \tag{10.15}$$

(Δ is the Laplace operator),

$$x(t) \Big|_{t \in \partial \Omega} = 0. \tag{10.16}$$

If $\alpha \neq n_1^2 + n_2^2$ for any integers n_1 and n_2, then the linear equation $\Delta x + \alpha x = 0$ has no nontrivial solutions subject to the boundary condition (10.16).We next consider the operator

$$Au(t) = \sum_{n=1}^{\infty} \sum_{m=1}^{\infty} (\alpha - n^2 - m^2)^{-1} \left\{ \int_0^\pi \int_0^\pi e_n(x) e_m(y) u(x, y) dx dy \right\} e_n(x) e_m(y), \tag{10.17}$$

$$u(t) \equiv u(x, y) \in L_2(R),$$

where $e_n(\tau) = \sqrt{\frac{2}{\pi}} \sin n\tau$. The operator (10.17) associates to any function $u(t) \in L_2(\Omega)$ a solution $x(t) = Au(t)$ of the equation $\Delta x + \alpha x = u(t)$. The eigenfunctions $e_{n,m}(t) = e_n(x) e_m(y)$ $(0 \leqslant x, y \leqslant \pi)$ of the operator (10.17) correspond to the eigenvalues $\lambda_{n,m} = (\alpha - n^2 - m^2)^{-1}$. We introduce the notation

$$\Lambda = \sup_{n,m \in \mathbb{N}} |\lambda_{n,m}|.$$

Let us now consider the subspace

$$E_0 = \{e(t) : e(t) \in L_2(\Omega), \ \|Ae(t)\| = \Lambda\|e(t)\|\}. \tag{10.18}$$

If $\dim E_0 = 1$, then

$$c_2\delta \ln\frac{1}{\delta} \leqslant \chi(\delta; e) \leqslant c_1\delta \ln\frac{1}{\delta}, \quad 0 \leqslant \delta \leqslant \delta_0; \ e \in E_0, \ \|e\| = 1. \tag{10.19}$$

If $\dim E_0 = 2$, then

$$c_2\sqrt{\delta} \leqslant \sup_{e\in E_0; \ \|e\|=1} \chi(\delta; e) \leqslant c_1\sqrt{\delta}, \quad 0 \leqslant \delta \leqslant \delta_0.$$

The dimension of the subspace (10.18) can be even greater. For instance, if $\alpha = 63$, then $\Lambda = \frac{1}{2}$ and $\dim E_0 = 6$; specifically, E_0 contains the functions:

$$\sin t_1 \sin 8t_2, \quad \sin 8t_1 \sin t_2, \quad \sin 7t_1 \sin 4t_2,$$
$$\sin 4t_1 \sin 7t_2, \quad \sin 5t_1 \sin 6t_2, \quad \sin 6t_1 \sin 5t_2.$$

If $2\alpha < 95$ is not a natural number, then either $\dim E_0 = 1$, or $\dim E_0 = 2$.

We assume that the function $\Psi(u)$ $(u \geqslant u_0 > 0)$ is positive, continuous, and nonincreasing. Moreover, let $\Psi(t, u)$ be a function that is bounded for $t \in \Omega$ and $u \geqslant 0$, and, for $u \geqslant u_0$, is given by

$$\Psi(t, u) = \begin{cases} \Psi(u) & \text{if } t \in \Omega_0, \\ 0 & \text{if } t \in \Omega \setminus \Omega_0, \end{cases} \tag{10.20}$$

where $\Omega_0 \subset \Omega$ and $\mu\Omega_0 > 0$. The function $u\Psi(t, u)$ lies in the class $\mathfrak{N}(u_0)$.

THEOREM 10.3. *Assume that, either* $\dim E_0 = 1$ *and*

$$\lim_{u\to\infty} \frac{u^2\Psi(u)}{\ln u} = \infty,$$

or, $\dim E_0 = 2$ *and*

$$\lim_{u\to\infty} u\sqrt{u}\Psi(u) = \infty.$$

Then there exists $\varepsilon_0 > 0$ *such that, whenever*

$$|f(t, x)| \leqslant \left(\frac{1}{\Lambda} + \varepsilon_0\right)|x| - \Psi(t, |x|), \quad t \in \Omega; \ x \in \mathbb{R}^1, \tag{10.21}$$

the existence of at least one solution of the Dirichlet problem (10.15)–(10.16) *follows.*

Theorem 10.3 follows from Theorem 8.1.

We proceed with the study of our second example. Let $\Omega \subset \mathbb{R}^2$ be a bounded domain with smooth boundary, and consider the Dirichlet problem

$$\Delta x = f(t, x), \tag{10.22}$$

with the boundary condition (10.16). We let A be the operator that associates to each function $u(t) \in L_2(\Omega)$ the solution $x(t) = Au(t)$ of the equation $\Delta x = u(t)$, subject to condition (10.16). Let us introduce the notation $\Lambda = \sup\{\lambda : \lambda \in \sigma(A)\}$. We also consider the subspace (10.18). The equality $\dim E_0 = 1$, and the estimates

$$c_2\delta \leqslant \chi(\delta; e) \leqslant c_1\delta, \quad 0 \leqslant \delta \leqslant \delta_0; \ e \in E_0, \ \|e\| = 1$$

hold true.

We assume that the function $\Psi(u)$ is positive, continuous, and nonincreasing for $u \geqslant u_0$. Moreover, let $\Psi(t, u)$ be a function that is bounded for $t \in \Omega$ and $u \geqslant 0$, and, for $u \geqslant u_0$, is given by (10.20), where $\Omega_0 \subset \Omega$ and $\mu\Omega_0 > 0$. The function $u\Psi(t, u)$ belongs to the class $\mathfrak{N}(u_0)$.

THEOREM 10.4. *Suppose that $\Omega = \Omega_0$ and*

$$\int\limits_{u_0}^{\infty} u\Psi(u)\mathrm{d}u = \infty,$$

or $\Omega \neq \Omega_0$ and

$$\lim_{u\to\infty} u^2\Psi(u) = \infty.$$

Then there exists $\varepsilon > 0$ such that, whenever the estimate (10.21) holds true, the existence of at least one solution of the Dirichlet problem (10.22), (10.16) follows.

Theorem 10.4 also follows from Theorem 8.1.

§11. Forced oscillations in control systems

11.1. CLOSED SINGLE-LOOP SYSTEMS. We consider a closed control system consisting of a linear link with the fractional-rational transfer function

$$W(p) = \frac{M(p)}{L(p)} \tag{11.1}$$

($M(p)$ and $L(p)$ are the polynomials given by (7.20) and (7.21), respectively), and a functional nonlinear link.

Its dynamics is described by the equation

$$L\left(\frac{\mathrm{d}}{\mathrm{d}t}\right) x(t) = M\left(\frac{\mathrm{d}}{\mathrm{d}t}\right) f(t, x). \tag{11.2}$$

The function $f(t, x)$ is assumed to be measurable and T-periodic in the variable t and continuous in the variable x. We will be concerned with problems related to forced T-periodic oscillations for this system, i.e., to T-periodic solutions of equation (11.2).

Assume that the numbers $\omega_k \mathrm{i}$ (where ω_k are given by (7.25)) are not roots of the polynomial $L(p)$, for any integer k. Then the operator A of the periodic problem for the linear link with the transfer function (11.1) is well defined (see Subsection 7.4).

Any T-periodic solution of equation (11.2), restricted to the interval $[0, T]$, is a solution for the operator equation

$$x(t) = Af[t, x(t)]. \tag{11.3}$$

Equation (11.3) is considered in the space L_2, and all the solutions $x(t) \in L_2$ of this equation also belong to the space C^{l-m-1}. Any solution of equation (11.3), extended by periodicity with period T, is a T-periodic solution of equation (11.2).

In this section we will give conditions for the existence of T-periodic oscillations for the closed single-loop system described above, as well as conditions for the applicability of the harmonic balance method to obtain T-periodic solutions of equation (11.2).

11.2. HARMONIC BALANCE METHOD (HBM). Recall that by $e_k(t)$ and $g_k(t)$ we denote the functions given by (7.26), and that P_k stand for the orthogonal projections onto the subspaces Π_k defined by (7.27). Given an arbitrary number N, we introduce the notations

$$E(N) = \bigoplus_{k=0}^{N} \Pi_k; \quad P(N) = \sum_{k=0}^{N} P_k.$$

Obviously, $P(N)$ is the orthogonal projection onto the $(2N+1)$-dimensional subspace $E(N)$ of all trigonometric polynomials of the form

$$x_N(t) = \xi_0 + \sum_{k=1}^{N} (\xi_k \cos \omega_k t + \eta_k \sin \omega_k t). \tag{11.4}$$

The projection $P(N)$ associates to every function $z(t) \in L_2$ the partial sum $P(N)z(t)$ in the Fourier series of the function $z(t)$.

The approximate T-periodic solution $x_N(t)$ provided by HBM has the form (11.4). The unknown coefficients

$$\xi_0, \xi_1, \eta_1, \ldots, \xi_N, \eta_N \tag{11.5}$$

in the trigonometric polynomial (11.4) are defined such that $x_N(t)$ satisfies the equality

$$L\left(\frac{d}{dt}\right) x_N(t) = M\left(\frac{d}{dt}\right) P(N)f[t, x_N(t)]. \tag{11.6}$$

Both the left and the right hand sides of relation (11.6) are trigonometric polynomials of degree N. By identifying the coefficients we get a system of $2N+1$ scalar equations — the HBM system of equations — with the $2N+1$ unknowns indicated in (11.5). If the numbers $\omega_k i$ are not roots of the polynomial $L(p)$, then equation (11.6) can be rewritten in the operator form

$$x_N(t) = AP(N)f[t, x_N(t)]. \tag{11.7}$$

Equation (11.7) may be equally well considered either in the finite-dimensional subspace $E(N)$, or in the whole space L_2 (or in $C([0, T])$); the set of its solutions remains the same.

11.3. REALIZABILITY AND CONVERGENCE OF THE HARMONIC BALANCE METHOD. The harmonic balance method is called *realizable* if equation (11.7) has solutions for any N. If the HBM equation is solvable only for all sufficiently large N, then we say that the method is *realizable for sufficiently large N*.

We denote by $\mathcal{F}$ the set of all T-periodic solutions of equation (11.2) restricted to $[0, T]$. In addition, we let $\mathcal{F}_N$ denote the set of approximate T-periodic solutions provided by HBM, that is, the solutions of equation (11.6). Both the sets $\mathcal{F}$ and $\mathcal{F}_N$ are included in L_2. Let us suppose that these sets are also included in a certain space E of functions $x(t) : [0, T] \to \mathbb{R}^1$, and let $\| \cdot \|_E$ denote the norm on E. We define the *Hausdorff deviation* of the set $\mathcal{F}_N$ from the set $\mathcal{F}$ by

$$\theta(\mathcal{F}_N, \mathcal{F}; E) = \sup_{z \in \mathcal{F}_N} \rho(z, \mathcal{F}; E) = \sup_{z \in \mathcal{F}_N} \inf_{x \in \mathcal{F}} \|x - z\| \tag{11.8}$$

Assume that the set $\mathcal{F}$ is nonempty and HBM is realizable (at least for large values of N). If

$$\lim_{N \to \infty} \theta(\mathcal{F}_N, \mathcal{F}; E) = 0, \tag{11.9}$$

then the HBM is called *convergent with respect to the metric on the space E*.

Let K be a subset of the space $E \subset L_2$. The harmonic balance method is said to be *convergent in the set* K (with respect to the metric on the space E) if the intersections $\mathcal{F} \cap K$ and $\mathcal{F}_N \cap K$ are nonempty, and

$$\lim_{N \to \infty} \theta(\mathcal{F}_N \cap K, \mathcal{F} \cap K; E) = 0. \tag{11.10}$$

Suppose next that HBM is realizable and each of the sets $\mathcal{F}_N$, $N = 1, 2, \ldots$, and $\mathcal{F}$ consists of a single element, i.e., the equations (11.7) ($N = 1, 2, \ldots$) and (11.3) have unique solutions. Then the definition (11.9) of the convergence of harmonic balance method has the very simple form

$$\lim_{N \to \infty} \|x - x_N\|_E = 0.$$

Let

$$\mathcal{B} = \{x(t) : \|x(t)\|_{C^{l-m-1}} \leqslant \rho\}.$$

From the general theory of the projection procedure (see, for instance, [Krasnoselskii, M. A., *et al.*, 1969]) we get:

LEMMA 11.1. *Assume that the sets* $\mathcal{F} \cap \mathcal{B}$ *and* $\mathcal{F}_N \cap \mathcal{B}$ *are nonempty. Then HBM converges in the ball* $\mathcal{B}$ *with respect to the metric on the space* C^{l-m-1}.

11.4. APPLICABILITY CONDITIONS FOR HBM. In this subsection we establish conditions of solvability for equation (11.3), i.e., conditions for the existence of forced T-periodic oscillations in a closed single-loop system, as well as conditions for the realizability and convergence of HBM.

Assume that the numbers $\omega_k i$ are not roots of the polynomial $L(p)$, for any integer k. Then, the operator A of the periodic problem for the linear link with the transfer function (11.1) is well defined. Recall that $w(T)$ denotes the norm of the operator A in the space L_2; this norm was defined by equality (7.29).

THEOREM 11.1. *Suppose that the nonlinearity* $f(t, x)$ *satisfies the estimate*

$$|f(t, x)| \leqslant k|x| + b(t), \quad 0 \leqslant t \leqslant T, \ x \in \mathbb{R}^1; \ b(t) \in L_2, \tag{11.11}$$

with

$$k \cdot w(T) < 1. \tag{11.12}$$

Then equation (11.3) has at least one T-*periodic solution* $x(t) \in C^{l-m-1}$, *and HBM is realizable and converges with respect to the norm of the space* C^{l-m-1}.

Proof. The solvability of equation (11.3) follows from conditions (11.11) and (11.12), in view of Theorem 7.1. The realizability of HBM is a consequence of Theorem 7.1, too; it is enough to notice that for every N, the estimate

$$\|AP(N)\|_{L_2 \to L_2} = \sup_{k=0,\pm 1,\ldots,\pm N} |W(\omega_k i)| \leqslant w(T)$$

holds. Since under the assumptions in Theorem 11.1 we have the general estimate

$$\|x\|_{L_2}, \|x_N\|_{L_2} \leqslant \|b(t)\|_{L_2} \cdot w(T)[1 - kw(T)]^{-1}$$

for both the solution $x(t)$ of equation (11.3) and the approximate solution $x_N(t)$ of degree N provided by HBM, the convergence of HBM follows from Lemma 11.1. Theorem 11.1 is proved.

Let us replace estimate (11.11) by the quite similar inequality

$$|f(t,x)| \leqslant k|x| - \Psi(t,|x|), \quad 0 \leqslant t \leqslant T, \ x \in \mathbb{R}^1, \tag{11.13}$$

where $\Psi(t,u)$ is bounded for $u \geqslant 0$ and $t \in [0,T]$, and nonnegative and continuous in u for $u \geqslant u_0 > 0$. Moreover, assume that for $t \in \Omega_0 \subset [0,T]$ (mes $\Omega_0 > 0$), and for $u \geqslant u_0$, the function $\Psi(t,u)$ does not depend on t, i.e.,

$$\Psi(t,u) \equiv \Psi(u), \quad t \in \Omega_0, \ u \geqslant u_0,$$

and that the function $u\Psi(u)$ is nonincreasing for $u \geqslant u_0$. This assumption means that $u\Psi(u) \in \mathfrak{N}(u_0)$.

Recall that l denotes the degree of the polynomial $L(p)$.

THEOREM 11.2. *Suppose that, if $l > 1$, then the function $\Psi(u)$ satisfies the condition*

$$\lim_{u \to \infty} u^{1+\kappa} \Psi(u) = \infty, \tag{11.14}$$

where

$$\kappa = \begin{cases} (2l-1)^{-1} & \text{if } l \text{ is even,} \\ (2l-2)^{-1} & \text{if } l \text{ is odd.} \end{cases} \tag{11.15}$$

Then the next assertions are true:

a) There exists a number $k_0 > [w(T)]^{-1}$ such that, for $k \leqslant k_0$, the estimate (11.13) implies the existence of at least one T-periodic solution $x(t) \in C^{l-m-1}$ of equation (11.3). In addition, HBM is realizable and converges in some ball

$$\left\{ x(t) : x(t) \in C^{l-m-1}, \ \|x\|_{C^{l-m-1}} \leqslant d \right\}$$

with respect to the norm of the space C^{l-m-1}.

b) *If $k \leqslant [w(T)]^{-1}$, then the estimate (11.13) implies a general a priori estimate*

$$\|x\|_{C^{l-m-1}}, \|x_N\|_{C^{l-m-1}} \leqslant \text{const} < \infty, \tag{11.16}$$

for all T-periodic solutions $x(t)$ of equation (11.3) and all the approximate solutions $x_N(t)$ provided by HBM. Therefore, HBM is realizable and converges with respect to the norm of the space C^{l-m-1}.

Theorem 11.2 will be proved in the following subsections of Section 11.

If a more detailed information on the polynomials $L(p)$ and $M(p)$ is available, then condition (11.14) can be weakened.

Let α denote the number of distinct roots p of the form $\omega_k i$ ($k = 0, \pm 1, \pm 2, \ldots$) of the equation $|M(p)| = w(T)|L(P)|$, i.e., the number of those k for which the maximum value in the right hand side of definition (7.31) of the number $w(T)$ is attained. Clearly, $1 \leqslant \alpha \leqslant 2l$.

If $\alpha = 1$, then condition (11.14) is no longer required. If $\alpha > 1$, then in (11.14) we can take $\kappa = (\alpha - 1)^{-1}$. Suppose that $\alpha = 2$ and the function $\Psi(t, u)$ does not depend on t for $u \geqslant u_0$ on the whole interval $[0, T]$ (i.e., $\Omega_0 = [0, T]$). Then, condition (11.14), which in this case becomes

$$\lim_{u \to \infty} u^2 \Psi(u) = \infty, \tag{11.17}$$

can be replaced by the less restrictive condition

$$\int_{u_0}^{\infty} u\Psi(u)\mathrm{d}u = \infty. \tag{11.18}$$

Theorem 11.2 can be complemented as follows. Suppose that $l > 1$ is odd, and $w(T) \neq |W(0)|$. Then in condition (11.14) we can set $\kappa = (2l - 3)^{-1}$.

11.5. DISTRIBUTIONS OF TRIGONOMETRIC POLYNOMIALS. Let us denote by K_{2s+1} ($s = 0, 1, \ldots$) the set of all trigonometric polynomials on the interval $[0, T]$ of the form

$$\xi_0 + \sum_{j=1}^{s} [\xi_j \cos \omega_{k(j)} t + \eta_j \sin \omega_{k(j)} t],$$

and by K_{2s} ($s = 1, 2, \ldots$) the set of trigonometric polynomials of the form

$$\sum_{j=1}^{s} [\xi_j \cos \omega_{k(j)} t + \eta_j \sin \omega_{k(j)} t],$$

where $0 < k(1) < k(2) < \cdots < k(s)$ are some fixed natural numbers. $\mathfrak{F}_r$ will denote the family

$$\mathfrak{F}_r = \mathfrak{F}_r[k(1), \ldots, k(s)] = \{e(t) : e(t) \in K_r, \|e\| = 1\},$$

where $\|\cdot\|$ stands for a certain norm on the finite dimensional space K_r.

LEMMA 11.2. *The functions* (1.5) *and* (1.6) *corresponding to the family* $\mathfrak{F}_r$ *of trigono-metric polynomials satisfy the next properties:*

a) *for* $r = 1$, *the functions* (1.5) *and* (1.6) *coincide and equal zero, for all sufficiently small values of* δ;

b) *for each* $r = 2, 3, \ldots$, *and any* $\delta_0 > 0$, *the inequalities*

$$c_1 \delta^{\frac{1}{r-1}} \leqslant \chi_U(\delta; \mathfrak{F}_r) \leqslant c_2^{\frac{1}{r-1}}, \quad 0 \leqslant \delta \leqslant \delta_0. \tag{11.19}$$

hold true, for certain $c_1, c_2 > 0$;

c) *for each odd integer* $r > 1$, *there exists* $\delta_0 > 0$ *such that*

$$\chi_L(\delta; \mathfrak{F}_r) = 0, \quad 0 \leqslant \delta \leqslant \delta_0; \tag{11.20}$$

d) *for each even integer* r *and any* $\delta_0 > 0$, *the inequalities*

$$c_1 \delta \leqslant \chi_L(\delta; \mathfrak{F}_r) \leqslant c_2 \delta, \quad 0 \leqslant \delta \leqslant \delta_0, \tag{11.21}$$

hold true, for certain $c_1, c_2 > 0$.

Proof. We start with a simple remark. The set K_{2s+1} coincides with the space of all solutions of the differential equation

$$\frac{d}{dt} \circ \left(\frac{d^2}{dt^2} + \omega_{k(1)}^2 \right) \circ \cdots \circ \left(\frac{d^2}{dt^2} + \omega_{k(s)}^2 \right) x(t) = 0$$

of order $2s + 1$, with constant coefficients. The set K_{2s} coincides with the space of all solutions of the differential equation

$$\left(\frac{d^2}{dt^2} + \omega_{k(1)}^2 \right) \circ \cdots \circ \left(\frac{d^2}{dt^2} + \omega_{k(s)}^2 \right) x(t) = 0$$

of order $2s$. Both these assertions have straightforward proofs. Thus, the set K_r is the space of solutions of an ordinary linear differential equation of order r, with constant coefficients.

We proceed with the proof of our lemma.

Assertion a) in Lemma 11.2 is obvious.

Assertion b) (inequality (11.19)) follows from Theorem 1.1 in view of the previous remark.

Assertion c) follows from the fact that $e_0(t) \in \mathfrak{F}_r$ (in case r is odd, $e_0(t)$ is the first function listed in (7.26)).

We prove now the last assertion — assertion d) — in Lemma 11.2. Assume that $e(t) \in \mathfrak{F}_r$, where r is even. Then obviously we have

$$\int\limits_0^T e(t)\mathrm{d}t = 0.$$

Therefore, any fixed function $e(t) \in \mathfrak{F}_r$ vanishes at least at one point $t_0 \in [0, T]$. All the norms on K_r are equivalent (since $\dim K_r = r < \infty$), hence

$$|e'(t)| \leqslant c, \quad 0 \leqslant t \leqslant T, \ e(t) \in \mathfrak{F}_r,$$

where $c > 0$ is a constant, the same one for all $e(t) \in \mathfrak{F}_r$. Consequently, from $e(t_0) = 0$ we get the estimate $|e(t)| \leqslant c|t - t_0|$ which, in its turn, implies that the left hand side in estimate (11.21) holds true.

To prove the estimate in the right hand side of (11.21) it suffices to consider the function $g_{k(1)}(t) \in \mathfrak{F}_r$. A direct computation shows that

$$\chi(\delta; g_{k(1)}) = \frac{2T}{\pi} \arcsin \sqrt{\frac{T}{2}} \delta,$$

and, since $\chi_{\mathrm{L}}(\delta; \mathfrak{F}_r) \leqslant \chi(\delta; g_{k(1)})$, the right hand side in (11.21) clearly follows.

Lemma 11.2 is completely proved.

Lemma 11.2 asserts, in particular, that the estimates of the functions (1.5) and (1.6) associated with the family $\mathfrak{F}_r = \mathfrak{F}_r[k(1), \ldots, k(s)]$ are determined (up to a factor) by the dimension r of the space K_r.

In the sequel we will need a simple algebraic result on the number of integers that may be roots of polynomials.

LEMMA 11.3. *Let $Q(k)$ be a polynomial with real coefficients of degree $\deg Q(k) = 2q$ that contains terms of even degree only, and takes nonnegative values when k is an integer. Assume, in addition, that q is odd. Then the number of integers that are roots of the polynomial Q does not exceed $2q - 1$.*

Proof. Since the number of all roots equals the degree of Q, to prove the lemma it is enough to show that, under its assumptions, the polynomial Q can not have $2q$

distinct integers as its roots. Suppose, on the contrary, that $Q(k)$ has $2q$ distinct integers as roots. Then these roots are given by $\pm k_1, \ldots, \pm k_q$, and the polynomial Q can be written as

$$Q(k) = a \left(k^2 - k_1^2 \right) \cdot \cdots \cdot \left(k^2 - k_q^2 \right).$$

Since the number q is odd, we get the equalities $\operatorname{sign} Q(k) = -\operatorname{sign} a$, for $k = 0$, and $\operatorname{sign} Q(k) = \operatorname{sign} a$, for $k \gg \max |k_s|$, i.e., the polynomial $Q(k)$ takes values of different signs when restricted to the set of integers. The just obtained contradiction proves the lemma.

11.6. DIMENSIONS OF INVARIANT SUBSPACES. We return to the study of the operator A of the periodic problem for the link with the transfer function (11.1). Let us denote by E_0 the set

$$E_0 = \{ x(t) : x(t) \in L_2, \; \|Ax(t)\| = w(T)\|x(t)\| \}. \tag{11.22}$$

LEMMA 11.4. *Let $l = \deg L(p) > 1$. Then there exists $c > 0$ such that*

$$\chi_U(\delta; E_0) \leqslant c \cdot \delta^\kappa \quad (\delta \geqslant 0), \tag{11.23}$$

where κ is the number given by (11.15).

Proof. Based on Lemma 11.2 (assertions a) and b)), in order to prove Lemma 11.4 it will be enough to establish the estimate $\dim E_0 \leqslant 2l$, if l is even, and the estimate $\dim E_0 \leqslant 2l - 1$, if l is odd. From the equalities

$$E_0 = \left\{ x(t) : x(t) \in L_2, \; A \sum_{k=0}^{\infty} P_k x(t)^2 = w^2(T) \sum_{k=0}^{\infty} P_k x(t)^2 \right\} =$$

$$= \left\{ x(t) : x(t) \in L_2, \; \sum_{k=0}^{\infty} |W(\omega_k \mathrm{i})|^2 P_k x(t)^2 = w^2(T) \sum_{k=0}^{\infty} P_k x(t)^2 \right\} =$$

$$= \left\{ x(t) : x(t) \in L_2, \; A \sum_{k=0}^{\infty} \left[|W(\omega_k \mathrm{i})|^2 - w^2(T) \right] P_k x(t)^2 = 0 \right\},$$

it follows that the set E_0 consists of functions $x(t)$ of the form

$$x(t) = \sum_{\substack{k=0,1,2,\ldots; \\ w(T)=|W(\omega_k \mathrm{i})|}} \xi_k P_k x(t).$$

Consequently, $\dim E_0$ coincide with the number of distinct integers k that are roots of the equation

$$w(T) = |W(\omega_k \mathrm{i})|, \tag{11.24}$$

which, in view of our further purposes, can be more conveniently rewritten as

$$Q(k) \stackrel{\text{def}}{=} w^2(T)|L(\omega_k i)|^2 - |M(\omega_k i)|^2 \equiv$$

$$\equiv w^2(T) \cdot \left[a_l - \left(2\frac{\pi}{T}\right)^2 a_{l-2}k^2 + \left(2\frac{\pi}{T}\right)^4 a_{l-4}k^4 - \cdots \right]^2 +$$

$$+ w^2(T) \left(2\frac{\pi}{T}\right)^2 k^2 \cdot \left[a_{l-1} - \left(2\frac{\pi}{T}\right)^2 a_{l-3}k^2 + \left(2\frac{\pi}{T}\right)^4 a_{l-5}k^4 - \cdots \right]^2 +$$

$$+ w^2(T) \left[b_m - \left(2\frac{\pi}{T}\right)^2 b_{m-2}k^2 + \left(2\frac{\pi}{T}\right)^4 b_{m-4}k^4 - \cdots \right]^2 +$$

$$+ w^2(T) \left(2\frac{\pi}{T}\right)^2 k^2 \left[b_{m-1} - \left(2\frac{\pi}{T}\right)^2 b_{m-3}k^2 + \left(2\frac{\pi}{T}\right)^4 b_{m-5}k^4 - \cdots \right]^2 = 0.$$

The polynomial $Q(k)$ has degree $2l$ and satisfies all the conditions in Lemma 11.3. Therefore, the number of its distinct roots in the set of integers does not exceed $2l$, and, in case $l > 1$ is odd, that number does not exceed $2l - 1$. Lemma 11.4 is proved.

It is possible to find examples (for l either even, or odd) of polynomials $L(p)$ and $M(p)$, such that $\dim E_0$ equals $2l$ (when $l = 2q$) or $2l-1$ (when $l = 2q+1$). These examples show that, generally speaking, the estimate (11.23) can not be improved.

Assume that $l = 2q$, and set

$$L(p) = \left(p^2 + \omega_1^2\right) \cdots \cdots \left(p^2 + \omega_q^2\right) + \left(p^2 + \omega_{q+1}^2\right) \cdots \cdots \left(p^2 + \omega_{2q}^2\right),$$

$$M(p) = \left(p^2 + \omega_1^2\right) \cdots \cdots \left(p^2 + \omega_q^2\right) - \left(p^2 + \omega_{q+1}^2\right) \cdots \cdots \left(p^2 + \omega_{2q}^2\right).$$

Then $L(p) \neq 0$ for $p = \omega_k i$ ($k = 0, 1, 2, \ldots$), $l = \deg L(p)$, $m = \deg M(p) = l - 2$, $w(T) = 1$, and equation (11.24) has $2l$ roots all of them being integers: $\pm 1, \pm 2, \ldots, \pm l$. In this case $\dim E_0 = 2l$.

Assume next that $l = 2q + 1$, and set

$$L(p) = p\left(p^2 + \omega_1^2\right) \cdots \cdots \left(p^2 + \omega_q^2\right) + p\left(p^2 + \omega_{q+1}^2\right) \cdots \cdots \left(p^2 + \omega_{2q}^2\right) + 1,$$

$$M(p) = p\left(p^2 + \omega_1^2\right) \cdots \cdots \left(p^2 + \omega_q^2\right) - p\left(p^2 + \omega_{q+1}^2\right) \cdots \cdots \left(p^2 + \omega_{2q}^2\right) + 1.$$

In this case again $L(p) \neq 0$, for $p = \omega_k i$ ($k = 0, 1, 2, \ldots$), $l = \deg L(p)$, $m = \deg M(p) = l - 2$, $w(T) = 1$. Equation (11.24) has $2l - 1$ distinct integers as roots: $0, \pm 1, \pm 2, \ldots \ldots, \pm(l - 1)$. In this case $\dim E_0 = 2l - 1$.

11.7. PROOF OF THEOREM 11.2. We will first prove the existence of at least one T-periodic solution $x(t) \in C^{l-m-1}$ of equation (11.3). To this end we use Theorem 8.1.

The subspace E_0 is invariant for the operator A of the periodic problem. Let P denote the orthogonal projection of L_2 onto E_0, and let $Q = I - P$. We have the inequalities

$$\|APx\| \leqslant w(T)\|Px\|, \quad \|AQx\| \leqslant w_1(T)\|Qx\|, \quad x \in L_2, \tag{11.25}$$

where

$$w_1(T) = \sup_{\substack{k=0,1,2,\ldots; \\ w(T)\neq|W(\omega_k i)|}} |W(\omega_k i)| < w(T). \tag{11.26}$$

Inequalities (11.25) are, in fact, the conditions (8.3) in Theorem 8.1.

From condition (11.14) it follows that

$$\lim_{\delta \to 0} \frac{\delta^\kappa}{(u_* + R_1\delta^{-1})\, \Psi\,(u_* + R_1\delta^{-1})} = 0,$$

for any $R_1 > 0$ and $u_* \geqslant u_0$. Therefore, for any $l > 0$, in view of Lemma 11.4 and assertion a) in Lemma 11.2, we get (again for any $R_1 > 0$ and $u_* \geqslant u_0$)

$$\lim_{\delta \to 0} \frac{\chi_U(\delta;\mathfrak{F})}{(u_* + R_1\delta^{-1})\, \Psi\,(u_* + R_1\delta^{-1})} = 0 \quad (\mathfrak{F} = \{e : e \in E_0, \|e\| = 1\}),$$

which implies that

$$\lim_{\delta \to 0} \sup_{e(t)\in E_0;\ \|e\|=1} \frac{\chi(\delta;e)}{\displaystyle\int_0^T \left[u_* + R_1\delta^{-1}|e(t)|\right] \Psi \left[u_* + R_1\delta^{-1}|e(t)|\right]\, dt} = 0,$$

where

$$R = R_1 \left[\sup_{e(t)\in E_0;\ \|e\|=1} |e(t)| \right]^{-1}.$$

The last conclusion stands for condition (8.7) in Theorem 8.1.

By Theorem 8.1 it follows that there exists a number $k_0 > [w(T)]^{-1}$, such that whenever $k \leqslant k_0$, the estimate (11.13) implies the existence of at least one solution $x(t) \in L_2$ of equation (11.3) (we also notice that if (11.13) holds true for $k \leqslant [w(T)]^{-1}$, then an a priori norm estimate of that solution is available).

The first assertion in Theorem 11.2 follows from the previous remarks and from the properties of the operator of the periodic problem.

The realizability of the harmonic balance method can be proved analogously. The convergence of HBM follows from Lemma 11.1.

11.8. REMARKS. Assume that the nonlinearity $f(t, x)$ satisfies the condition

$$-b(t) + \beta|x| \leqslant f(t, x)\operatorname{sign} x \leqslant \alpha|x| + b(t), \quad t \in [0, T], \ x \in \mathbb{R}^1, \tag{11.27}$$

where $\alpha > \beta$, and $b(t) \in L_2$ is a nonnegative function. If $\alpha = -\beta = k$, then condition (11.27) coincides with (11.11).

In addition, suppose that the polynomial $L(p) - \dfrac{\alpha + \beta}{2} M(p)$ does not have roots of the form $\omega_k \mathrm{i}$. Denote by $w = w[T, \alpha, \beta]$ the number given by

$$w = \max_{k=0, \pm 1, \cdots} \left| \frac{M(\omega_k \mathrm{i})}{L(\omega_k \mathrm{i}) - \dfrac{\alpha + \beta}{2} M(\omega_k \mathrm{i})} \right|.$$

If $\alpha = -\beta = k$, then clearly $w = w(T)$.

THEOREM 11.3. *Assume that $w \cdot (\alpha - \beta) < 2$. Then equation (11.3) has at least one T-periodic solution $x(t) \in C^{l-m-1}$, and HBM is realizable and converges with respect to the norm of the space C^{l-m-1}.*

Theorem 11.3 follows by applying Theorem 11.1 to the equation

$$\left[L\left(\frac{\mathrm{d}}{\mathrm{d}t} \right) - \frac{\alpha + \beta}{2} M\left(\frac{\mathrm{d}}{\mathrm{d}t} \right) \right] x(t) = M\left(\frac{\mathrm{d}}{\mathrm{d}t} \right) \left[f(t, x) - \frac{\alpha + \beta}{2} x \right],$$

which is equivalent to (11.3). Condition (11.27) has to be rewritten as

$$\left| f(t, x) - \frac{\alpha + \beta}{2} x \right| \leqslant \frac{\alpha - \beta}{2} |x| + b(t), \quad t \in [0, T], \ x \in \mathbb{R}^1.$$

In the same way we can obtain an analog of Theorem 11.2 in case

$$\beta|x| + \Psi(t, |x|) \leqslant f(t, x)\operatorname{sign} x \leqslant \alpha|x| - \Psi(t, |x|), \quad t \in [0, T], \ x \in \mathbb{R}^1.$$

Chapter 2 consists mainly of results obtained in [Krasnoselskii, A.M., 1980b; Krasnoselskii, A.M., 1986; Krasnoselskii, A.M., 1987].

Chapter 3

The use of arguments
of leading eigenvalues

In this chapter we continue the study of the nonlinear operator equation $x = A\mathfrak{f}x$, where A is a normal completely continuous linear operator acting on a real Hilbert space H, and $\mathfrak{f}$ is a nonlinear operator.

To start with, let us suppose that the nonlinear operator $\mathfrak{f}$ satisfies the estimate $\|\mathfrak{f}x\| \leqslant k\|x\| + b$ (this is the case, for instance, when $H = L_2$ and $\mathfrak{f}$ is the superposition operator with a nonlinearity $f(t,x)$ subject to the condition $|f(t,x)| \leqslant k|x| + b(t)$). If $k\|A\| < 1$, then for the equation $x = A\mathfrak{f}x$ we may invoke the Schauder Principle to conclude that there exists at least one solution $x \in H$ of this equation. If $k\|A\| \geqslant 1$, then there are examples of equations $x = A\mathfrak{f}x$ without solutions.

However, if all the leading eigenvalues of the operator A have nonzero imaginary parts, then it is still possible to establish the solvability of equation $x = A\mathfrak{f}x$ in case an estimate $k\|A\| < \mu$ holds true, for a certain $\mu > 1$.

If the nonlinearity $\mathfrak{f}x$ satisfies the estimate $\|\mathfrak{f}x\| \leqslant \|B_1 x\| + \|B_2 x\| + b$, where B_1 and B_2 are bounded linear operators, then the simplest condition for the solvability of equation $x = A\mathfrak{f}x$ is given by $\|A\|(\|B_1\| + \|B_2\|) < 1$. It turns out that this condition may also be considerably weakened in many different situations.

The methods discussed in the present chapter are applied to the forced oscillation problem for various intricate systems, as well as to the two-point boundary value problem. These applications lead to new conditions for the existence and uniqueness of solutions. We also exhibit conditions that allow the use of harmonic balance method for complicated systems with delay, and for systems with control by derivatives. The method is quite general and it can be also used for some other nonlinear problems.

§12. Use of the arguments principle

12.1. AN AUXILIARY ALGEBRAIC EQUATION. Let $q_1 \in [0, 1)$ and $q_2 \in (0, 1)$ be two fixed numbers. Later on we will use the fourth degree equation

$$q_1^2 z^2 (z^2 - 1) - (1 - q_1^2 z^2)(1 - q_2 z)^2 = 0. \tag{12.1}$$

Since the left side of equality (12.1) is negative for $z = 1$ and positive for $z = q_2^{-1}$, equation (12.1) has at least one root in the interval $(1, q_2^{-1})$. For every such root μ we clearly have $\mu q_1 < 1$ and $\mu q_2 < 1$. On the other hand, for $zq_1 < 1$ and $zq_2 < 1$, the left side of equality (12.1) is a strictly increasing function, therefore the root μ of equation (12.1) in the interval $(1, q_2^{-1})$ is unique. We will denote it by $\mu(q_1, q_2)$. Recall that

$$q_1 \mu(q_1, q_2) < 1, \quad q_2 \mu(q_1, q_2) < 1. \tag{12.2}$$

Table 12.1 contains values of $\mu(q_1, q_2)$ for a few values of q_1 and q_2.

Table 12.1

q_1 q_2	.100	.200	.300	.400	.500	.600	.700	.800	.900
.100	2.739	2.373	2.064	1.808	1.597	1.424	1.281	1.163	1.067
.200	2.041	1.860	1.693	1.544	1.412	1.297	1.197	1.112	1.042
.300	1.710	1.598	1.491	1.391	1.299	1.216	1.142	1.078	1.027
.400	1.506	1.431	1.357	1.286	1.220	1.158	1.103	1.055	1.018
.500	1.363	1.312	1.260	1.209	1.161	1.115	1.074	1.038	1.012
.600	1.257	1.222	1.185	1.150	1.115	1.081	1.051	1.026	1.008
.700	1.173	1.150	1.126	1.102	1.078	1.055	1.034	1.017	1.005
.800	1.105	1.091	1.077	1.062	1.047	1.033	1.020	1.010	1.003
.900	1.048	1.042	1.035	1.029	1.022	1.015	1.009	1.004	1.001

12.2. BASIC RESULTS. We proceed with the study of the equation

$$x = A\mathfrak{f}x, \tag{12.3}$$

in the space $L_2 = L_2(\Omega, \mathbb{R}^m)$. We will assume that the linear operator A on the space

L_2 is normal and completely continuous, with the spectral decomposition

$$Ax(t) = \sum_{n=0}^{\infty} |\lambda_n| U_n P_n x(t) \tag{12.4}$$

(see Subsection 7.3). The nonlinear operator $\mathfrak{f}$ is the superposition operator

$$\mathfrak{f}x(t) = f[t, x(t)], \tag{12.5}$$

corresponding to a function $f(t,x) : \Omega \times \mathbb{R}^m \to \mathbb{R}^m$ that is measurable in t and continuous in x, and satisfies the inequality

$$|f(t,x)| \leqslant k|x| + b(t), \quad t \in \Omega, \ x \in \mathbb{R}^m; \ b(t) \in L_2. \tag{12.6}$$

We next choose $\eta \in (0, \|A\|]$ and consider the finite dimensional subspace

$$E_0 = \left(\sum_{\substack{n=0,1,\dots; \\ |\lambda_n| \geqslant \eta}} P_n \right) L_2. \tag{12.7}$$

In addition, set

$$q_1 = \|A\|^{-1} \max_{|\lambda_n| < \eta} |\lambda_n|, \tag{12.8}$$

$$q_2 = \|A\|^{-1} \max_{e(t) \in E_0; \ \|e\|=1} [|A^* x(t)|, |x(t)|], \tag{12.9}$$

where A^* denotes, as usual, the adjoint on L_2 of the operator A. The inequalities $q_1 < 1$ and $q_2 \leqslant 1$ are always true.

THEOREM 12.1. *Assume that $q_2 < 1$, and*

$$k\|A\| < \mu(q_1, q_2). \tag{12.10}$$

Then equation (12.3) has at least one solution in the space L_2.

Theorem 12.1 will be proved in Subsection 12.4.

If $q_2 < 1$, then, obviously, among the eigenvalues λ of the operator (12.4) satisfying $|\lambda| \geqslant \eta$ there are no real eigenvalues. The converse is not true in the general case. However, we have:

LEMMA 12.1. *Let $\Omega = [a, b]$ and consider equation (12.3) in the space of scalar-valued functions $L_2 = L_2(\Omega, \mathbb{R}^1)$. Suppose that all the functions in the subspace*

(12.7) *are analytic, and that among the leading eigenvalues of the normal completely continuous linear operator A there are no real eigenvalues. Then the number q_2 defined by* (12.9) *satisfies the inequality $q_2 < 1$.*

Lemma 12.1 will be proved in Subsection 12.5.

From Theorem 12.1 and Lemma 12.1 we get:

THEOREM 12.2. *Assume that all the conditions in Lemma 12.1 are satisfied. Then inequality* (12.6), *with a coefficient k subject to condition* (12.10), *implies the existence of at least one solution of equation* (12.3) *in L_2.*

Theorems 12.1 and 12.2 can be used to get various criteria of solvability for equation (12.3), choosing different values of η in the previous definitions of the subspace (12.7) and the quantities (12.8) and (12.9). Below, in the study of a few specific nonlinear boundary value problems, we will use Theorems 12.1 and 12.2 for $\eta = \|A\|$.

Under the assumptions of Theorems 12.1 and 12.2, the completely continuous vector field $x - A\mathfrak{f}x$ is everywhere different from zero, on any sphere $\|x\| = \rho$ with a sufficiently large radius ρ — as it will follow from our subsequent proof —, and its rotation on such spheres equals 1.

We also notice that condition (12.10) is equivalent to the algebraic condition

$$q_1^2 k^2 \|A\|^2 (k^2 \|A\|^2 - 1) - (1 - q_1^2 k^2 \|A\|^2) \cdot (1 - q_2 k \|A\|)^2 < 0.$$

12.3. A THEOREM ON A PRIORI ESTIMATES. We next consider the equation

$$x = \xi A \mathcal{F} x, \quad 0 \leqslant \xi \leqslant 1, \tag{12.11}$$

where A and $\mathcal{F}$ are a linear and a nonlinear operator, respectively, acting on a real Hilbert space H with norm $\| \cdot \|$ and inner product $[\cdot, \cdot]$. The operator $\mathcal{F}$ is not necessarily supposed to be a superposition operator.

Further on we will assume that the space H is the othogonal direct sum of two invariant subspaces E_0 and E_1 of the linear operator A. Let P and Q $(P+Q=I)$ denote the orthogonal projections onto the subspaces E_0 and E_1, respectively. Clearly the operator A commutes with the projections P and Q.

Let $q_1 \in [0, 1)$, $q_2 \in (0, 1)$ and $k, \lambda, c_1, c_2 \geqslant 1$ be some fixed numbers.

Finally, let $\mathcal{B} = \mathcal{B}(q_1, q_2, k, \lambda, c_1, c_2)$ denote the family of all equations (12.11), such that the operators A and $\mathcal{F}$ satisfy the next four estimates, for any $x \in H$:

$$\|APx\| \leqslant \lambda \|Px\|, \tag{12.12}$$

$$\|AQx\| \leqslant q_1 \lambda \|Qx\|, \tag{12.13}$$

$$\|\mathcal{F}x\| \leqslant k\|x\| + c_1, \tag{12.14}$$

$$|[Px, A\mathcal{F}x]| \leqslant \|Px\|[c_2 + \lambda k(q_2\|Px\| + \|Qx\|)]. \tag{12.15}$$

As a value of λ it is convenient to choose $\|A\|$. We notice that the first three estimates (12.12)–(12.14) always imply an estimate slightly different from (12.15), namely,

$$|[Px, A\mathcal{F}x]| \leqslant \|Px\|[\lambda c_1 + \lambda k(\|Px\| + \|Qx\|)]. \tag{12.16}$$

Inequality (12.15) with $c_2 = \lambda c_1$ differs from (12.16) merely by the coefficient $q_2 < 1$ in its right hand side.

Recall that $\mu(q_1, q_2)$ stands for the unique root of equation (12.1) in the interval $(1, q_2^{-1})$.

THEOREM 12.3. *Let k, λ, q_1 and q_2 be such that*

$$\lambda k < \mu(q_1, q_2). \tag{12.17}$$

Then the norm $\|x\|$ of each solution x of any equation (12.11) in the family $\mathcal{B}$ satisfies the general a priori estimate

$$\|x\| \leqslant c = c(k, \lambda, q_1, q_2)(c_1 + c_2). \tag{12.18}$$

The constant $c(k, \lambda, q_1, q_2)$ in (12.18) above can be explicitly computed. Precise formulas will follow from the proof below.

Proof. Since the next chain of inequalities

$$\|\xi A\mathcal{F}x\| \leqslant \|A\mathcal{F}x\| \leqslant \lambda \|\mathcal{F}x\| \leqslant \lambda k\|x\| + \lambda c_1$$

clearly holds true, the inequality $\lambda k < 1$ leads to the estimate $\|x\| \leqslant \lambda(1 - \lambda k)^{-1} c_1$ for all solutions x of any equation (12.11) in the family $\mathcal{B}$. Therefore, if $\lambda k < 1$ the theorem is proved. For the rest of the proof we assume that $1 \leqslant k\lambda < \mu(q_1, q_2)$. In view of (12.2) we have $q_1 \lambda k < 1$ and $q_2 \lambda k < 1$, hence

$$q_1^2 \lambda^2 k^2 (\lambda^2 k^2 - 1) < (1 - q_1^2 \lambda^2 k^2)(1 - q_2 \lambda k)^2.$$

Let

$$\beta = \frac{1}{2}\left[q_1 \sqrt{\frac{\lambda^2 k^2 - 1}{1 - q_1^2 \lambda^2 k^2}} + \frac{1 - q_2 \lambda k}{\lambda k}\right], \tag{12.19}$$

and observe that

$$q_1 \sqrt{\frac{\lambda^2 k^2 - 1}{1 - q_1^2 \lambda^2 k^2}} < \beta < \frac{1 - q_2 \lambda k}{\lambda k}. \tag{12.20}$$

Using β, let us consider the sets

$$G_1 = \{x \in H,\ \beta \|Px\| \leqslant \|Qx\|\}, \tag{12.21}$$

$$G_2 = \{x \in H,\ \beta \|Px\| > \|Qx\|\}. \tag{12.22}$$

The union of the sets (12.21) and (12.22) coincides with the whole space H. In order to prove estimate (12.18) we will follow different approaches corresponding to the cases when the solution x under consideration lies in the set (12.21) or (12.22), respectively.

To start with, assume that $x = \xi A \mathcal{F} x$ $(0 \leqslant \xi \leqslant 1)$ and $x \in G_1$. Then $\|Qx\| = \|\xi Q A \mathcal{F} x\| = \xi \|AQ\mathcal{F}x\|$, and by (12.13) we find

$$\|Qx\|^2 \leqslant \xi^2 q_1^2 \lambda^2 \|Q\mathcal{F}x\|^2 = \xi^2 q_1^2 \lambda^2 (\|\mathcal{F}x\|^2 - \|P\mathcal{F}x\|^2).$$

In view of (12.12) we have

$$\xi q_1 \lambda \|P\mathcal{F}x\| \geqslant \xi q_1 \|AP\mathcal{F}x\| = q_1 \|Px\|,$$

and therefore,

$$\|Qx\|^2 \leqslant q_1^2 \lambda^2 \|\mathcal{F}x\|^2 - q_1^2 \|Px\|^2. \tag{12.23}$$

From (12.14) it follows that $\|\mathcal{F}x\|^2 \leqslant c_1^2 + 2kc_1 \|x\| + k^2 \|x\|^2$. Consequently, by (12.23) we get the relation

$$\|Qx\|^2 \leqslant q_1^2 \lambda^2 c_1^2 + 2q_1^2 \lambda^2 kc_1 \|x\| + q_1^2 k^2 \|x\|^2 - q_1^2 \|Px\|^2,$$

which can be rewritten as

$$\|Qx\|^2 \leqslant q_1^2 \lambda^2 c_1^2 + 2q_1^2 \lambda^2 kc_1 \|x\| + q_1^2 (\lambda^2 k^2 - 1)\|Px\|^2 + q_1^2 \lambda^2 k^2 \|Qx\|^2.$$

On the other hand, since for all $x \in G_1$ we have the estimates $\|x\| \leqslant \beta^{-1}\sqrt{1 + \beta^2}\|Qx\|$ and $\|Px\| \leqslant \beta^{-1}\|Qx\|$, and because we also assume that $\lambda k \geqslant 1$, the last relation implies the inequality

$$\|Qx\|^2 \leqslant q_1^2 \lambda^2 c_1^2 + 2q_1^2 \lambda^2 kc_1 \beta^{-1}\sqrt{1 + \beta^2}\|Qx\| + $$
$$+ \left[q_1^2(\lambda^2 k^2 - 1)\beta^{-2} + q_1^2 \lambda^2 k^2\right]\|Qx\|^2. \tag{12.24}$$

We next introduce the notation

$$\alpha_1 = 1 - q_1^2(\lambda^2 k^2 - 1)\beta^{-2} - q_1^2\lambda^2 k^2, \quad \alpha_2 = q_1^2\lambda^2 kc_1\beta^{-1}\sqrt{1 + \beta^2}, \quad \alpha_3 = q_1^2\lambda^2,$$

and observe that (12.24) coincides with a quadratic inequality in $\|Qx\|$, namely,

$$\alpha_1\|Qx\|^2 - 2\alpha_2 c_1\|Qx\| - \alpha_3 c_1^2 \leqslant 0. \tag{12.25}$$

Clearly α_1 is positive in view of the left side of (12.20), and both α_2 and α_3 are nonnegative, therefore (12.25) leads to the estimate

$$\|Qx\| \leqslant \frac{\alpha_2 + \sqrt{\alpha_2^2 + \alpha_1\alpha_3}}{\alpha_1}c_1,$$

which in its turn implies

$$\|x\| \leqslant \beta^{-1}\sqrt{1 + \beta^2}\frac{\alpha_2 + \sqrt{\alpha_2^2 + \alpha_1\alpha_3}}{\alpha_1}c_1.$$

Since the numbers α_1, α_2, α_3 and β depend only on q_1, q_2, k and λ and are independent of c_1 (all the more on c_2), the last relation provides the a priori estimate (12.18) of any solution $x \in G_1$ of an equation (12.11) in the family $\mathcal{B}$.

We next proceed with the case when $x \in \xi A\mathcal{F}x$ ($0 \leqslant \xi \leqslant 1$) and $x \in G_2$. In this case we have $\|Px\|^2 = |[Px, \xi PA\mathcal{F}x]| \leqslant |[Px, A\mathcal{F}x]|$, and from (12.15) we obtain

$$\|Px\|^2 \leqslant \|Px\|[c_2 + \lambda k(q_2\|Px\| + \|Qx\|)].$$

Since $x \in G_2$ the last inequality implies

$$\|Px\| \leqslant c_2 + \lambda k(q_2 + \beta)\|Px\|. \tag{12.26}$$

From the right side of estimate (12.20) it follows that $1 - \lambda k(q_2 + \beta) > 0$, therefore (12.26) can be rewritten as

$$\|Px\| \leqslant [1 - \lambda k(q_2 + \beta)]^{-1}c_2.$$

Thus we get

$$\|x\| \leqslant \sqrt{1 + \beta^2}[1 - \lambda k(q_2 + \beta)]^{-1}c_2,$$

i.e., the solutions $x \in G_2$ of any equation in $\mathcal{B}$ also satisfy the estimate (12.18).

Theorem 12.3 is completely proved.

12.4. PROOF OF THEOREM 12.1. Under the assumptions of Theorem 12.1 the operator $A\mathcal{F}$ acts and is completely continuous on the space L_2. Based on Leray-Schauder Principle, to conclude the proof of Theorem 12.1 it will be enough to show that the norm of all solutions of any equation (12.11) is subject to the general a priori estimate $\|x\| \leqslant \mathrm{const} < \infty$. To this end, in view of Theorem 12.3, it suffices to check the estimates (12.12)–(12.15) for all $x \in L_2$, where P stands for the orthogonal projection onto the subspace (12.7), $Q = I - P$, $\lambda = \|A\|$, and the numbers q_1 and q_2 are defined by equalities (12.8) and (12.9), respectively.

The estimate (12.12) for $\lambda = \|A\|$ is obvious. The estimate (12.13) follows from definition (12.8) of q_1. The estimate (12.14) is obtained from (12.6) setting $c_1 = \|b(t)\|$. Finally, the estimate (12.15) is a consequence of the next chain of relations:

$$|[Px, A\mathcal{F}x]| = |[A^*Px, \mathcal{F}x]| \leqslant [|A^*Px|, |\mathcal{F}x|] \leqslant [|A^*Px|, k|x| + |b(t)|] \leqslant$$
$$\leqslant k[|A^*Px|, |Px|] + k[|A^*Px|, |Qx|] + [|A^*Px|, |b(t)|] \leqslant k\lambda q_2\|Px\|^2 +$$
$$+k\lambda\|Px\|\,\|Qx\| + \lambda\|Px\|\,\|b(t)\| = \|Px\|[\lambda\|b(t)\| + k\lambda(q_2\|Px\| + \|Qx\|)].$$

In some of the relations above we used the equalities $\|A^*\| = \|A\| = \lambda$. Theorem 12.1 is proved.

12.5. PROOF OF LEMMA 12.1. Assume that all the conditions in Lemma 12.1 are fulfilled and $q_2 = 1$. Since the subspace E_0 is finite dimensional, there exists a function $e(t) \in E_0$ with $\|e\| = 1$ such that

$$\int_\Omega |A^*e(t)e(t)|\mathrm{d}\mu = \|A\|.$$

Consequently, we first get the equality $\|A^*e(t)\| = \|A\|$, and, second, it follows that the functions $|A^*e(t)|$ and $\|A\| \cdot |e(t)|$ coincide. Since both the functions $A^*e(t)$ and $\|A\|e(t)$ are analytic, we conclude that either $A^*e(t) = \|A\|e(t)$, or $A^*e(t) = -\|A\|e(t)$, i.e., the operator A^* has at least one real eigenvalue, $\|A\|$ or $-\|A\|$. Clearly the last conclusion contradicts one of our assumptions, namely, the fact that the operator A has only nonreal leading eigenvalues. The proof of Lemma 12.1 is complete.

§13. Joint norms of operators

13.1. BASIC DEFINITIONS. Throughout this section we assume that H is a real Hilbert space equipped with a given modulus. More specifically, by a *modulus* on H we mean an isometric positively homogeneous nonlinear operator $M : H \to H$ satisfying the weak triangle inequality

$$(M(x + y), Mz) \leqslant (Mx + My, Mz), \quad x, y, z \in H. \tag{13.1}$$

In addition, we suppose that (13.1) becomes an equality for all $z \in H$ if and only if $Mx + My = M(x + y)$.

If H is a finite dimensional space with a fixed orthonormal basis, then as an example of a modulus M on H we may consider the operator that sends any vector $\{x_1, \ldots, x_n\}$ into the vector $\{|x_1|, \ldots, |x_n|\}$. If H is a Hilbert space consisting of scalar-valued functions (for instance, L_2), then we can take as a modulus M on H the operator that sends any function $x(t)$ into $|x(t)|$. Inequality (13.1) is obvious in both the previous examples.

In what follows we will assume that the operator M is fixed.

Let A and B be two bounded linear operators acting on H.

DEFINITION. The *joint norm* of two operators A and B on a Hilbert space H with a modulus M is the number

$$N(A, B) = N(A, B; H, M) = \sup_{\|x\|=1} \|MAx + MBx\|. \tag{13.2}$$

Let us next assume that $\mathfrak{f}(x_1, x_2) : H \times H \to H$ is a given nonlinear operator satisfying the estimate

$$\|\mathfrak{f}(x_1, x_2)\| \leqslant \|k_1 Mx_1 + k_2 Mx_2\| + c. \tag{13.3}$$

Such an estimate holds true, for instance, when $\mathfrak{f}$ is the superposition operator

$$\mathfrak{f}(x_1, x_2) = f[t, x_1(t), x_2(t)],$$

acting on the space L_2 with the standard modulus introduced above, and

$$|f(t, x_1, x_2)| \leqslant k_1|x_1| + k_2|x_2| + c(t). \tag{13.4}$$

If the joint norm $N(k_1 A, k_2 B)$ of the operators $k_1 A$ and $k_2 B$ is less than 1, then in view of (13.3) the operator $x \mapsto \mathfrak{f}(Ax, Bx)$ maps the ball

$$\mathfrak{B}(\rho) = \{x : x \in H, \ \|x\| \leqslant \rho\}$$

with radius $\rho = c[1 - N(k_1 A, k_2 B)]^{-1}$ into itself. If, in addition, the operator $\mathfrak{f}(Ax, Bx)$ is completely continuous on H, then based on the Schauder Principle we conclude that equation $x = \mathfrak{f}(Ax, Bx)$ has at least one solution in the ball $\mathfrak{B}(\rho)$.

The preceding remarks show that the knack of computing (or well estimating) the joint norm of two operators can lead to criteria for the existence of solutions of various nonlinear equations.

A simple and obvious estimate of the joint norm (13.2) is given by

$$N(A, B) \leqslant \|A\| + \|B\|. \tag{13.5}$$

There are cases when this estimate yields the only conclusion we could reach.

For instance, if both the operators A and B are completely continuous, then the estimate (13.5) may not be improved, if and only if there exists $x \in H$ with $\|x\| = 1$, such that $\|Ax\| = \|A\|$, $\|Bx\| = \|B\|$, and the vectors MAx and MBx are collinear.

An exact computation of the joint norm of two operators is, of course, impossible without some additional assumptions. Even in the simple case when H is a two-dimensional space, the computation of the joint norm $N(A, B)$ of two arbitrarily chosen operators A and B is practically possible only with a computer aid.

In the following subsections we will be concerned with some specific situations when a quite good estimate of the joint norm is available.

13.2. REMARKS. **A.** There is a natural way of defining the joint norm of several operators. More precisely, let A_i $(i = 1, \ldots, n; \ n > 2)$ be some bounded linear operators acting on a Hilbert space H with a modulus M. Their joint norm is defined by

$$N(A_1, \ldots, A_n) = \sup_{\|x\|=1} \left\| \sum_{i=1}^{n} MA_i x \right\|, \tag{13.6}$$

and it can be used in the study of nonlinear equations of the form

$$x = \mathfrak{f}(A_1 x, \ldots, A_n x).$$

B. We may introduce, in a quite analogous manner, the joint norm of two bounded operators A and B that do not act on a Hilbert space, but merely on a Banach space. For instance, if A and B are operators on L_p $(p \neq 2)$ or on C, then we can use the number

$$N(k_1 A, k_2 B) = \sup_{\|x\|=1} \| k_1 |Ax(t)| + k_2 |Bx(t)| \|$$

to investigate the equation $x = \mathfrak{f}(Ax, Bx)$ with a nonlinear superposition operator $\mathfrak{f}$ subject to condition (13.4).

C. In what follows we restrict our attention to the case of two operators on a Hilbert space. Instead of the joint norm (13.2) we will estimate the quantity

$$\mathfrak{J} = \mathfrak{J}(q_1, q_2, q_3; A, B) = \sup_{\|x\|=1} \left[q_1 \|Ax\|^2 + 2q_2(MAx, MBx) + q_3 \|Bx\|^2 \right]. \quad (13.7)$$

If $q_1 = q_2 = q_3 = 1$, then clearly $\mathfrak{J} = [N(A, B)]^2$; for $q_1 = k_1^2$, $q_2 = k_1 k_2$ and $q_3 = k_2^2$ we get $\mathfrak{J} = [N(k_1 A, k_2 B)]^2$.

D. If $N(k_1 A, k_2 B) = 1$, then it is possible to investigate the nonlinear equation $x = \mathfrak{f}(Ax, Bx)$ in L_2, along the lines and using the methods developed in Chapter 2. In this respect, it will be more convenient to replace condition (13.4) above by

$$|f(t, x_1, x_2)| \leqslant k_1 |x_1| + k_2 |x_2| - \varphi(|x_1|),$$

or by some other analogous conditions (see Section 18).

13.3. EXAMPLES OF EXACT COMPUTATION OF THE JOINT NORM. We first assume that $H = \mathbb{R}^n$ is a finite dimensional Euclidean space with the modulus $M\{x_1, \ldots, x_n\} = \{|x_1|, \ldots, |x_n|\}$. In this case quantity (13.7) corresponding to the diagonal operators

$$A = \begin{bmatrix} a_1 & & 0 \\ & \ddots & \\ 0 & & a_n \end{bmatrix}, \quad B = \begin{bmatrix} b_1 & & 0 \\ & \ddots & \\ 0 & & b_n \end{bmatrix} \quad (13.8)$$

equals

$$\mathfrak{J}(q_1, q_2, q_3; A, B) = \max_{i=1,\ldots,n} \left[q_1 a_i^2 + 2q_2 |a_i b_i| + q_3 b_i^2 \right]. \quad (13.9)$$

In particular,

$$N(A, B) = \max_{i=1,\ldots,n} (|a_i| + |b_i|). \quad (13.10)$$

Equality (13.9) follows from the identity

$$q_1 \|Ax\|^2 + 2q_2(MAx, MBx) + q_3 \|Bx\|^2 = \sum_{i=1}^{n} \left[q_1 a_i^2 + 2q_2 |a_i b_i| + q_3 b_i^2 \right] x_i^2,$$

combined with the general fact that the maximum value of a positive definite quadratic form $\sum c_i x_i^2$ on the unit sphere $\sum x_i^2 = 1$ equals $\max_i c_i$.

We next suppose that H is an infinite dimensional space and let A and B be two self-adjoint completely continuous operators acting on H, such that $AB = BA$.

Then there exists an orthonormal basis $\{e_i\}$ for H consisting of joint eigenvectors e_i of the operators A and B. In addition, we assume that the modulus M on H is defined in terms of the basis $\{e_i\}$ by the formula

$$Mx = \sum_{i=1}^{\infty} |(x, e_i)| e_i.$$

Then the quantity (13.7) is given by

$$\mathfrak{I}(q_1, q_2, q_3; A, B) = \max_{i=1,\ldots,\infty} \left[q_1 a_i^2 + 2q_2 |a_i b_i| + q_3 b_i^2 \right],$$

a formula which is analogous to (13.9) above, where a_i and b_i denote the eigenvalues of A and B corresponding to the eigenvector e_i, respectively.

Unfortunately, the modulus M is seldomly related so nicely to a basis $\{e_i\}$ consisting of joint eigenvectors.

We proceed with yet another example. Consider once again a finite dimensional space with a fixed orthonormal basis and let M be the standard modulus. If A and B are two matrices with nonnegative entries, then $N(A, B) = \|A + B\|$. An analogous result is also true for integral operators on the space L_2 with nonnegative kernels.

13.4. ESTIMATES INDEPENDENT OF MODULUS. In this subsection our main concern will be to estimate quantity (13.7) independently of the choice of a modulus on the space H. Actually, instead of (13.7) we will estimate the quantity

$$\mathfrak{I}_1(q_1, q_2, q_3; A, B) = \sup_{\|x\|=1} \left[q_1 \|Ax\|^2 + 2q_2 \|Ax\| \cdot \|Bx\| + q_3 \|Bx\|^2 \right]. \tag{13.11}$$

Since any modulus is an isometry, and therefore $(MAx, MBx) \leqslant \|Ax\| \cdot \|Bx\|$, we clearly have

$$\mathfrak{I}(q_1, q_2, q_3; A, B) \leqslant \mathfrak{I}_1(q_1, q_2, q_3; A, B).$$

Let us also notice that for the operators (13.8) on a finite dimensional Euclidean space, the quantity (13.11) can be computed exactly.

In what follows we let $\mathfrak{R}(q_1, q_2, q_3) = \mathfrak{R}(q)$ denote the set of all ordered quadruples $\{r_1, r_2, s_1, s_2\}$ of nonnegative numbers satisfying the inequalities $r_1 > r_2$, $s_1 < s_2$, and such that the largest positive root $\zeta = \zeta(q, r_1, r_2, r_3, r_4)$ of the quadratic equation

$$\left(\frac{\Delta s}{\Delta} \zeta - q_1 \right) \left(\frac{\Delta r}{\Delta} \zeta - q_3 \right) = q_2^2 \quad (\Delta s = s_2 - s_1, \ \Delta r = r_1 - r_2, \ \Delta = r_1 s_2 - r_2 s_1) \tag{13.12}$$

satisfies the inequalities

$$\frac{\Delta}{\Delta r}\left[q_3 + q_2\left(\frac{r_2}{s_2}\right)^{\frac{1}{2}}\right] < \zeta < \frac{\Delta}{\Delta r}\left[q_3 + q_2\left(\frac{r_1}{s_1}\right)^{\frac{1}{2}}\right]. \tag{13.13}$$

Obviously any equation (13.12) has at least one positive root, and for the largest root ζ we have the estimates

$$\zeta\Delta s > q_1\Delta, \quad \zeta\Delta r > q_3\Delta. \tag{13.14}$$

Relations (13.13) satisfied by ζ can be equally well stated in a more symmetric form, namely,

$$\left(\zeta\frac{\Delta s}{\Delta} - q_1\right)\frac{r_2}{s_2} < \left(\zeta\frac{\Delta r}{\Delta} - q_3\right) < \left(\zeta\frac{\Delta s}{\Delta} - q_1\right)\frac{r_1}{s_1}. \tag{13.15}$$

Let us now consider an important example. If

$$q_1 q_3 = q_2^2, \tag{13.16}$$

then $\zeta = \dfrac{\Delta}{\Delta r \Delta s}(q_1\Delta r + q_3\Delta s)$. The estimates (13.15) above can be rewritten in this case as

$$\frac{r_2}{s_2} < \frac{q_1}{q_3}\left(\frac{\Delta r}{\Delta s}\right)^2 < \frac{r_1}{s_1}.$$

We also notice that in this case, if we start with arbitrary r_i and s_i ($r_1 > r_2 \geqslant 0$, $s_2 > s_1 \geqslant 0$), then the relation $\{r_1, r_2, s_1, s_2\} \in \mathfrak{R}(q_1, q_2, q_3)$ is completely described in terms of the ratio of the coefficients q_1 and q_3.

Assuming that condition (13.16) is fulfilled, then

$$\zeta \geqslant q_1 r_i + 2q_2\sqrt{r_i s_i} + q_3 s_i, \quad i = 1, 2.$$

THEOREM 13.1. *We have the equality*

$$\mathfrak{J}(q_1, q_2, q_3; A, B) = \max\{J_1, J_2\},$$

where A and B are the operators defined by (13.8),

$$J_1 = \max_{i=1,\ldots,n}\left[q_1 a_i^2 + 2q_2|a_i b_i| + q_3 b_i^2\right] \tag{13.17}$$

and

$$J_2 = \max_{i,j:\left\{a_i^2, a_j^2, b_i^2, b_j^2\right\}\in\mathfrak{R}(q)} \zeta\left(q, a_i^2, a_j^2, b_i^2, b_j^2\right). \tag{13.18}$$

The maximum value in (13.11) is always attained at a vector $\{x_1, \ldots, x_n\}$ with no more than two different from zero components; if $\mathfrak{J}_1 = J_1$, then that maximum value is attained at a vector with only one nonzero component.

Theorem 13.1 will be proved in the next subsection.

In spite of the cumbersome form of the numbers J_1 and J_2, Theorem 13.1 could quite easily be used in specific situations.

Let, for instance, $n > 2$, $a_i = [(i-1)^2 + 1]^{-1}$, $b_i = (i-1)a_i$, and assume that the numbers q_1, q_2, q_3 are related as in equality (13.16). Then

$$J_1 = \max \left\{ q_1, \frac{1}{4} \left(\sqrt{q_1} + \sqrt{q_3} \right)^2 \right\},$$

$$J_2 = q_1 + \frac{1}{3} q_3 \quad (q_3 < 9q_1).$$

Consequently, if $q_3 < 9q_1$, then $\mathfrak{J}_1 = \frac{1}{3}q_3 + q_1$, and whenever $q_3 \geqslant 9q_1$, then $\mathfrak{J}_1 = \frac{1}{4}(\sqrt{q_1} + \sqrt{q_3})^2$. In this case

$$N(k_1 A, k_2 B) \leqslant \begin{cases} \sqrt{k_1^2 + \frac{1}{3}k_2^2} & \text{if } k_2 < 3k_1, \\ \frac{1}{2}(k_1 + k_2) & \text{if } k_2 \geqslant 3k_1. \end{cases}$$

Theorem 13.1 provides the best upper estimate of the expression

$$q_1 \sum_{i=1}^{n} a_i^2 x_i^2 + 2q_2 \sqrt{\left(\sum_{i=1}^{n} a_i^2 x_i^2 \right) \left(\sum_{i=1}^{n} b_i^2 x_i^2 \right)} + q_3 \sum_{i=1}^{n} b_i^2 x_i^2.$$

From this theorem it follows that for any operators A and B satisfying the estimates

$$\|Ax\|^2 \leqslant \sum_{i=1}^{n} a_i^2 x_i^2, \quad \|Bx\|^2 \leqslant \sum_{i=1}^{n} b_i^2 x_i^2,$$

we have the inequalities

$$\mathfrak{J}(q_1, q_2, q_3; A, B) \leqslant \mathfrak{J}_1(q_1, q_2, q_3; A, B) \leqslant \max(J_1, J_2),$$

where J_1 and J_2 are defined according to (13.17) and (13.18).

From Theorem 13.1 we also draw estimates of quantity (13.11) for operators on infinite dimensional spaces. To be more specific, assume that $\{e_i\}$ is a given orthonormal basis for H. Let A and B be two operators on H such that

$$\|Ax\|^2 \leqslant \sum_{i=1}^{\infty} a_i^2 x_i^2, \quad \|Bx\|^2 \leqslant \sum_{i=1}^{\infty} b_i^2 x_i^2 \quad (a_i, b_i \to 0), \tag{13.19}$$

for any $x \in H$, $x = \sum x_i e_i$. Such estimates are fulfilled (with $=$ instead of $\leqslant$), for instance, when A and B are completely continuous normal commuting operators. In this case we have the decompositions

$$Ax = \sum_{i=1}^{\infty} |\lambda_i^A| U_i^A P_i x, \quad Bx = \sum_{i=1}^{\infty} |\lambda_i^B| U_i^B P_i x, \tag{13.20}$$

with mutually orthogonal projections $P_i x$; estimates (13.19) follow from (13.20).

We next consider two sequences A_n and B_n of operators satisfying the estimates

$$\|A_n x\|^2 \leqslant \sum_{i=1}^{n} a_i^2 x_i^2, \quad \|B_n x\| \leqslant \sum_{i=1}^{n} b_i^2 x_i^2. \tag{13.21}$$

If the decompositions (13.20) hold true, then as operators A_n and B_n we naturally choose the operators

$$A_n x = \sum_{i=1}^{n} |\lambda_i^A| U_i^A P x_i, \quad B_n x = \sum_{i=1}^{n} |\lambda_i^B| U_i^B P x_i. \tag{13.22}$$

Since $a_i, b_i \to 0$, then for sufficiently large values of n we have

$$\mathfrak{J}_1(q_1, q_2, q_3; A, B) = \mathfrak{J}_1(q_1, q_2, q_3; A_n, B_n);$$

therefore, in view of Theorem 13.1, for the operators A and B we have the estimates

$$\mathfrak{J}(q_1, q_2, q_3; A, B) \leqslant \mathfrak{J}_1(q_1, q_2, q_3; A_n, B_n) \leqslant \max(J_1, J_2), \tag{13.23}$$

with J_1 and J_2 defined by formulas (13.17) and (13.18), where now the indices i and j take arbitrary values in the set of natural numbers. For the sake of later references we summarize the just proved result as a distinctive statement.

THEOREM 13.2. *Assume that the operators A and B acting on a Hilbert space H are completely continuous and satisfy the estimates (13.19) with respect to the orthonormal basis $\{e_i\}$. Then inequality (13.23) holds true.*

REMARK. Statements analogous to Theorems 13.1 and 13.2 can be obtained in estimating the quantity

$$\max_{\|x\|=1} \sum_{i,j=1}^{m} q_{ij} \|A_i x\| \cdot \|A_j x\|.$$

If all the operators A_i are diagonal with respect to a given basis, then quantity (13.24) may be computed exactly. The maximum value (13.24) is attained at a vector x with no more than m nonzero components.

13.5. PROOF OF THEOREM 13.1. To prove our theorem we will investigate the extrema of the function

$$\mathbb{J}(x) = q_1 \sum_{l=1}^{n} \xi_l x_l^2 + 2q_2 \sqrt{\left(\sum_{l=1}^{n} \xi_l x_l^2\right)\left(\sum_{l=1}^{n} \eta_l x_l^2\right)} + q_3 \sum_{l=1}^{n} \eta_l x_l^2 \qquad (13.25)$$

with nonnegative coefficients ξ_l and η_l on the sphere

$$\sum_{l=1}^{n} x_l^2 = 1. \qquad (13.26)$$

Specifically, we will use the well-known Lagrange multiplier method. We introduce the Lagrangean

$$L(x, \lambda) = \mathbb{J}(x) - \lambda \left(\sum_{l=1}^{n} x_l^2 - 1\right),$$

and then equate with zero its partials:

$$\frac{\partial L}{\partial \lambda} = 0, \quad \frac{\partial L}{\partial x_l} = 0, \qquad l = 1, \ldots, n.$$

Equality $\dfrac{\partial L}{\partial \lambda} = 0$ is equivalent to (13.26); equalities $\dfrac{\partial L}{\partial x_l} = 0$ have the form

$$x_l \left\{ q_1 \xi_l + q_2 \left[\xi_l \frac{u}{v} + \eta_l \frac{v}{u} \right] + q_3 \eta_l = \lambda \right\} = 0, \quad l = 1, \ldots, n. \qquad (13.27)$$

In (13.27) above, as well as for the rest of this proof, we use the notation

$$u = \sqrt{\sum_{l=1}^{n} \xi_l x_l^2}, \quad v = \sqrt{\sum_{l=1}^{n} \eta_l x_l^2}.$$

If each of the equalities (13.27) is multiplied by x_l and all the resulting equations are added, then we get

$$q_1 \sum_{l=1}^{n} \xi_l x_l^2 + 2q_2 \sqrt{\left(\sum_{l=1}^{n} \xi_l x_l^2\right)\left(\sum_{l=1}^{n} \eta_l x_l^2\right)} + q_3 \sum_{l=1}^{n} \eta_l x_l^2 - \lambda \sum_{l=1}^{n} x_l^2 = 0,$$

i.e., $\lambda = \mathbb{J}(x)$. Thus, the value of function (13.25) at any point of extrema (x, λ) equals λ.

We now proceed with a more detailed analysis of equation (13.27). For every l, at least one of the numbers x_l or

$$\tau_l = q_1\xi_l + q_2\left[\xi_l\frac{u}{v} + \eta_l\frac{v}{u}\right] + q_3\eta_l - \lambda$$

equals zero. If all the numbers x_l but one are zero, and $x_m = 1$, then

$$\mathbb{J}(x) = q_1\xi_m + 2q_2\sqrt{\xi_m\eta_m} + q_3\eta_m. \tag{13.28}$$

If a couple of components x_i and x_j $(i \neq j)$ are nonzero, then $\tau_i = \tau_j = 0$, therefore the numbers

$$z_1 = q_1 + q_2\frac{u}{v}, \quad z_2 = q_3 + q_2\frac{v}{u}$$

are solutions for the system of two linear equations

$$\begin{cases} z_1\xi_i + z_2\eta_i = \lambda, \\ z_1\xi_j + z_2\eta_j = \lambda. \end{cases} \tag{13.29}$$

The case when

$$\Delta = \det\begin{bmatrix} \xi_i & \eta_i \\ \xi_j & \eta_j \end{bmatrix} = 0,$$

may not be considered. If $\Delta \neq 0$, then from (13.29) we find the equalities

$$z_1 = \frac{\lambda}{\Delta}\Delta\eta, \quad z_2 = \frac{\lambda}{\Delta}\Delta\xi \quad (\Delta\xi = \xi_i - \xi_j, \ \Delta\eta = \eta_j - \eta_i). \tag{13.30}$$

We choose the numbers i and j such that Δ is positive (if $\Delta < 0$, then we simply change the roles of i and j). The positivity of the numbers z_1 and z_2 implies the inequalities $\xi_i > \xi_j$ and $\eta_j > \eta_i$. In view of the definition of z_1 and z_2 it follows that

$$q_2\frac{u}{v} = \frac{\lambda\Delta\xi}{\Delta} - q_3, \quad q_2\frac{v}{u} = \frac{\lambda\Delta\eta}{\Delta} - q_1, \tag{13.31}$$

therefore the number λ is the largest positive root of the equation

$$\left(\frac{\Delta\eta}{\Delta}\lambda - q_1\right) \cdot \left(\frac{\Delta\xi}{\Delta}\lambda - q_3\right) = q_2^2,$$

an equation that coincides with (13.12). Thus we proved that any extreme value λ is the largest root of equation (13.12) (or λ equals one of the numbers (13.28)).

We next show that the maximum value of function (13.25) is attained at a vector with no more than two nonzero components.

Assume that x_i, x_j and x_k are different from zero. Then $\tau_i = \tau_j = \tau_k = 0$, i.e.,

$$\begin{cases} z_1\xi_i + z_2\eta_i = \lambda, \\ z_1\xi_j + z_2\eta_j = \lambda, \\ z_1\xi_k + z_2\eta_k = \lambda. \end{cases}$$

Therefore, there are some positive numbers ε_1 and ε_2 ($\varepsilon_1 + \varepsilon_2 = 1$) such that one of the following three systems of equations is true:

$$\begin{cases} \xi_i = \varepsilon_1\xi_j + \varepsilon_2\xi_k \\ \eta_i = \varepsilon_1\eta_j + \varepsilon_2\eta_k \end{cases} \quad \begin{cases} \xi_j = \varepsilon_1\xi_k + \varepsilon_2\xi_i \\ \eta_j = \varepsilon_1\eta_k + \varepsilon_2\eta_i \end{cases} \quad \begin{cases} \xi_k = \varepsilon_1\xi_i + \varepsilon_2\xi_j \\ \eta_k = \varepsilon_1\eta_i + \varepsilon_2\eta_j \end{cases}. \tag{13.32}$$

Without any loss of generality we will suppose that the last system in (13.32) is true. Then we can set instead of x_i, x_j and x_k in $\{\lambda, x_1, \ldots, x_n\}$ the new values

$$x_i' = \sqrt{x_i^2 + \varepsilon_1 x_k^2}, \quad x_j' = \sqrt{x_j^2 + \varepsilon_2 x_k^2}, \quad x_k' = 0.$$

The new vector $\{\lambda, x_1, \ldots, x_i', \ldots, x_j', \ldots, x_k', \ldots, x_n\}$ corresponds to an extreme value, too. In fact the value of function (13.25) is preserved, but the number of nonzero components of this new vector is less by 1 than in the initial vector.

Thus, the extreme value λ corresponds to a vector with two nonzero components x_i and x_j. In order to conclude the proof of our theorem it remains to prove that whenever a vector $\{\lambda, 0, \ldots, x_i, \ldots, 0, \ldots, x_j, \ldots, 0\}$ provides an extrema, then $\{\xi_i, \xi_j, \eta_i, \eta_j\} \in \mathfrak{R}(q)$. From the identities

$$\xi_i x_i^2 + \xi_j x_j^2 = u^2, \quad \eta_i x_i^2 + \eta_j x_j^2 = v^2,$$

considered as linear equations in x_i^2 and x_j^2, we get the relations

$$\frac{\xi_j}{\eta_j} < \left(\frac{u}{v}\right)^2 < \frac{\xi_i}{\eta_i},$$

which coincide (in view of (13.31)) with conditions (13.13) required for $\{\xi_i, \xi_j, \eta_i, \eta_j\} \in \mathfrak{R}(q)$. Theorem 13.1 is proved.

13.6. ESTIMATES DEPENDING ON MODULUS. In view of our next purposes we will assume that there exist some numbers $\lambda_A \geqslant \mu_A \geqslant 0$, $\lambda_B \geqslant \mu_B \geqslant 1$ and $\gamma < 1$ and two orthogonal subspaces E_0 and E_1, such that

$$\|Ax_0\| \leqslant \lambda_A\|x_0\|, \quad \|Bx_0\| \leqslant \lambda_B\|x_0\| \qquad (x_0 \in E_0); \tag{13.33}$$

$$\|Ax_1\| \leqslant \mu_A\|x_1\|, \quad \|Bx_1\| \leqslant \mu_B\|x_1\| \qquad (x_1 \in E_1); \tag{13.34}$$

$$(M(Ax_0), M(Bx_0)) \leqslant \gamma \lambda_A \lambda_B \|x_0\|^2 \quad (x_0 \in E_0); \tag{13.35}$$

$$(Ax_0, Ax_1) = (Bx_0, Bx_1) = 0 \quad (x_0 \in E_0, x_1 \in E_1). \tag{13.36}$$

If the estimates (13.33) hold true, then (13.35) is always fulfilled with $\gamma = 1$. The constants λ_A and λ_B can coincide with the norms $\|A\|$ and $\|B\|$ of the operators A and B, respectively.

We also consider the function

$$D = D(q_1, q_2, q_3, \lambda_A, \lambda_B, \mu_A, \mu_B, \gamma) \stackrel{\text{def}}{=}$$

$$= \sqrt{[q_1 (\lambda_A^2 - \mu_A^2) + q_3 (\lambda_B^2 - \mu_B^2) + 2q_2(\gamma\lambda_A\lambda_B - \mu_A\mu_B)]^2 + 4q_2^2(\lambda_A\mu_B + \mu_A\lambda_B)^2} +$$

$$+ q_1 (\lambda_A^2 + \mu_A^2) + q_3 (\lambda_B^2 + \mu_B^2) + 2q_2(\gamma\lambda_A\lambda_B + \mu_A\mu_B). \tag{13.37}$$

THEOREM. 13.3. *Under the previous assumptions we have*

$$\mathfrak{J}(q_1, q_2, q_3; A, B) \leqslant \frac{1}{2}D(q_1, q_2, q_3, \lambda_A, \lambda_B, \mu_A, \mu_B, \gamma). \tag{13.38}$$

Theorem 13.3 does not always provide an estimate of quantity (13.7) better than $\mathfrak{J}(q_1, q_2, q_3; A, B) \leqslant q_1\lambda_A^2 + 2q_2\lambda_A\lambda_B + q_3\lambda_B^2$. However, if $\varepsilon > 0$ is given, then for sufficiently small values of μ_A and μ_B we get

$$D \leqslant 2\left(q_1\lambda_A^2 + 2q_2\gamma\lambda_A\lambda_B + q_3\lambda_B^2\right) + \varepsilon.$$

Since clearly $\sqrt{a+b} \leqslant \sqrt{a} + \sqrt{b}$ for any $a, b > 0$, we always have

$$D \leqslant 2\left(q_1\lambda_A^2 + 2q_2\gamma\lambda_A\lambda_B + q_3\lambda_B^2 + q_2(\lambda_A\mu_B + \lambda_B\mu_A)\right).$$

The proof of Theorem 13.3 is preceded by a simple lemma.

LEMMA 13.1 *The maximum value of the function*

$$J(u, v) = 2au^2 + 2buv + 2cv^2, \quad a, b, c \geqslant 0, \tag{13.39}$$

on the circle $u^2 + v^2 = 1$ *equals*

$$a + b + \sqrt{(a - c)^2 + b^2}. \tag{13.40}$$

Proof of Lemma 13.1. Since $u^2 + v^2 = 1$ and the numbers u and v can be assumed nonnegative, there is $t \in \left[0, \frac{\pi}{2}\right]$ such that $u = \sin t$ and $v = \cos t$. Consequently

$$J(u, v) = 2a \sin^2 t + 2b \sin t \cos t + 2c \cos^2 t =$$

$$= \left[a + b + \sqrt{(a-c)^2 + b^2}\right] \cos \left[2t - \arccos \frac{c-a}{\sqrt{(a-c)^2 + b^2}}\right],$$

i.e., $J(u, v)$ attains its maximum value for

$$u = \sin \left[\frac{1}{2} \arcsin \frac{c-a}{\sqrt{(a-c)^2 + b^2}}\right], \quad v = \cos \left[\frac{1}{2} \arccos \frac{c-a}{\sqrt{(a-c)^2 + b^2}}\right].$$

Clearly that value is given by (13.40) above. The lemma is proved.

The number (13.40) coincides with the spectral radius of the matrix

$$\begin{bmatrix} 2a & b \\ b & 2c \end{bmatrix}.$$

Proof of Theorem 13.3. Let us denote

$$u = \frac{\|Px\|}{\|x\|}, \quad v = \frac{\|Qx\|}{\|x\|},$$

where $x \in H$, $x \neq 0$, is a given vector and P and Q are the orthogonal projections onto the subspaces E_0 and E_1, respectively. Clearly $u^2 + v^2 = 1$.

From (13.36) it follows that

$$\|Ax\|^2 = \|APx\|^2 + \|AQx\|^2, \quad \|Bx\|^2 = \|BPx\|^2 + \|BQx\|^2.$$

Therefore, by (13.33) and (13.34) we find the estimates

$$\|Ax\|^2 = \left(\lambda_A^2 u^2 + \mu_A^2 v^2\right) \|x\|^2, \quad \|Bx\|^2 = \left(\lambda_B^2 u^2 + \mu_B^2 v^2\right) \|x\|^2. \tag{13.41}$$

In order to estimate (MAx, MBx) we will use assumption (13.35) and the triangle inequality (13.1). The conclusion of Theorem 13.3 follows from (13.41) and the next chain of relations

$$(MAx, MBx) \leqslant$$

$$\leqslant (MAPx, MBPx) + (MAQx, MBPx) + (MAPx, MBQx) + (MAQx, MBQx) \leqslant$$

$$\leqslant \gamma \lambda_A \lambda_B \|Px\|^2 + \mu_A \lambda_B \|Qx\| \cdot \|Px\| + \lambda_A \mu_B \|Px\| \cdot \|Qx\| + \mu_A \mu_B \|Qx\|^2 =$$

$$= \left[\gamma \lambda_A \lambda_B u^2 + (\mu_A \lambda_B + \lambda_A \mu_B) uv + \mu_A \mu_B v^2\right] \cdot \|x\|^2.$$

The proof is complete.

Results analogous to Theorem 13.3 can be established for more general quadratic forms, namely,

$$\sum_{i,j=1}^{n} q_{ij}(MA_i x, MA_j x).$$

More precisely, if we assume that

$$\|A_i Px\| \leqslant \lambda_i \|Px\|, \quad \|A_i Qx\| \leqslant \mu_i \|Qx\|, \qquad (A_i Px, A_i Qx) = 0,$$

$$(MA_i Px, MA_j Px) \leqslant \gamma_{ij}\lambda_i\lambda_j \|Px\|^2 \quad (i \neq j),$$

for all $i, j = 1, \ldots, n$ and all $x \in H$, where $\gamma_{ij} < 1$, then

$$\sum_{i,j=1}^{n} q_{ij}(MA_i x, MA_j x) \leqslant r\|x\|^2,$$

where r is the spectral radius of the matrix

$$\begin{bmatrix} \displaystyle\sum_{i=1}^{n} q_{ii}\lambda_i^2 + \sum_{\substack{i,j=1 \\ i\neq j}}^{n} q_{ij}\lambda_i\lambda_j\gamma_{ij} & \displaystyle\sum_{\substack{i,j=1 \\ i\neq j}}^{n} q_{ij}\lambda_i\mu_j \\[3ex] \displaystyle\sum_{\substack{i,j=1 \\ i\neq j}}^{n} q_{ij}\lambda_i\mu_j & \displaystyle\sum_{i=1}^{n} q_{ii}\mu_i^2 + \sum_{\substack{i,j=1 \\ i\neq j}}^{n} q_{ij}\mu_i\mu_j \end{bmatrix}.$$

If the information on operators A and B is more detailed than that provided by conditions (13.33)–(13.36), then quantity (13.7) can be estimated with a better accuracy than in (13.38).

Suppose that the space H is represented as a direct sum of mutually orthogonal subspaces E_i ($i = 0, \ldots, n$, $n \geqslant 1$). In addition, assume that

a) $\qquad \|Ax_i\| \leqslant \lambda_{Ai}\|x_i\|, \quad \|Bx_i\| \leqslant \lambda_{Bi}\|x_i\| \qquad (x_i \in E_i; \; i = 0, \ldots, n),$

where $\lambda_{A0} \geqslant \lambda_{A1} \geqslant \cdots \geqslant \lambda_{An} \geqslant 0$, $\lambda_{B0} \geqslant \lambda_{B1} \geqslant \cdots \geqslant \lambda_{Bn} \geqslant 0$;

b) $\qquad (MAx_i, MBx_j) \leqslant \gamma_{ij}\lambda_{Ai}\lambda_{Bj}\|x_i\| \cdot \|x_j\| \quad (x_i \in E_i, \; x_j \in E_j),$

where all the constants γ_{ij} are strictly less than 1, with the possible exception of those γ_{ij} for which one of the indices i or j equals n;

c) $\qquad (Ax_i, Ax_j) = (Bx_i, Bx_j) = 0 \quad (x_i \in E_i, \; x_j \in E_j; \; i \neq j).$

Then $\mathfrak{J}(q_1, q_2, q_3; A, B) \leqslant r_1$, where r_1 is the spectral radius of the matrix $A_1 + 2q_2 A_2$, and

$$A_1 = \begin{bmatrix} q_1 \lambda_{A0}^2 + q_3 \lambda_{B0}^2 & & 0 \\ & \ddots & \\ 0 & & q_1 \lambda_{An}^2 + q_3 \lambda_{Bn}^2 \end{bmatrix},$$

$$A_2 = \begin{bmatrix} \gamma_{00} \lambda_{A0} \lambda_{B0} & \cdots & \gamma_{0n} \lambda_{A0} \lambda_{Bn} \\ \cdots & \cdots & \cdots \\ \gamma_{n0} \lambda_{An} \lambda_{B0} & \cdots & \gamma_{nn} \lambda_{An} \lambda_{Bn} \end{bmatrix} + \begin{bmatrix} \gamma_{00} \lambda_{A0} \lambda_{B0} & \cdots & \gamma_{0n} \lambda_{An} \lambda_{B0} \\ \cdots & \cdots & \cdots \\ \gamma_{n0} \lambda_{A0} \lambda_{Bn} & \cdots & \gamma_{nn} \lambda_{An} \lambda_{Bn} \end{bmatrix}.$$

The constant r_1 is not necessary less than $q_1 \lambda_{A0}^2 + 2q_2 \lambda_{A0} \lambda_{B0} + q_3 \lambda_{B0}^2$ or $\dfrac{D}{2}$. However, if λ_{Ai} and λ_{Bi} $(i > 0)$ as well as γ_{ij} $(i, j = 0, \ldots, n-1)$ are sufficiently small, then the estimates

$$r_1 < \frac{D}{2}, \quad r_1 < q_1 \lambda_{0A}^2 + 2q_2 \lambda_{A0} \lambda_{B0} + q_3 \lambda_{B0}^2$$

hold true.

Theorem 13.3 is a particular case (when $n = 1$) of the just mentioned result.

13.7. UNIQUENESS OF SOLUTIONS. Assume that instead of estimate (13.3) we have the Lipschitz-type condition

$$\|\mathfrak{f}(x_1, x_2) - \mathfrak{f}(y_1, y_2)\| \leqslant \|k_1 M(x_1 - y_1) + k_2 M(x_2 - y_2)\|. \tag{13.42}$$

This condition is fulfilled for any superposition operator $\mathfrak{f}x = f(t, x_1(t), x_2(t))$ on the spaces L_p or C, provided that

$$|f(t, x_1, x_2) - f(t, y_1, y_2)| \leqslant k_1 |x_1 - y_1| + k_2 |x_2 - y_2|, \quad t \in \Omega, \ x_1, x_2, y_1, y_2 \in \mathbb{R}^n.$$

THEOREM 13.4. *Suppose that condition (13.42) holds true and*

$$N = N(k_1 A, k_2 B) < 1.$$

Then the equation $x = \mathfrak{f}(Ax, Bx)$ has no more but one solution $x \in H$.

Proof. The proof is very simple. Assume that $x_i = \mathfrak{f}(Ax_i, Bx_i)$ $(i = 1, 2)$. Then

$$\|x_1 - x_2\| \leqslant \|k_1 M A(x_1 - x_2) + M k_2 B(x_1 - x_2)\| \leqslant N \|x_1 - x_2\|,$$

and, since $N < 1$, the conclusion follows.

§14. Two-point boundary value problems (the nonquasilinear case)

14.1. SETTING THE PROBLEM. In this section we will continue the study initiated in Chapter 2 of the equation

$$Lz \overset{\text{def}}{=} z'' + p(t)z' + q(t)z = f(t, z, z'), \tag{14.1}$$

with boundary conditions

$$z(0) = z(\pi) = 0. \tag{14.2}$$

Equation (14.1) is considered for $p(t) \equiv 0$ and $q(t) \equiv \alpha > 1$, i.e., it has the form

$$z'' + \alpha z = f(t, z, z'). \tag{14.3}$$

The nonlinearity $f(t, z, y) : [0, \pi] \times \mathbb{R}^1 \times \mathbb{R}^1 \to \mathbb{R}^1$ is assumed (for simplicity) to be jointly continuous and to satisfy the estimate

$$|f(t, z, y)| \leqslant k_1|z| + k_2|y| + c, \quad -\infty < z, y < \infty, \ 0 \leqslant t \leqslant \pi. \tag{14.4}$$

If $\alpha \neq n^2$ (where n is a natural number), then the boundary value problem (14.3)–(14.2) is equivalent to the operator equation

$$x(t) = f[t, Ax(t), A_1x(t)], \tag{14.5}$$

where by A and A_1 we denote the completely continuous operators acting on $L_2 = L_2(0, \pi)$ defined by

$$Ax(t) = -\frac{2}{\pi} \sum_{n=1}^{\infty} \frac{1}{n^2 - \alpha}(x(t), \sin nt) \sin nt, \tag{14.6}$$

$$A_1x(t) = -\frac{2}{\pi} \sum_{n=1}^{\infty} \frac{n}{n^2 - \alpha}(x(t), \sin nt) \cos nt. \tag{14.7}$$

Specifically, by the equivalence between problem (14.3)–(14.2) and equation (14.5) we mean that any solution $x(t) \in L_2$ of the operator equation (14.5) yields the twice continuously differentiable solution $z(t) = Ax(t)$ of problem (14.3)–(14.2), and, conversely, any twice continuously differentiable solution $z(t)$ of problem (14.3)–(14.2) provides the solution $x(t) = f(t, z, z')$ of equation (14.5).

Every solution $x(t) \in L_2$ of the operator equation (14.5) is a continuous function. The values of the operators (14.6) and (14.7) are related by the equality
$$\frac{d}{dt} Ax(t) \equiv A_1 x(t).$$

THEOREM 14.1. *Suppose that*

$$N(k_1 A, k_2 A_1) = N(k_1 A, k_2 A_1; L_2, |\cdot|) < 1.$$

Then the two-point boundary value problem (14.3)–(14.2) has at least one twice continuously differentiable solution.

Theorem 14.1 is an easy consequence of the Schauder Principle. All we need is to observe that the operator $x(t) \mapsto f[t, Ax(t), A_1 x(t)]$ is completely continuous on L_2 and sends the ball

$$\mathfrak{B} = \{\|x\| \leqslant [1 - N(k_1 A, k_2 A_1)]^{-1} c\sqrt{\pi}\} \in L_2$$

into itself.

It is possible to find examples of nonlinearities $f(t, z, y)$ that satisfy estimate (14.4) but $N(k_1 A, k_2 A_1) = 1$, and problem (14.3)–(14.2) has no absolutely continuous differentiable solutions.

Since the norms of operators (14.6) and (14.7) on the space L_2 are given by

$$\|A\| = \max_{n \in \mathbb{N}} |n^2 - \alpha|^{-1}, \quad \|A_1\| = \max_{n \in \mathbb{N}} n|n^2 - \alpha|^{-1},$$

the conclusion of Theorem 14.1 remains true if instead of inequality $N(k_1 A, k_2 A_1) < 1$ we consider the stronger constraint

$$k_1 \max_{n \in \mathbb{N}} |n^2 - \alpha|^{-1} + k_2 \max_{n \in \mathbb{N}} n|n^2 - \alpha|^{-1} < 1. \tag{14.8}$$

In particular, for $1 < \alpha < 2$ both the maximum values in (14.8) are attained at $n = 1$; therefore, in this case condition (14.8) becomes

$$k_1 + k_2 < \alpha - 1. \tag{14.9}$$

If $2 \leqslant \alpha \leqslant 2\frac{1}{2}$, then inequality (14.8) is equivalent to

$$\frac{k_1}{\alpha - 1} + \frac{2k_2}{4 - \alpha} < 1. \tag{14.10}$$

Actually, it is easy to rewrite condition (14.8) for all possible values of α. More precisely, there are three different cases:

a) if $n^2 < \alpha \leqslant n^2 + n$, then

$$k_1\|A\| + k_2\|A_1\| = \frac{k_1 + k_2 n}{\alpha - n^2};$$

b) if $n^2 + n + \frac{1}{2} < \alpha < (n+1)^2$, then

$$k_1\|A\| + k_2\|A_1\| = \frac{k_1 + k_2(n+1)}{(n+1)^2 - \alpha};$$

c) if $n^2 + n < \alpha \leqslant n^2 + n + \frac{1}{2}$, then

$$k_1\|A\| + k_2\|A_1\| = \frac{k_1}{\alpha - n^2} + \frac{k_2(n+1)}{(n+1)^2 - \alpha}.$$

In the following subsections we will discuss examples of estimates of the joint norm $N(k_1 A, k_2 A_1)$ that will provide solvability conditions for problem (14.3)–(14.2) better than (14.8). As a matter of fact, we will separately consider the cases $1 < \alpha < 2$ and $2 < \alpha < 2\frac{1}{2}$. An analogous approach can be developed for values of α greater than $2\frac{1}{2}$. For instance, the case $n^2 + n < \alpha \leqslant n^2 + n + \frac{1}{2}$ may be handled in the same way as the case $2 < \alpha < 2\frac{1}{2}$, and the cases $n^2 + n < \alpha \leqslant n^2 + n$ or $n^2 + n + \frac{1}{2} < \alpha < (n+1)^2$ can be treated similarly to the case $1 < \alpha < 2$. In the subsequent study of the cases $1 < \alpha < 2$ and $2 < \alpha < 2\frac{1}{2}$ we will use different results from Section 13.

14.2. THE CASE $1 < \alpha < 2$. In this case the solvability condition for problem (14.3)–(14.2) stated by inequality (14.8) above reduces to (14.9). Theorem 13.3 leads to a less restrictive condition.

Represent the space L_2 as a direct sum of the subspaces

$$E_0 = \left\{ x_0(t) : x_0(t) = \xi \sin t, \ \xi \in \mathbb{R}^1 \right\}$$

and

$$E_1 = \{ x_1(t) : (x_1(t), \sin t) = 0 \}.$$

The operators A and A_1 satisfy relations (13.33)–(13.36) for

$$\lambda_A = \lambda_{A_1} = \frac{1}{\alpha - 1}, \quad \mu_A = \frac{1}{4 - \alpha}, \quad \mu_{A_1} = \frac{2}{4 - \alpha}, \quad \gamma = \frac{2}{\pi}.$$

Therefore, the joint norm $N(k_1 A, k_2 A_1)$ has the estimate

$$2N^2(k_1 A, k_2 A_1) \leqslant D \overset{\text{def}}{=}$$

$$= D\left(k_1^2, k_1k_2, k_2^2, (\alpha-1)^{-1}, (\alpha-1)^{-1}, 2(4-\alpha)^{-1}, 2(4-\alpha)^{-1}, 2\pi^{-1}\right) =$$

$$= \left\{ k_1^2\left[\frac{1}{(\alpha-1)^2} + \frac{1}{(4-\alpha)^2}\right] + k_2^2\left[\frac{1}{(\alpha-1)^2} + \frac{4}{(4-\alpha)^2}\right] + 2k_1k_2\left[\frac{2}{\pi}\frac{1}{(\alpha-1)^2} + \frac{2}{(4-\alpha)^2}\right]\right\} +$$

$$+ \left(4k_1^2k_2^2\frac{3}{(\alpha-1)^2(4-\alpha)^2} + \left\{k_1^2k_1^2\left[\frac{1}{(\alpha-1)^2} - \frac{1}{(4-\alpha)^2}\right] + \right.\right.$$

$$\left.\left. + k_2^2k_1^2\left[\frac{1}{(\alpha-1)^2} - \frac{4}{(4-\alpha)^2}\right] + 2k_1k_2k_1^2\left[\frac{2}{\pi}\frac{1}{(\alpha-1)^2} - \frac{3}{(4-\alpha)^2}\right]\right\}^2\right)^{1/2},$$

where D is the function defined by (13.37). Condition (14.8) for the solvability of problem (14.3)–(14.2) can be replaced by condition $D < 2$. If $1 < \alpha < \alpha_0$ (where α_0 is a well-determined number in the interval $(1, 2)$), then the latter condition is better than condition (14.8), for any k_1 and k_2. In case $\alpha_0 < \alpha < 2$, condition $D < 2$ is better than condition (14.8) only for sufficiently large values of k_2/k_1.

For instance, assume that $\alpha = 1\frac{1}{2}$ and the nonlinearity $f(t, z, y)$ satisfies the estimate (14.4) for $k_1 = k_2 = k$. Then, condition (14.8) asserts that $k < \frac{1}{4}$, whereas condition $D < 2$ provides the less restrictive relation $k <\approx .27$.

We mention yet another example. If $\alpha = 2 - \varepsilon$ (with $\varepsilon > 0$ very small) and $k_2 < .36k_1$, then condition $D < 2$ is better than condition (14.8). For a different $\alpha > \alpha_0$, the admissible values of k_2/k_1 must be greater than .36.

14.3. THE CASE $2 < \alpha < 2\frac{1}{2}$. In this case the operators A and A_1 satisfy the estimates

$$\|Ax\|^2 \leqslant \sum_{n=1}^{\infty} \frac{1}{|n^2 - \alpha|}x_n^2, \quad \|A_1x\|^2 \leqslant \sum_{n=1}^{\infty} \frac{n}{|n^2 - \alpha|}x_n^2,$$

where x_n are the coefficients in the decomposition of x with respect to the orthonormal basis for L_2 provided by the functions $e_n(t) = \sqrt{2\pi^{-1}}\sin nt$. In order to find a solvability condition for the two-point boundary value problem (14.3)–(14.2) less restrictive than (14.8), we will apply Theorem 13.2 in case $q_1 = k_1^2$, $q_2 = k_1k_2$, $q_3 = k_2^2$ ($k_2 \neq 0$). Since for $2 < \alpha < 2\frac{1}{2}$ we have the inequalities

$$(\alpha - 1)^{-1} > (4 - \alpha)^{-1} > \cdots > (n^2 - \alpha)^{-1} > \cdots,$$

as well as

$$2(4 - \alpha)^{-1} > (\alpha - 1)^{-1} > 3(9 - \alpha)^{-1} > \cdots > n(n^2 - \alpha)^{-1} > \cdots,$$

it follows that

$$J_1^2 = \max\left\{\frac{1}{(\alpha - 1)^2}(k_1 + k_2)^2, \frac{1}{(4 - \alpha)^2}(k_1 + 2k_2)^2\right\}.$$

If
$$\frac{k_1}{k_2} > \frac{3(\alpha - 2)}{5 - 2\alpha},$$

then
$$J_1 = \frac{k_1 + k_2}{\alpha - 1},$$

whereas
$$\frac{k_1}{k_2} \leqslant \frac{3(\alpha - 2)}{5 - 2\alpha}$$

implies
$$J_1 = \frac{k_1 + 2k_2}{4 - \alpha}.$$

We go on finding estimates for J_2. If
$$\frac{\alpha^2 - 4}{2(5 - 2\alpha)} < \frac{k_1}{k_2} < \frac{\alpha^2 - 4}{5 - 2\alpha},$$

then we get
$$\{(\alpha - 1)^{-2}, (4 - \alpha)^{-2}, (\alpha - 1)^{-2}, 4(4 - \alpha)^{-2}\} \in \mathfrak{R}\left(k_1^2, k_1 k_2, k_2^2\right),$$

and, therefore, we obtain
$$J_2^2 = \frac{k_1^2}{\alpha^2 - 4} + \frac{k_2^2}{5 - 2\alpha}.$$

If inequalities (14.11) fail, we have $J_1 \geqslant J_2$. Since the double inequality
$$\frac{\alpha^2 - 4}{2(5 - 2\alpha)} < \frac{3(\alpha - 2)}{5 - 2\alpha} < \frac{\alpha^2 - 4}{5 - 2\alpha}$$

holds true for all $\alpha \in \left(2, 2\frac{1}{2}\right)$, we conclude that

$$N(k_1 A, k_2 A_1) \leqslant \begin{cases} \dfrac{k_1 + 2k_2}{4 - \alpha} & \text{if } \dfrac{k_1}{k_2} \leqslant \dfrac{\alpha^2 - 4}{2(5 - 2\alpha)}, \\[3mm] \sqrt{\dfrac{k_1^2}{\alpha^2 - 4} + \dfrac{k_2^2}{5 - 2\alpha}} & \text{if } \dfrac{\alpha^2 - 4}{2(5 - 2\alpha)} < \dfrac{k_1}{k_2} < \dfrac{\alpha^2 - 4}{5 - 2\alpha}, \\[3mm] \dfrac{k_1 + k_2}{\alpha - 1} & \text{if } \dfrac{\alpha^2 - 4}{5 - 2\alpha} \leqslant \dfrac{k_1}{k_2}. \end{cases}$$

14.4. EQUATIONS CONTAINING THE DERIVATIVE IN THEIR LINEAR PART. We next consider the two-point boundary value problem for the equation

$$z'' - 2bz' + \alpha z = f(t, z, z'). \tag{14.12}$$

If $\alpha - b^2$ is not the square of a natural number, then problem (14.12)–(14.2) is equivalent, in the usual sense, to the operator equation (14.5), where now the operators A and A_1 are acting on the space L_2^* of all square-integrable functions with respect to the inner product

$$\langle x, y \rangle \stackrel{\text{def}}{=} \int_0^\pi \exp\{-2bt\} x(t) y(t) \mathrm{d}t, \tag{14.13}$$

and are defined by the formulas

$$A x(t) = \sum_{n=1}^\infty \frac{1}{\alpha - (n^2 + b^2)} \langle x, e_n \rangle e_n(t), \tag{14.14}$$

and

$$A_1 x(t) = \sum_{n=1}^\infty \frac{\sqrt{n^2 + b^2}}{\alpha - (n^2 + b^2)} \langle x, e_n \rangle g_n(t). \tag{14.15}$$

In formulas (14.14) and (14.15) above we denoted by $\{e_n(t)\}$ the orthonormal basis on L_2^* consisting of the functions $e_n(t) = \sqrt{2/\pi} \exp\{bt\} \sin nt$ $(n = 1, 2, \ldots)$, and $g_n(t)$ stand for the functions $g_n(t) = e_n'(t) \sqrt{n^2 + b^2}$ which are unit vectors in L_2^*, too.

From (14.14) and (14.15) it is clear that in order to find estimates of the joint norm of the operators $k_1 A$ and $k_2 A_1$ (A is the operator (14.14) and A_1 is the operator (14.15)) we can still use the methods developed in Section 13. For instance, if $b = 1$ and $\alpha = (2, \sqrt{10})$, then

$$\|A\| = [(\alpha - (1 + b^2))]^{-1}, \quad \|A_1\| = \sqrt{1 + b^2}[\alpha - (1 + b^2)]^{-1}, \quad \langle |e_1|, |g_1| \rangle = \frac{2}{\pi} < 1.$$

§15. Forced oscillations in quasilinear systems

15.1. SETTING THE PROBLEM. In this section we are concerned with finding existence and uniqueness conditions for forced periodic oscillations in systems whose dynamics is described by the equation

$$L\left(\frac{\mathrm{d}}{\mathrm{d}t}\right) x = M\left(\frac{\mathrm{d}}{\mathrm{d}t}\right) f(t, x) \tag{15.1}$$

(for notations and terminology see Sections 7 and 11). Recall that the operator of the periodic problem, denoted by A, acts on the space $L_2 = L_2([0, T], \mathbb{R}^1)$ according to the formula

$$A x(t) = \sum_{n=0}^\infty |W(\omega_n \mathrm{i})| U_n P_n x(t) \tag{15.2}$$

Its norm is given by

$$\|A\| = w(T) = \max_{n=0,\pm1,\ldots} |W(\omega_n \mathrm{i})|,$$

and in what follows we will only consider the case when all the leading eigenvalues of the operator A are nonreal. By $W(p)$ we denote the fractional rational transfer function of the linear link of the system, and $\omega_n = \dfrac{2n\pi}{T}$ $(n = 0, \pm1, \pm2, \ldots)$. Throughout this section we assume that $\omega_n \mathrm{i}$ are not roots of the polynomial L.

As it was mentioned above, the main assumption along this section is that equality $w(T) = |W(\omega_n \mathrm{i})|$ (or any other analogous relation) is possible but for values of n satisfying $\operatorname{Im} W(\omega_n \mathrm{i}) \neq 0$.

We also suppose, as usual, that the nonlinearity $f(t, x)$ satisfies the Caratheodory condition and is T-periodic in t, i.e.,

$$f(t, x) \equiv f(t + T, x), \quad 0 \leqslant t \leqslant T, \ -\infty < x < \infty.$$

If the estimate

$$|f(t, x)| \leqslant k|x| + b(t), \quad 0 \leqslant t \leqslant T, \ -\infty < x < \infty; \ b(t) \in L_2, \tag{15.3}$$

holds true and $kw(T) < 1$, then (see Theorem 11.1) equation (15.1) has at least one solution $x(t)$ that is continuous together with all its derivatives up to the order $l - m - 1$ ($l = \deg L$, $m = \deg M$).

Using the results proved in Section 12, the basic condition $kw(T) < 1$ can be successfully and significantly improved.

15.2. THE NUMBER $q_2(\eta)$. Let η be a fixed number in the interval $(0, w(T)]$, and let $\mathbb{N}(\eta)$ denote the set of those nonnegative integers n for which $|W(\omega_n \mathrm{i})| \geqslant \eta$. We will always assume that

$$\operatorname{Im} W(\omega_n \mathrm{i}) \neq 0, \quad n \in \mathbb{N}(\eta). \tag{15.4}$$

In addition, let us consider the subspace

$$E[\eta] = \left(\sum_{n \in \mathbb{N}(\eta)} P_n \right) L_2 \equiv \bigoplus_{n \in \mathbb{N}(\eta)} \Pi_n \tag{15.5}$$

(for the definition of the subspaces Π_n, see Subsection 7.4). Since $\operatorname{Im} W(0) = 0$, we clearly have $0 \notin \mathbb{N}(\eta)$, hence the dimension of the subspace (15.5) is even. If $\eta = w(T)$, then the subspace (15.5) coincides with the subspace (11.22). In case $\dim E[\eta] = 2$

(this case will be considered below in more detail) there exists a natural number s such that

$$E[\eta] = \left\{ e(t) : e(t) = \xi \cos \omega_s t + \zeta \sin \omega_s t; \ \xi, \zeta \in \mathbb{R}^1 \right\}.$$

We next introduce the number

$$q_2(\eta) \stackrel{\text{def}}{=} [w(T)]^{-1} \max_{e(t) \in E[\eta]; \ \|e\|=1} [|Ae(t)|, |e(t)|]. \tag{15.6}$$

From assumption (15.4) and Lemma 12.1 it follows that

$$q_2 < 1. \tag{15.7}$$

The quantity (15.6) can be equally well defined as

$$q_2(\eta) = [w(T)]^{-1} \max_{e(t) \in E[\eta]; \ \|e\|=1} [|A^*e(t)|, |e(t)|].$$

Our specific purpose in this section is to compute the number (15.6) when $\dim E[\eta] = 2$. Under this additional assumption we clearly have $w(T) = |W(\omega_s i)|$, where s is a natural number. Moreover, $E[\eta] = E[w(T)]$ and

$$|W(\omega_n i)| < w(T) \quad (n \neq s).$$

Any normalized function $e(t) \in E_0 \stackrel{\text{def}}{=} E[w(T)]$ has the form

$$e(t) = \sqrt{\frac{2}{T}} \sin(\omega_s t + \delta) \tag{15.8}$$

for an initial phase δ. For a function $e(t)$ as in (15.8) we have the equality (see (7.26))

$$Ae(t) = \sqrt{\frac{2}{T}} w(T) \sin(\omega_s t + \delta + \tau), \tag{15.9}$$

where $\tau = \arg W(\omega_s i)$. By assumption (15.4) we know that $\tau \neq 0$ and $\tau \neq \pi$. From (15.8) and (15.9) it follows that

$$[|Ae(t)|, |e(t)|] = \int_0^T |Ae(t)e(t)| dt = \frac{2}{T} w(T) \int_0^T |\sin(\omega_s t + \delta + \tau) \sin(\omega_s t + \delta)| dt.$$

But

$$\frac{2}{T} \int_0^T |\sin(\omega_s t + \delta + \tau) \sin(\omega_s t + \delta)| dt = \frac{2}{\omega_s T} \int_\delta^{\omega_s T + \delta} |\sin(\theta) \sin(\theta + \tau)| d\theta =$$

$$= \frac{1}{s\pi} \int_0^{2s\pi} |\sin(\theta) \sin(\theta + \tau)| d\theta = \frac{2}{\pi} \int_0^\pi |\sin(\theta) \sin(\theta + \tau)| d\theta,$$

hence
$$[\|Ae(t)\|, |e(t)|] = w(T)Q(\tau), \quad e(t) \in E_0; \ \|e\| = 1, \tag{15.10}$$

where by $Q(\tau)$ we denote the function

$$Q(\tau) \stackrel{\text{def}}{=} \frac{2}{\pi} \int\limits_0^\pi |\sin(\theta)\sin(\theta+\tau)| d\theta. \tag{15.11}$$

If τ is in the interval $[0, \pi]$, then $Q(\tau)$ is given by

$$Q(\tau) = \pi^{-1}[(\pi - 2\tau)\cos\tau + 2\sin\tau]. \tag{15.12}$$

We notice that the function (15.11) is periodic with period π and even. In addition, $Q(0) = Q(\pi) = 1$ and $Q\left(\frac{\pi}{2}\right) = \frac{2}{\pi} = \inf Q(\tau)$. Below we indicate a few values of this function:

$\tau\pi^{-1}$	.000	.100	.200	.300	.400	.500	.600	.700	.800	.900	1.00
$Q(\tau)$	1.000	.958	.860	.750	.667	.637	.667	.750	.860	.958	1.00

Thus, we conclude that if $\dim E[\eta] = 2$, then the number (15.6) is defined by the equality

$$q_2(\eta) = Q[\arg W(\omega_s i)], \tag{15.13}$$

where s is the natural number satisfying $w(T) = |W(\omega_s i)|$.

15.3. MAIN THEOREMS. Besides the previous notations we set

$$q_1(\eta) = [w(T)]^{-1} \max_{n=0,\pm1,\pm2,\dots;\ |W(\omega_s i)|<\eta} |W(\omega_n i)|. \tag{15.14}$$

If $\eta = w(T)$, then the number (15.14) is given by

$$q_1 = \frac{w_1(T)}{w(T)},$$

where $w_1(T)$ is defined by formula (11.26).

In both the theorems stated below we will use the root $\mu(q_1, q_2)$ of equation (12.1) as it was defined in Subsection 12.1.

THEOREM 15.1. *Let $\eta \in (0, w(T)]$ be such that condition (15.4) holds true. Assume that the nonlinearity $f(t,x)$ in equation (15.1) satisfies estimate (15.3) with a coefficient k subject to the condition*

$$kw(T) < \mu[q_1(\eta), q_2(\eta)].$$

Then equation (15.1) has at least one T-periodic solution $x(t) \in C^{l-m-1}$.

Theorem 15.1 follows straightforwardly from Theorem 12.1.

In case dim $E[\eta] = 2$, Theorem 15.1 can be stated as follows:

THEOREM 15.2. *Assume that the maximum value in the equality*

$$w(T) = \max_{n=0,1,\dots} |W(\omega_n i)|$$

is attained for a unique number $n = s > 0$, *such that* $\operatorname{Im} W(\omega_n i) \neq 0$. *In addition, suppose that the coefficient k in (15.3) satisfies the inequality*

$$kw(T) \leqslant \mu \left[\frac{w_1(T)}{w(T)}, Q(\arg W(\omega_s i)) \right]. \tag{15.15}$$

Then condition (15.3) implies the existence of at least one T-periodic solution $x(t) \in C^{l-m-1}$ *of equation (15.1).*

15.4. A UNIQUENESS THEOREM. If together with condition (15.3) the nonlinearity $f(t, x)$ satisfies the Lipschitz condition

$$|f(t, x) - f(t, y)| \leqslant |x - y|, \quad 0 \leqslant t \leqslant T, \quad -\infty < x, y < \infty, \tag{15.16}$$

then Theorems 15.1 and 15.2 above can be complemented with a uniqueness result. For a smooth function $f(t, x)$ condition (15.16) is equivalent to condition $|f'_x(t, x)| \leqslant k$. If $f(t, 0) \in L_2$, then condition (15.16) implies inequality (15.3).

THEOREM 15.3. *Suppose that the constraint (15.16) holds true and all the conditions in Theorem 15.1 (or 15.2) are fulfilled. Then the solution* $x(t) \in C^{l-m-1}$ *of equation (15.1) is unique.*

Proof. Let $x_1(t), x_2(t) \in C^{l-m-1}$ be two T-periodic solutions of equation (15.1) and assume that the conditions of Theorem 15.1 as well as condition (15.16) are fulfilled. For the functions $x_1(t)$ and $x_2(t)$ we have the equalities

$$x_1(t) = Af[t, x_1(t)], \quad x_2(t) = Af[t, x_2(t)],$$

where A is the operator (15.2). Therefore, the difference $z_0(t) = x_1(t) - x_2(t)$ is a solution of the equation

$$z(t) = Ag[t, z(t)],$$

where the function

$$g(t, x) \stackrel{\text{def}}{=} f[t, z + x_2(t)] - f[t, x_2(t)]$$

satisfies, in view of (15.16), the condition

$$|g(t, z)| \leqslant k|z|, \quad 0 \leqslant t \leqslant T, \quad -\infty < z < \infty. \tag{15.18}$$

Further, let us consider the superposition operator

$$Fz(t) = g[t, z(t)]. \tag{15.19}$$

From (15.18) we get that the operator (15.19) satisfies condition (12.14), where $q_1 = q_1(\eta)$ and $c_1 = 0$. Since

$$|[Pz, AFz]| \leqslant [|A^*Pz|, k|z|] \leqslant \|Pz\|kw(T)\{q_2(\eta)\|Pz\| + \|Qz\|\},$$

the estimate (12.15) is also true, with $c_2 = 0$. Therefore, based on Theorem 12.3, we conclude that the solution $z_0(t)$ of equation (15.17) satisfies the estimate $\|z_0\| = 0$, i.e., $z_0(t) \equiv 0$. Theorem 15.3 is proved.

15.5. A REMARK ON THE CONVERGENCE OF APPROXIMATION METHODS. Recall that the harmonic balance method provides a specific construction of approximations to T-periodic solutions of equation (15.1). Under the conditions in Theorem 15.1 the harmonic balance method is realizable and converges. Moreover, under the conditions in Theorem 15.3 the approximate solution of order N is unique, for each $N = 1, 2, \ldots$.

In this subsection we first describe the method of simple iterations for approximating the T-periodic solutions of equation (15.1), and then we compare the convergence condition for this method with conditions of applicability of the harmonic balance method.

Given an initial approximation $x_0(t) \in L_2$ we define a sequence $x_n(t)$ of approximate T-periodic solutions by formula

$$L\left(\frac{\mathrm{d}}{\mathrm{d}t}\right) x_{n+1} = M\left(\frac{\mathrm{d}}{\mathrm{d}t}\right) f[t, x_n(t)], \quad n = 0, 1, 2, \ldots. \tag{15.20}$$

In order to construct recurrently the approximation sequence it is enough to find a T-periodic solution x_{n+1} of the linear equation (15.20) with constant coefficients. If the Lipschitz condition (15.16) is satisfied and $kw(T) < 1$, then the sequence $x_n(t)$ converges in C^{l-m-1} to the unique solution $x_*(t)$ of equation (15.1) for any choice of the initial approximation $x_0(t)$. If $kw(T) > 1$, then the sequence $x_n(t)$ does not necessarily converge.

Consequently, if k in (15.16) lies in the interval $(1, \mu[q_1(\eta), q_2(\eta)])$, then, generally speaking, the method of simple iterations is not applicable, whereas the harmonic balance method is realizable and converges.

15.6. A SECOND ORDER EQUATION. Let us next consider the equation

$$x'' + ax' + bx = f(t, x) \tag{15.21}$$

with the periodic conditions

$$x(0) = x(2\pi), \quad x'(0) = x'(2\pi). \tag{15.22}$$

We assume that $a \neq 0$ and that the function $f(t, x)$ satisfies the Caratheodory condition and is periodic in t with period 2π. Let

$$\kappa \stackrel{\text{def}}{=} 2b - a^2,$$

and suppose that $\kappa > 1$ and

$$\kappa \neq n^2 + (n+1)^2, \quad n = 1, 2, \ldots .$$

Then there exists a natural number s such that

$$s^2 + (s-1)^2 < \kappa < s^2 + (s+1)^2.$$

Let λ_1 and λ_2 denote the numbers defined by

$$\lambda_1 = \frac{1}{\sqrt{(b - s^2)^2 + a^2 s^2}},$$

$$\lambda_2 = \begin{cases} \dfrac{1}{\sqrt{[b - (s-1)^2]^2 + a^2(s-1)^2}} & \text{if } \kappa \leqslant 2s^2 + 2, \\[2ex] \dfrac{1}{\sqrt{[b - (s+1)^2]^2 + a^2(s+1)^2}} & \text{if } \kappa > 2s^2 + 2. \end{cases}$$

It is easy to check that $\lambda_1 > \lambda_2$.

THEOREM 15.4. *Assume that the number k satisfies the condition*

$$k < \lambda_1^{-1} \mu \left[\frac{\lambda_2}{\lambda_1}, Q \left(\arctan \frac{as}{b - s^2} \right) \right].$$

If condition (15.3) is satisfied, then equation (15.21) has at least one solution subject to condition (15.22); under the additional condition (15.13), that solution is unique.

§16. Forced oscillations in systems with delay

16.1. EQUIVALENT OPERATOR EQUATIONS. In this section we are concerned with the periodic regime in systems with a dynamics described by equations with delays:

$$L\left(\frac{\mathrm{d}}{\mathrm{d}t}\right)x(t) = M\left(\frac{\mathrm{d}}{\mathrm{d}t}\right)f[t, x(t - h_1), \ldots, x(t - h_n)]. \tag{16.1}$$

In equation (16.1) above we let $L(p)$ and $M(p)$ denote, once again, the coprime polynomials (7.20) and (7.21), $l = \deg L(p)$ and $m = \deg M(p)$, respectively. The nonlinearity $f(t, x_1, \ldots, x_n)$ is jointly continuous in the variables $x_1, \ldots, x_n$, and measurable and T-periodic in the variable t. Without any significant loss of generality we assume that the delays satisfy the inequalities

$$0 \leqslant h_1 < h_2 < \cdots < h_n < T. \tag{16.2}$$

We also suppose that the polynomial $L(p)$ has no roots of the form $\omega_s \mathrm{i}$ $(s = 0, \pm 1, \ldots)$. Then the operator A (see, for instance, (15.2)) of the periodic problem for the linear link with the fractional-rational transfer function $W(p)$ is well defined. Following [Krasnoselskii, M. A., 1960] we introduce the operators

$$S(h)x(t) = \begin{cases} x(t - h) & \text{if } h \leqslant t \leqslant T, \\ x(t - h + T) & \text{if } 0 \leqslant t < h. \end{cases} \tag{16.3}$$

Any operator (16.3) will act on functions defined on the interval $[0, T]$; the values of the operators (16.3) are functions defined on the same interval. The operators (16.3) are linear and isometric on any space L_p $(1 \leqslant p \leqslant \infty)$. For every $h \in [0, T]$, the operator $S(h)$ sends a function of the form $\xi \cos \omega_s t + \eta \sin \omega_s t$ $(s = 0, 1, \ldots)$ into a function of the same form. Therefore, all the operators $S(h)$ commute with the operator of periodic problem.

Let us consider the operator equations

$$x(t) = Af[t, S(h_1)x, \ldots, S(h_n)x] \tag{16.4}$$

and

$$y(t) = f[t, AS(h_1)y, \ldots, AS(h_n)y]. \tag{16.5}$$

Each solution $x(t)$ of equation (16.4) corresponds to a solution

$$y(t) = f[t, S(h_1)x, \ldots, S(h_n)x]$$

of equation (16.5); conversely each solution $y(t)$ of equation (16.5) corresponds to the solution

$$x(t) = Ay(t)$$

of equation (16.4). A solution $x(t)$ of equation (16.4) is a T-periodic solution of equation (16.1) (in the same sense in which a solution of equation (11.3) is a solution of equation (11.2)).

16.2. THE SIMPLEST SOLVABILITY CONDITIONS. Throughout this subsection we assume that the nonlinearity $f(t, x_1, \ldots, x_n)$ satisfies the constraint

$$|f(t, x_1, \ldots, x_n)| \leqslant \sum_{j=1}^{n} k_j |x_j| + b(t),$$

$$(16.6)$$

$$0 \leqslant t \leqslant T, \quad -\infty < x_j < \infty \ (j = 1, \ldots, n); \quad b(t) \in L_2.$$

In this case the superposition operator

$$\mathfrak{f}(x_1, \ldots, x_n) = f[t, x_1(t), \ldots, x_n(t)] : (L_2)^n \to L_2,$$

is well-defined and continuous. The properties of the operator of the periodic problem (see Section 7) imply that the operator

$$Bx(t) = f[t, AS(h_1)x, \ldots, AS(h_n)x]$$

$$(16.7)$$

is a completely continuous one on the space L_2. For the values of this operator we have the estimate

$$\|Bx\| \leqslant w(T) \left(\sum_{j=1}^{n} k_j \right) \|x\| + \|b(t)\|.$$

$$(16.8)$$

In (16.8) above, as well as below, $\| \cdot \|$ denotes the norm on the space L_2.

According to the Schauder Fixed Point Principle we get the the next result.

THEOREM 16.1. *Assume that*

$$w(T) \cdot \sum_{j=1}^{n} k_j < 1.$$

$$(16.9)$$

Then equation (16.1) has at least one solution $x(t) \in C^{l-m-1}$ *with period T.*

Proof. It is enough to notice that under the conditions in Theorem 13.1 and in view of estimate (1.8), the operator (16.7) sends the ball $\mathfrak{B}_\rho = \{x(t) \in L_2 : \|x\| \leqslant \rho\}$ with radius

$$\rho = \|b(t)\| \left(1 - w(T) \sum_{j=1}^{n} k_j \right)^{-1}$$

into itself. The theorem is proved.

The constructions presented further in this section are aimed to weaken the main restriction (16.9) in Theorem 16.1.

16.3. USING THE JOINT NORM. Following the constructions developed in Section 13, we denote by

$$w = w(M, L, T; h_1, \ldots, h_n; k_1, \ldots, k_1)$$

the joint norm of the operators $k_1 AS(h_1), \ldots, k_n AS(h_n)$. The quantity w can be computed using the equality

$$w = \sup_{x \in L_2; \ \|x\|=1} \sum_{j=1}^{n} k_j |AS(h_j)x(t)|. \tag{16.10}$$

The next improvement of Theorem 16.1 is obvious.

Assume that quantity w is strictly less than 1. Then equation (16.1) has at least one solution $x(t) \in C^{l-m-1}$ with period T.

In order to take full advantage of this result in the case of specified delays and polynomials $L(p)$ and $M(p)$ we can use Theorem 13.3 or its analogues (for $n \geqslant 2$). We give a few details for the case $n = 2$.

Specifically, assume that $n = 2$, $h_1 = 0$ and $h_2 = h$, i.e., equation (16.1) has the form

$$\left(\frac{d}{dt}\right) x(t) = M\left(\frac{d}{dt}\right) f[t, x(t), x(t-h)]. \tag{16.11}$$

In addition, suppose that the maximum value in the definition

$$w(T) = \max_{s=0,1,2,\ldots} |W(\omega_s i)| \quad \left(W(p) = \frac{M(p)}{L(p)}\right). \tag{16.12}$$

of the norm $w(T)$ of the operator of the periodic problem in the space L_2 is attained for but one value of s, and $s > 0$. Then the quantity

$$w = N[k_1 A, k_2 S(h)A]$$

admits the estimate

$$2w^2 \leqslant D \overset{\text{def}}{=} D(k_1^2, k_1 k_2, k_2^2, w(T), w(T), w_1(T), w_1(T), \gamma) =$$
$$= \left[w^2(T) + w_1^2(T)\right] \left(k_1^2 + k_2^2\right) + 2 \left[\gamma w^2(T) + w_1^2(T)\right) k_1 k_2 +$$
$$+ \sqrt{\left\{\left[w^2(T) - w_1^2(T)\right] \left(k_1^2 + k_2^2\right) + 2 \left[\gamma w^2(T) - w_1^2(T)\right] k_1 k_2\right\}^2 + 16 w^2(T) w_1^2(T) k_1^2 k_2^2}.$$

In the formula above, $w_1(T)$ denotes the quantity (11.28) and $\gamma = Q[\arg W(\omega_s i) - \omega_s h]$, where $Q(\tau)$ is the function (15.11). If the inequality

$$2k_1 k_2 w_1^2(T) < (1 - \gamma)[w^2(T) - w_1^2(T)](k_1 + k_2)^2 \tag{16.13}$$

holds true, then condition $D < 2$ is less restrictive than condition (16.9). If $(2 - 2\gamma)w^2(T) \geqslant (3 - 2\gamma)w_1^2(T)$, then inequality (16.13) is fulfilled for any k_1 and k_2. In case $(2 - 2\gamma)w^2(T) < (3 - 2\gamma)w_1^2(T)$, inequality (16.13) is true only for sufficiently large values of the ratios k_1/k_2 or k_2/k_1.

16.4. THE MAIN SOLVABILITY CONDITION. Let us, once more, assume that the maximum value in (16.10) is attained for a unique positive integer s. Given the nonnegative numbers $k_1, \ldots, k_n$ we define the quantities

$$k_* = \sum_{j=1}^{n} k_j, \quad k_{**} = \sum_{j=1}^{n} \{k_j \cdot Q[\arg W(\omega_s i) - \omega_s h]\}, \tag{16.14}$$

and the number

$$\mu_* = \mu\left(\frac{w_1(T)}{w(T)}, \frac{k_{**}}{k_*}\right) \tag{16.15}$$

(recall that $\mu(q_1, q_2)$ denotes the unique root of equation (12.1) in the interval $(1, q_2^{-1})$). The numbers (16.14) clearly satisfy the inequality $k_{**} \leqslant k_*$, the equality $k_{**} = k_*$ being possible if and only if the following n equalities

$$k_j \cdot \sin[\arg W(\omega_s i) - \omega_s h] = 0, \quad j = 1, \ldots, n.$$

hold true.

THEOREM 16.2. *Suppose that $k_{**} < k_*$ and*

$$k_* \cdot w(T) < \mu_*. \tag{16.16}$$

Then the estimate (16.6) implies the existence of at least one T-periodic solution $x(t) \in C^{l-m-1}$ of equation (16.1).

16.5. PROOF OF THEOREM 16.2. We first consider the operator

$$Fx(t) = f[t, S(h_1)x, \ldots, S(h_n)x]. \tag{16.17}$$

Since each operator $S(h)$ is isometric on the space L_2, from (16.6) it follows that the operator (16.17) acts continuously on L_2 and the norms of its values on any ball

$\mathfrak{B} \subset L_2$ are bounded. Consequently, the operator AF is completely continuous on L_2. If we manage to show that the norms of all solutions $x(t)$ of all equations

$$x(t) = \xi AFx(t), \quad 0 \leqslant \xi \leqslant 1, \tag{16.18}$$

admit a general a priori estimate $\|x(t)\| \leqslant \text{const} < \infty$, then the Leray-Schauder Principle will imply the existence of fixed points for the operator AF. Thus, all we need in order to prove our theorem is to establish an a priori estimate in L_2 of the solutions of equation (16.18).

We let $E_0 \subset L_2$ denote the two dimensional subspace consisting of functions $e(t)$ of the form $\eta \sin(\omega_s t + \delta)$ $(\eta, \delta \in \mathbb{R}^1)$, and let E_1 be the orthogonal complement of E_0 in L_2. In Section 7 the subspace E_0 was denoted by Π_s. Let P and Q be the orthogonal projections onto E_0 and E_1, respectively. In view of Theorem 12.3, the existence of an a priori estimate suitable for our purposes will follow as soon as we will prove that under the assumptions of Theorem 16.2 equation (16.18) is in the class

$$\mathcal{B} = \mathcal{B}\left(\frac{w_1(T)}{w(T)}, \frac{k_{**}}{k_*}, k_*, w(T), c_1, c_2\right),$$

for some positive constants c_1 and c_2. To accomplish this goal, we have to check the following four estimates:

$$\|APx\| \leqslant w(T)\|Px\|, \tag{16.19}$$

$$\|AQx\| \leqslant w_1(T)\|Qx\|, \tag{16.20}$$

$$\|Fx\| \leqslant k_*\|x\| + c_1, \tag{16.21}$$

$$|[Px, AFx]| \leqslant \|Px\|[c_2 + w(T)k_{**}\|Px\| + w(T)k_*\|Qx\|]. \tag{16.22}$$

Inequalities (16.19) and (16.20) are obvious. Inequality (16.21) follows from (16.6) if we choose $c_1 = \|b(t)\|$. It remains to prove the last inequality (16.22). Since any function $Px(t) \in E_0$ has the form

$$Px(t) = \|Px\|\sqrt{\frac{2}{T}} \sin(\omega_s t + \delta)$$

and

$$A^*Px(t) = w(T)\|Px\|\sqrt{\frac{2}{T}} \sin[\omega_s t + \delta - \arg W(\omega_s \mathrm{i})],$$

then

$$[|A^*Px|, |S(h_s)Px|] =$$

$$= \frac{2}{T}w(T)\|Px\|^2 \int\limits_0^\pi \sin[\omega_s t + \delta - \arg W(\omega_s \mathrm{i})] \sin[\omega_s(t - h) + \delta]\mathrm{d}t,$$

that is,

$$[|A^*Px|, |S(h_s)Px|] = w(T)\|Px\|^2 Q[\arg W(\omega_s i) - \omega_s h]. \qquad (16.23)$$

The inequality

$$\|[Px, AFx]\| \leqslant [|A^*Px|, |f[t, S(h_1)x, \ldots, S(h_n)x]|]$$

together with (16.6) leads to the relations

$$\|[Px, AFx]\| \leqslant \left[|A^*Px|, \left(b(t) + \sum_{j=1}^{n} k_j |S(h_j)x(t)| \right) \right] \leqslant$$

$$\leqslant \|A^*Px\| \cdot \|b(t)\| + \sum_{j=1}^{n} k_j [|A^*Px|, |S(h_j)x|] \leqslant$$

$$\leqslant c_2\|Px\| + \sum_{j=1}^{n} k_j [|A^*Px|, |S(h_j)Px|] + \sum_{j=1}^{n} k_j [|A^*Px|, |S(h_j)Qx|].$$

In their turn, these relations and (16.23) show that the estimate (16.22) holds true for $c_2 = w(T)\|b(t)\|$. The proof of Theorem 16.2 is complete.

REMARK. In the study of specific equations of the form (16.1) it is possible to establish different existence conditions for forced T-periodic oscillations by using joint norms and Theorem 16.2. In this respect it will be interesting to compare qualitatively these existence conditions.

16.6. UNIQUENESS OF T-PERIODIC SOLUTIONS. If both condition (16.6) and Lipschitz condition

$$|f(t, x_1, \ldots, x_n) - f(t, y_1, \ldots, y_n)| \leqslant \sum_{j=1}^{n} k_j |x_j - y_j|,$$

$$0 \leqslant t \leqslant T, \quad -\infty < x_j, y_j < \infty \ (j = 1, \ldots, n); \ b(t) \in L_2, \qquad (16.24)$$

are imposed, then the existence criteria discussed above can be complemented with uniqueness criteria.

THEOREM 16.3. *Assume that condition (16.24) and all the conditions in Theorem 16.2 are fulfilled. Then equation (16.1) has a unique T-periodic solution.*

Proof. Let $x(t)$ and $y(t)$ be two T-periodic solutions of equation (16.1). Then the function $z(t) = x(t) - y(t)$ is a T-periodic solution of the equation

$$L\left(\frac{\mathrm{d}}{\mathrm{d}t}\right)z(t) = M\left(\frac{\mathrm{d}}{\mathrm{d}t}\right)Gz(t), \qquad (16.25)$$

where
$$Gz(t) = f[t, z(t - h_1) + y(t - h_1), \ldots, z(t - h_n) + y(t - h_n)] -$$
$$- f[t, y(t - h_1), \ldots, y(t - h_n)].$$

The restriction of the function $z(t)$ to the interval $[0, T]$ (also denoted by $z(t)$) is a solution of the operator equation

$$z = AGz. \tag{16.26}$$

As in the previous subsection, let us consider the subspaces E_0 and E_1 and the projections P and Q. We claim that under the assumptions of our theorem equation (16.26) is in the class

$$\mathcal{B} = \mathcal{B}\left(\frac{w_1(T)}{w(T)}, \frac{k_{**}}{k_*}, k_*, w(T), 0, 0\right) \tag{16.27}$$

$(c_1 = c_2 = 0!)$. This claim will clearly conclude the proof of Theorem 16.3, since in view of Theorem 12.3 it implies the equality $z(t) = 0$.

Thus all we need is to prove our claim. Based on estimates (16.19) and (16.20) we first observe that it is enough to prove the relations

$$\|Gz\| \leqslant k_* \|z\| \tag{16.28}$$

and

$$|[Pz, AGz]| \leqslant w(T)\|Pz\|[k_{**}\|Pz\| + k_*\|Qz\|]. \tag{16.29}$$

Next we notice that inequality (16.28) follows from (16.24), whereas inequality (16.29) follows from (16.23) and the following chain of relations:

$$|[A^*Pz, Gz]| \leqslant \left[|A^*Pz|, \sum_{j=1}^{n} k_j |S(h_j)z|\right] \leqslant$$

$$\leqslant \sum_{j=1}^{n} k_j[|A^*Pz|, (|S(h_j)Pz| + |S(h_j)Qz|)] \leqslant$$

$$\leqslant w(T)k_{**}\|Pz\|^2 + w(T)\|Pz\|k_8\|Qz\|.$$

Theorem 16.3 is proved.

§17. Remarks on forced oscillations in systems with control by derivatives

17.1. THE GENERAL SETTING. In this section we consider a system whose dynamics is described by the equation

$$L\left(\frac{\mathrm{d}}{\mathrm{d}t}\right)x(t) = M\left(\frac{\mathrm{d}}{\mathrm{d}t}\right)f(t,x,x'). \qquad (17.1)$$

We assume that the polynomials $L(p)$ and $M(p)$ and the nonlinearity $f(t,x_1,x_2)$ satisfy the usual restrictions; in addition, we suppose that

$$\deg L(p) > \deg M(p) + 1.$$

Equations similar to (17.1) above (or of a more general form) have been considered by many authors (see, for instance, [Babitskii, Krupenin, 1985], [Rosenwasser, 1969]). In the case of system (17.1), the problem of T-periodic oscillations may be reduced to the operator equation

$$x = f(t, Ax, A_1x), \qquad (17.2)$$

where A and A_1 stand for the operators of the T-periodic problems for linear links corresponding, respectively, to the transfer functions

$$W(p) = \frac{M(p)}{L(p)}, \quad W_1(p) = pW(p). \qquad (17.3)$$

To equation (17.2), considered in $L_2[0, T]$, we can apply all the results given in Section 13.

It makes sense, of course, to consider equation (17.2) in a more general setting, assuming that the transfer functions W and W_1 which determine the operators A and A_1 are defined by the formulas

$$W(p) = \frac{M(p)}{L(p)}, \quad W_1(p) = \frac{M_1(p)}{L_1(p)}, \qquad (17.4)$$

instead of formulas (17.3) above.

Throughout this section we let $N(B_1, B_2)$ denote the joint norm of two operators B_1 and B_2 on the space L_2 with the modulus $|\cdot|$.

THEOREM 17.1. *Assume that the nonlinearity $f(t,x,y)$ satisfies the estimate*

$$|f(t,x,y)| \leqslant k_1|x| + k_2|y| + b(t), \quad 0 \leqslant t \leqslant T, \ x,y \in \mathbb{R}^1, \qquad (17.5)$$

where the coefficients k_1 and k_2 are such that

$$N(k_1 A, k_2 A_1) < 1. \tag{17.6}$$

Then equation (17.2) has at least one solution.

In case of specific transfer functions W and W_1, the joint norm of the operators A and A_1 is uniquely determined by these functions and the period T, and it can be either explicitly computed or estimated. Below we discuss two particular examples of transfer functions.

The first example is dealing with equation (17.1) where

$$L(p) = p^2 + \alpha, \quad M(p) \equiv 1.$$

We will separately consider the cases $0 < \alpha < \frac{1}{2}$ and $\frac{1}{2} < \alpha < 1$. Analogous considerations are still possible for $\alpha = \frac{1}{2}$, but are more cumbersome.

The second example is devoted to equation (17.2) where the operators A and A_1 are determined by two transfer functions of a special form.

17.2. A SECOND ORDER EQUATION. In this subsection we consider a particular form of equation (17.1), namely,

$$x'' + \alpha x = f(t, x, x'). \tag{17.7}$$

The transfer functions (17.3) of this equation are given by

$$W(p) = \frac{1}{p^2 + \alpha}, \quad W_1(p) = \frac{p}{p^2 + \alpha}.$$

We assume that the function $f(t, x, y)$ is periodic in t with period 2π.

Let $0 < \alpha < \frac{1}{2}$. In this case the absolute values of the first two eigenvalues of the operators A and A_1 are equal to α^{-1} and $(1-\alpha)^{-1}$ (for A) and to 0 and $(1-\alpha)^{-1}$ (for A_1). All the other eigenvalues of A and A_1 have absolute values that are less than $(1-\alpha)^{-1}$, and, consequently, their corresponding invariant subspaces may not be considered. Thus we get

$$N(k_1 A, k_2 A_1) \leqslant \tilde{N} \stackrel{\text{def}}{=} \max_{x_1^2 + x_2^2 = 1} \left\{ k_1 \sqrt{\frac{x_1^2}{\alpha^2} + \frac{x_2^2}{(1-\alpha)^2}} + k_2 \frac{x_2}{1-\alpha} \right\}.$$

In view of Theorem 13.1, the quantity $\tilde{N}$ is given by

$$\tilde{N} = \begin{cases} \sqrt{\dfrac{k_1^2}{\alpha^2} + \dfrac{k_2^2}{1-2\alpha}} & \text{if } \dfrac{k_1}{k_2} \geqslant \dfrac{\alpha^2}{1-2\alpha}, \\[4mm] \dfrac{k_1 + k_2}{1-\alpha} & \text{if } \dfrac{k_1}{k_2} < \dfrac{\alpha^2}{1-2\alpha}. \end{cases}$$

Let $\frac{1}{2} < \alpha < 1$. In this case the norm of both the operators A and A_1 equals $(1-\alpha)^{-1}$ and is attained on the subspace E_0 of functions of the form $\xi \sin(t + \delta)$. We have the relations

$$\|APx\|, \|A_1 Px\| \leqslant \frac{1}{1-\alpha}\|Px\|,$$

$$\|AQx\| \leqslant \frac{1}{\alpha}\|Qx\|, \quad \|A_1 Qx\| \leqslant \frac{2}{4-\alpha}\|Qx\|, \quad \gamma = \frac{2}{\pi}, \tag{17.8}$$

where P is the orthogonal projection onto E_0 and $Q = I - P$. Inequalities (17.8) are specific counterparts of conditions (13.33)–(13.36). Based on Theorem 13.3 we can now find estimates of the joint norm $N(k_1 A, k_2 A_1)$. Unfortunately, for arbitrary values of α these estimates have a cumbersome form (see (13.37) and (13.38)). For instace, if $\alpha = .8$, then condition (17.6) is fulfilled as soon as

$$w^2(T)\left(k_1^2 + k_2^2 + k_1 k_2\left(\frac{2}{\pi} + \frac{3}{16}\right)\right) \leqslant 1 \quad (w(T) = 5).$$

17.3. AN EXAMPLE OF A FOURTH ORDER EQUATION. In this subsection we investigate equation (17.2), where the operators A and A_1 are determined by the transfer functions

$$W(p) = \frac{1}{p^2 + 3}, \quad W_1(p) = \frac{p^2}{p^4 + 2},$$

and $T = 2\pi$. In this case we have

$$N(k_1 A, k_2 A_1) \leqslant \tilde{N} \overset{\text{def}}{=} \max_{x_1^2 + x_2^2 = 1}\left\{k_1\sqrt{x_1^2 + \frac{1}{4}x_2^2} + k_2\sqrt{\frac{4}{81}x_1^2 + \frac{1}{9}x_2^2}\right\}.$$

Based on Theorem 13.1 we conclude that

$$\tilde{N} = \begin{cases} k_1 + \dfrac{2}{9}k_2 & \text{if } \dfrac{k_2}{k_1} \leqslant 2.7, \\[2ex] \sqrt{\dfrac{8}{5}k_1^2 + \dfrac{32}{243}k_2^2} & \text{if } 2.7 < \dfrac{k_2}{k_1} < 8.1, \\[2ex] \dfrac{1}{2}k_1 + \dfrac{1}{3}k_2 & \text{if } 8.1 \leqslant \dfrac{k_2}{k_1}. \end{cases}$$

The constructions developed in the present section can be successfully used in the study of equations with nonlinearities containing derivatives of different orders.

§18. Extensions of the joint norm method

18.1. PRELIMINARIES. In this section we are concerned with the natural and also challanging problem of improving the results of Section 13. In Section 13 we stated and carefully analysed a result that, for the sake of convenience, is summarized below.

Let A and B be two completely continuous linear operators on a Hilbert space H with modulus M, and let $\mathfrak{f}(x_1, x_2) : H \times H \to H$ be a nonlinear continuous operator, such that

$$\|\mathfrak{f}(x_1, x_2)\| \leqslant \|k_1 M x + k_2 M x_2\| + c, \quad x_1, x_2 \in H, \tag{18.1}$$

for some nonnegative coefficients k_1 and k_2. Assume that the joint norm $N(k_1 A, k_2 B)$ of the operators $k_1 A$ and $k_2 B$ is less than 1. Then the equation

$$x = \mathfrak{f}(Ax, Bx) \tag{18.2}$$

has at least one solution $x \in H$. The norm of any solution (18.2) satisfies the estimate

$$\|x\| \leqslant c[1 - N(k_1 A, k_2 B)]^{-1}.$$

The just mentioned result has a lot of important improvements. In the present section we will indicate two possible extensions.

The first improved version deals with the case when the joint norm of the operators A and B is attained for a common leading eigenvector of both the operators A and B. In this case we have $N(k_1 A, k_2 B) = k_1 \|A\| + k_2 \|B\|$. If $N(k_1 A, k_2 B) = 1$, then there are a lot of situations when equation (18.2) has no solutions at all. The use of weak nonlinearities (analogous to the approach developed in Section 8) will enable us to establish the existence of solutions in spite of the lack of a priori norm estimates.

The second improved version is related to the case when the joint norm is not attained for eigenvalues of the operators A and B. In this case it is possible to use some specific constructions with a noticeable geometric meaning, analogous to the constructions developed in Section 12.

18.2. CONDITIONS INVOLVING RESONANCE. Assume that $H = L_2(\Omega, \mathbb{R}^n)$ and $Mx(t) = |x(t)|$ (for notations see Section 13). Let A and B be two completely continuous operators such that the subspaces

$$E_A = \{e(t) : e(t) \in L_2, \|Ae(t)\| = \|A\| \cdot \|e\|\}$$

and

$$E_B = \{e(t) : e(t) \in L_2, \ \|Be(t)\| = \|B\| \cdot \|e\|\}$$

coincide, and denote $E_A = E_B = E_0$. We also assume that the subspace E_0, as well as its orthogonal complement in L_2, denoted by E_1, are invariant for both the operators A and B. Moreover, let $q < 1$ be such that the estimates

$$\|Ax(t)\| \leqslant q\|A\| \cdot \|x(t)\|, \quad \|Bx(t)\| \leqslant q\|B\| \, \|x(t)\|, \qquad x \in E_1, \tag{18.3}$$

hold true. If in addition to all the previous conditions there exists a function $e(t) \in E_0$ that is an eigenvector of both the operators A and B, i.e., $Ae = \pm\|A\|e$ and $Be = \pm\|B\|e$, then

$$N(k_1 A, k_2 B) = k_1\|A\| + k_2\|B\|,$$

for any nonnegative numbers k_1 and k_2.

Suppose next that $\mathfrak{f}$ is the superposition operator

$$\mathfrak{f}(x_1, x_2) = f[t, x_1(t), x_2(t)], \tag{18.4}$$

corresponding to a function $f(t, x_1, x_2) : \Omega \times \mathbb{R}^n \times \mathbb{R}^n \to \mathbb{R}^n$ that is jointly continuous in the variables x_1 and x_2, and measurable in the variable t. The solvability condition for equation (18.2) may be sought of the form

$$|f(t, x, y)| \leqslant k_1|x| + k_2|y| - \Psi(t, |x| + |y|), \quad t \in \Omega, \ x, y \in \mathbb{R}^n, \ k_1, k_2 > 0. \tag{18.5}$$

In (18.5) above, $\Psi(t, u)$ $(u \geqslant 0)$ denotes a bounded function subject to the condition that the function $\Phi(t, u) = u\Psi(t, u)$ belongs to one of the classes $\mathfrak{N}(u_0)$. Recall that the definition of these classes was given at the beginning of Section 2.

THEOREM 18.1. *Suppose that the function $\Phi(t, u)$ satisfies condition (8.7) for every $u_* \geqslant u_0$ and all $R > 0$. Then there exists $\varepsilon > 0$ such that condition (18.5), where $k_1\|A\| + k_2\|B\| < 1 + \varepsilon$, implies the existence of at least one solution $x(t) \in L_2$ of equation (18.2).*

The proof of Theorem 18.1 is quite cumberous. Since the difficulties involved in the proof may be successfully overcome in a completely analogous way to that used in the proof of Theorem 8.1, we concede ourselves to skip the proof.

Under the conditions of Theorem 18.1 there are no a priori norm estimates of the solutions of equation (18.2).

18.3. AN APPLICATION OF THE USE OF ARGUMENTS METHOD. We continue the study of equation (18.2) in the Hilbert space $H = L_2(\Omega, \mathbb{R}^n)$, with the nonlinearity defined as in (18.4). Specifically, we introduce the notations

$$E_0 = \{e(t) : e(t) \in H, \ \||k_1|Ae(t)| + k_2|Be(t)|\|| = N(k_1A, k_2B)\|e\|\},$$

$$E_1 = \{x(t) : x(t) \in H, \ x(t) \perp E_0\},$$

and assume that

$$\||k_1|Ax(t)| + k_2|B(t)|\|\| \leqslant q_1 N(k_1A, k_2B)\|x\|, \quad x \in E_1, \tag{18.6}$$

$$[|x|, k_1|Ax(t)| + k_2|Bx(t)|] \leqslant q_2 N(k_1A, k_2B)\|x\|^2, \quad x \in E_0, \tag{18.7}$$

for some positive constants $q_1, q_2 < 1$. It should be emphasized that all the subsequent arguments do not rely on any assumption concerning the invariance of the subspaces E_0 and E_1 with respect to the operators A and B.

Suppose that $q_1 < \frac{\sqrt{3}}{3}$ and $q_2 < (1 - 3q_1^2)(1 - q_1^2)^{-1}$. Then the system of equations

$$\begin{cases} y = \cos\alpha + q_1\sin\alpha \\ y = q_2 + q_1\operatorname{tg}\alpha \end{cases} \tag{18.8}$$

has a unique solution $\{y, \alpha\}$ with $\alpha \in \left(0, \frac{1}{2}\pi\right)$, and the first component $y = y(q_1, q_2)$ of this solution is less than 1.

THEOREM 18.2. *Assume that the estimates (18.6) and (18.7) hold true. Let the numbers k_1 and k_2 be such that*

$$N(k_1A, k_2B) < \frac{1}{y(q_1, q_2)}. \tag{18.9}$$

Then the condition

$$|f(t, x, y)| \leqslant k_1|x| + k_2|y| + b(t), \quad t \in \Omega, \ x, y \in \mathbb{R}^n; \ b(t) \in L_2 \tag{18.10}$$

implies the existence of at least one solution of equation (18.2).

Of course, the root $y(q_1, q_2)$ of (18.8) can be computed by quadratures. It satisfies the inequalities $q_1, q_2 < y(q_1, q_2)$.

18.4. PROOF OF THEOREM 18.2. Let us consider the equation

$$x = \xi f(Ax, Bx) \tag{18.11}$$

for $0 \leqslant \xi \leqslant 1$. Let $x_n(t) \in L_2$ be a sequence of solutions of equation (18.11) corresponding to some $\xi_n \in [0, 1]$, and assume that $\|x_n\| \to \infty$ as $n \to \infty$. We denote by P and Q the orthogonal projections onto the subspaces E_0 and E_1, respectively. The next chain of relations

$$[Px_n, Px_n] = [Px_n, \xi P\mathfrak{f}(Ax_n, Bx_n)] \leqslant [Px_n, P\mathfrak{f}(Ax_n, Bx_n)] =$$
$$= [Px_n, \mathfrak{f}(Ax_n, Bx_n)] \leqslant [|Px_n|, |f(t, Ax_n, Bx_n)|] \leqslant$$
$$\leqslant [|Px_n|, k_1|Ax_n| + k_2|Bx_n|] + c_1\|Px_n\| \leqslant$$
$$\leqslant [|Px_n|, k_1|APx_n| + k_2|BPx_n|] + [|Px_n|, k_1|AQx_n| + k_2|BQx_n|] + c_1\|Px_n\| \leqslant$$
$$\leqslant q_2 N(k_1 A, k_2 B)\|Px_n\|^2 + q_1 N(k_1 A, k_2 B)\|Px_n\| \, \|Qx_n\| + c_1\|Px_n\|$$

implies the inequality

$$\|Px_n\| \leqslant N(k_1 A, k_2 B)[q_2\|Px_n\| + q_1\|Qx_n\|] + c_1. \tag{18.12}$$

Let $\|Px_n\| = \cos\alpha_n \|x_n\|$ and $\|Qx_n\| = \sin\alpha_n \|x_n\|$, where $0 \leqslant \alpha_n \leqslant \frac{1}{2}\pi$. Then (18.12) can be rewritten in the form

$$\cos\alpha_n \leqslant N(k_1 A, k_2 B)[q_2\cos\alpha_n + q_1\sin\alpha_n] + \frac{c_1}{\|x_n\|}. \tag{18.13}$$

From the following new chain of relations

$$\|x_n\| = \|\mathfrak{f}(Ax_n, Bx_n)\| \leqslant \|k_1|Ax_n| + k_2|Bx_n|\| + c_2 \leqslant$$
$$\leqslant \|k_1|APx_n| + k_2|BPx_n|\| + \|k_2|AQx_n| + k_2|BQx_n|\| + c_2 \leqslant$$
$$\leqslant N(k_1 A, k_2 B)[\|Px_n\| + q_1\|Qx_n\|] + c_2$$

we get the estimate

$$1 \leqslant N(k_1 A, k_2 B)[\cos\alpha_n + q_1\sin\alpha_n] + \frac{c_2}{\|x_n\|}. \tag{18.14}$$

If

$$\lim_{n\to\infty} \cos\alpha_n = 0,$$

then (18.14) leads to the false inequality $1 \leqslant q_1 N(k_1 A, k_2 B)$. Therefore,

$$\lim_{n\to\infty} \cos\alpha_n > 0.$$

We next choose a subsequence of the bounded sequence α_n that converges to a number $\alpha_* = [0, \frac{1}{2}\pi)$. From (18.14) it follows that

$$1 \leqslant N(k_1 A, k_2 B)[\cos\alpha_* + q_1\sin\alpha_*],$$

and (18.13) implies

$$1 \leqslant N(k_1 A, k_2 B)[q_2 + q_1 \operatorname{tg} \alpha_*],$$

i.e.,

$$\frac{1}{N(k_1 A, k_2 B)} \leqslant \min\{\cos \alpha_* + q_1 \sin \alpha_*, q_2 + q_1 \operatorname{tg} \alpha_*\}. \tag{18.15}$$

But

$$y(q_1, q_2) = \max_{\alpha \in [0, \frac{1}{2}\pi]} \min\{\cos \alpha + q_1 \sin \alpha, q_2 + q_1 \operatorname{tg} \alpha\};$$

consequently, from (18.15) we get the inequality

$$\frac{1}{N(k_1 A, k_2 B)} \leqslant y(q_1, q_2),$$

that clearly contradicts assumption (18.9).

Thus we conclude that under the conditions of Theorem 18.2 there exists an a priori norm estimate in L_2 of all the solutions of all the equations (18.11) for $0 \leqslant \xi \leqslant 1$. Theorem 18.2 follows from the Leray-Schauder Principle.

The proof of Theorem 18.2 is complete.

Chapter 3 consists mainly of results obtained in [Krasnoselskii, A. M., 1984a], [Krasnoselskii, A. M., 1984b], [Krasnoselskii, A. M., 1991].

Chapter 4
Weak nonlinearities

§19. Equations with weak nonlinearities

19.1. PRELIMINARIES. The present chapter is concerned with the study of the nonlinear equation

$$x(t) = Af[t, x(t)] \qquad (19.1)$$

in various spaces of scalar-valued functions $x(t)$. We will assume that the function $f(t, x) : \Omega \times \mathbb{R}^1 \to \mathbb{R}^1$ has the form

$$f(t, x) = \alpha x + \varphi(t, x) + b(t), \qquad (19.2)$$

where the proper nonlinearity — the function $\varphi(t, x)$ — is locally bounded in x, i.e.,

$$\sup_{|x| \leqslant x_0} \sup_{t \in \Omega} |\varphi(t, x)| < \infty \qquad (19.3)$$

for any nonnegative x_0, and, in addition,

$$\lim_{|x| \to \infty} \sup_{t \in \Omega} x^{-1} |\varphi(t, x)| = 0. \qquad (19.4)$$

From now on such a function $\varphi(t, x)$ will be called a *weak nonlinearity*. Every weak nonlinearity $\varphi(t, x)$ satisfies the two-sided estimate

$$|\varphi(t, x)| \leqslant \varepsilon |x| + c, \quad t \in \Omega, \ x \in \mathbb{R}^1,$$

where $c = c(\varepsilon) > 0$, for any $\varepsilon > 0$.

Equations with nonlinearities as (19.1) above were considered by many authors. We would like to mention here, for instance, the results due to M. A. Krasnoselskii, Landesman and Lazer, Fučik, Mawhin. Some results in this chapter abut upon theorems due to Fučik and his colaborators (see [Fučik, 1974, 1980], [Fučik, Hess, 1979], [Fučik, Krbec, 1977]).

19.2. SIMPLE CONDITIONS FOR THE SOLVABILITY OF EQUATION (19.1).

THEOREM 19.1. *Assume A is a completely continuous operator acting on the space L_p and $b(t) \in L_p$. Moreover, suppose that $\alpha \neq 0$ and α^{-1} does not belong to the spectrum $\sigma(A)$ of the operator A. Then equation (19.1) having the nonlinear function $f(t,x)$ of the form (19.2) with a weak nonlinearity $\varphi(t,x)$, has at least one solution in the space L_p.*

Proof. Under the assumptions in our theorem, the operator $I - \alpha A$ is continuously invertible. Therefore, any solution $x(t)$ of the equation

$$x(t) = (I - \alpha A)^{-1} A[\varphi(t,x) + b(t)] \tag{19.6}$$

is a solution of equation (19.1). We next choose $c > 0$ such that the function $\varphi(t,x)$ satisfies the estimate (19.3) for $\varepsilon = \frac{1}{2}\|(I - \alpha A)^{-1}A\|_{L_p \to L_p}$. Then the completely continuous operator $(I - \alpha A)^{-1}Af(t,x)$ sends the ball $\{x(t) : x(t) \in L_p, \|x(t)\| \leqslant \rho\}$ of radius $\rho = 2\varepsilon\|b\|$ into itself. Based on the Schauder Fixed Point Principle, equation (19.6) has at least one solution. The theorem is proved.

Throughout Chapter 4 we will consider the case when $\alpha^{-1} \in \sigma(A)$.

19.3. FREDHOLM CONDITION. Suppose that $\varphi(t,x) \equiv 0$ $(t \in \Omega, x \in \mathbb{R}^1)$ and let A be a completely continuous operator acting on a Hilbert space H of functions $x(t) : \Omega \to \mathbb{R}^1$ (for instance, $H = L_2 = L_2(\Omega, \mathbb{R}^1)$). In this case the existence of solutions for the equation (19.1), where $f(t,x) = \alpha x + b(t)$, is completely answered by the classical Fredholm theorems. The key role is played by the *Fredholm condition*. Specifically, equation (19.1) has solutions if and only if the function $Ab(t)$ is orthogonal to the kernel of the adjoint $(I - \alpha A)^*$ of the operator $(I - \alpha A)$. For example, if $H = L_2$ and α^{-1} is a simple eigenvalue of the completely continuous operator A, then the Fredholm condition means that

$$\int_\Omega b(t)g(t)\mathrm{d}\mu = 0, \tag{19.7}$$

where $g(t)$ denotes an eigenfunction of the operator A^* corresponding to the eigenvalue α^{-1} of this operator.

We now return to the general case when the nonlinear function $\varphi(t, x)$ is no longer identically zero. As it follows, for instance, from Landesman's and Lazer's results, the Fredholm condition still plays an essential role in the case of nonzero nonlinearites $\varphi(t, x)$. To illustrate the point, we mention an example (see, for instance, [Fučik, Kufner, 1980]).

Let $\mathfrak{F}$ denote the class of all bounded nonlinearities $\varphi(t, x) \equiv \varphi(x)$ ($t \in [0, \pi]$, $x \in \mathbb{R}^1$) satisfying the conditions

$$\varphi^- \stackrel{\text{def}}{=} \lim_{x \to -\infty} \varphi(x) < \varphi(y) < \lim_{x \to +\infty} \varphi(x) \stackrel{\text{def}}{=} \varphi^+, \quad y \in \mathbb{R}^1, \tag{19.8}$$

with $\varphi^- < 0 < \varphi^+$. The two-point boundary value problem

$$-x'' - x + \varphi(x) = f(t), \quad x(0) = x(\pi) = 0 \tag{19.9}$$

has solutions for all $\varphi \in \mathfrak{F}$, if and only if the Fredholm condition, which in this case has the form

$$\int_0^\pi f(t) \sin t \, dt = 0, \tag{19.10}$$

is fulfilled.

Of course, the Fredholm condition alone does not guarantee the solvability of equation (19.1) for arbitrary nonlinearities $\varphi(t, x)$.

For example, problem (19.9) has no solution if the Fredholm condition (19.10) holds and the function $\varphi(x)$ is either positive or negative.

In the sequel we will prove a few new theorems relying on the Fredholm condition. Under some supplementary restrictions, the rotation at infinity of the vector field $x(t) - Af[t, x(t)]$ equals 0, if the Fredholm condition fails, and is different from 0, in case Fredholm condition holds.

The examples discussed in this section carry along a general feature. They are formulated in terms of problem (19.9) for the sake of simplicity only.

It should be mentioned that in many articles devoted to the above sketched topic (see, for instance, [Landesman, Lazer, 1970a, 1970b], [Dancer, 1977], [Nakao, 1972], [Kazdan, Werner, 1975], [Nečas, 1973], [Williams, 1970], [Hess, 1974], [Fučik, Nečas, Kučera, 1975]) the Fredholm condition is replaced by the Landesman-Lazer condition, the latter having a more general form.

The next statement provides a sample of a Landesman-Lazer type theorem.

Suppose the nonlinearity $\varphi(t, x)$ is continuous and satisfies conditions (19.8). Then problem (19.9) has solutions if and only if the "forcing term" $f(t)$ is subject to the Landesman-Lazer condition

$$\varphi^- < \frac{1}{2} \int\limits_0^\pi f(t) \sin t \mathrm{d}t < \varphi^+.$$

Section 24 below presents a series of theorems in which Landesman-Lazer type conditions or some of their analogues are used. We will not formulate conditions for the lack of solutions, giving instead similar ones in terms of the vanishing of rotation of some vector fields.

§20. Equations with normal operators

20.1. SOLVABILITY OF THE OPERATOR EQUATION. We next consider the operator equation

$$x = A[x + \varphi(t, x)] + b(t), \tag{20.1}$$

where A is a normal completely continuous linear operator acting on the space $L_2 = L_2(\Omega, \mathbb{R}^1)$, the nonlinearity $\varphi(t, x)$ is weak — i.e., $\varphi(t, x)$ satisfies conditions (19.3) and (19.4) —, and $b(t) \in L_2$.

In addition, we assume that 1 is an eigenvalue of the operator A, and let

$$E_0 = \{x(t) : x(t) \in L_2, \ Ax(t) = x(t)\} \tag{20.2}$$

denote the corresponding eigenspace. The Fredholm condition provides a necessary and sufficient condition for the solvability of the linearized at infinity equation

$$x = Ax + b(t). \tag{20.3}$$

Since the operator A on L_2 is normal, the Fredholm condition reduces to $b(t) \perp E_0$, i.e.,

$$\int\limits_\Omega b(t) e(t) \mathrm{d}\mu = 0, \quad e(t) \in E_0. \tag{20.4}$$

In the theorems stated below we will again use functions $\Phi(t, x)$ in the classes $\mathfrak{N}(u_*)$. The definition of these classes was introduced in Subsection 2.1. Recall that by $\chi(\delta; e)$ we denote the distribution $\chi(\delta; e) = \mu\{t : t \in \Omega, \ |e(t)| \leqslant \delta\}$ of the function $e(t)$ (see Section 1).

THEOREM 20.1. *Suppose that the weak nonlinearity $\varphi(t, x)$ satisfies either the estimate*

$$\varphi(t, x)\operatorname{sign} x \geqslant \Psi(t, |x|), \quad t \in \Omega, \ x \in \mathbb{R}^1, \tag{20.5}$$

or the estimate

$$\varphi(t,x) \cdot \operatorname{sign} x \leqslant -\Psi(t,|x|), \quad t \in \Omega, \ x \in \mathbb{R}^1, \tag{20.6}$$

where $\Psi(t,u)$ ($t \in \Omega$, $u \geqslant 0$) is a scalar-valued bounded function such that the function $\Phi(t,u) = u\Psi(t,u)$ belongs to one of the classes $\mathfrak{N}(u_*)$ and the equality

$$\lim_{\delta \to 0} \sup_{e(t) \in E_0, \ \|e\|=1} \frac{\chi(\delta;e)}{\displaystyle\int_\Omega \Phi\left[t, u_0 + R\delta^{-1}|e(t)|\right] \mathrm{d}\mu} = 0 \tag{20.7}$$

is fulfilled for all $u \geqslant u_*$.

Moreover, assume that the Fredholm condition (20.4) is satisfied and at least one solution in L_2 of equation (20.3) lies in the space $L_\infty = L_\infty(\Omega, \mathbb{R}^1)$, too.

Then equation (20.1) has at least one solution $x(t) \in L_2$.

The proof is delegated to the next subsection. Examples of sufficient conditions for equality (20.7) have been already presented in Section 2.

Natural analogues of Theorem 20.1 are true for the equation

$$x = A[\alpha x + \varphi(t,x)] + b(t). \tag{20.8}$$

Suppose that the equation $x = \alpha Ax$ has nontrivial solutions, i.e., $\alpha^{-1} \in \sigma(A)$. Let $\varphi(t,x)$ be a weak nonlinearity satisfying the conditions of Theorem 20.1, where this time E_0 is the subspace

$$E_0 = \{x(t) : x(t) \in L_2, \ \alpha Ax(t) = x(t)\}.$$

Moreover, assume that $b(t) \perp E_0$ and the linearized equation $x = \alpha Ax + b(t)$ has at least one solution in the space L_∞. Then equation (20.8) has at least one solution $x(t) \in L_2$.

The just stated result follows straightforwardly from Theorem 20.1.

We also observe that under the assumptions in Theorem 20.1, the equation

$$x = A[x - \varphi(t,x)] + b(t) \tag{20.9}$$

has at least one solution.

The hypothesis on the existence of bounded solution for equation (20.3) is satisfied, for instance, whenever the Fredholm condition holds and the operator A acts from L_2 into L_∞.

It can be proved that the assumptions in Theorem 20.1 imply the existence of an a priori estimate for the norm in L_2 of the solutions of equation (20.1).

20.2. Proof of theorem 20.1. The proof presented in the sequel consists of two steps. We first prove our theorem in the case $b(t) \equiv 0$, and then we show how the general case reduces to the case $b(t) \equiv 0$.

Step 1. Let d denote the distance from 1 to the set of all numbers λ^{-1}, where $\lambda \neq 1$ and $\lambda \in \sigma(A)$. In addition, set $d_0 = \min\{1, d\}$ and let γ be a real number satisfying the equality

$$|\gamma - 1| = \frac{1}{3}d_0. \tag{20.10}$$

Since either $\gamma = 1 - \frac{1}{3}d_0$, or $\gamma = 1 + \frac{1}{3}d_0$, clearly $\gamma^{-1} \notin \sigma(A)$, hence the operator $I - \gamma A$ is continuously invertible. Consequently, the equation

$$x = A[x + \varphi(t, x)] \tag{20.11}$$

is equivalent to the equation

$$x = A_\gamma[(1 - \gamma)x + \varphi(t, x)], \tag{20.12}$$

where

$$A_\gamma = (I - \gamma A)^{-1} A.$$

Equation (20.12) is an equation of type (8.1).

Let P denote the orthogonal projection on the space L_2 onto the subspace (20.2), and let $Q = I - P$ be the orthogonal projection onto the orthogonal complement E_1 of E_0 in L_2. Since

$$\|A_\gamma P x\| = \frac{1}{|1 - \gamma|}\|Px\|$$

and

$$\|A_\gamma Q x\| = \sup_{\lambda \neq 0,\ \lambda \in \sigma(A)} \frac{1}{|\lambda^{-1} - \gamma|}\|Qx\|, \quad x \in L_2,$$

from (20.10) we get the relations

$$\|A_\gamma P x\| = 3d_0^{-1}\|Px\|, \quad \|A_\gamma Q x\| \leqslant 1.5 d_0^{-1}\|Qx\|, \qquad x \in L_2. \tag{20.13}$$

Relations (20.13) play in our proof the role of conditions (8.3) in Theorem 8.1 applied to equation (20.12).

We also introduce the notation

$$d_1 = \sup_{t \in \Omega,\ u \geqslant 0} |\Psi(t, u)|$$

and let d_2 be a positive number such that the estimate (19.5) holds for $\varepsilon = \frac{1}{3}d_0$ with $c = d_2$, i.e.,

$$|\varphi(t,x)| \leqslant \frac{1}{3}d_0|x| + d_2, \quad t \in \Omega, \ x \in \mathbb{R}^1. \tag{20.14}$$

Let us next consider the function

$$\Psi_1(t,u) = \begin{cases} \Psi(t,u) & \text{if } u \geqslant \dfrac{3(d_1 + d_2)}{d_0}, \\ -\max\{d_1, d_2\} & \text{if } u < \dfrac{3(d_1 + d_2)}{d_0}. \end{cases} \tag{20.15}$$

It obviously satisfies the inequalities

$$\Psi_1(t,u) \leqslant \Psi(t,u), \quad t \in \Omega, \ u \geqslant 0 \tag{20.16}$$

and

$$d_2 - \frac{1}{3}d_0 u \leqslant -\Psi_1(t,u), \quad t \in \Omega, \ u \geqslant 0. \tag{20.17}$$

Assume now that condition (20.5) is fulfilled, and choose $\gamma = 1 + \frac{1}{3}d_0$. This number satisfies equality (20.10) as well as the estimates (20.13) obtained from (20.10). Since $\gamma > 1$, then — in view of (20.5) —

$$[(1-\gamma)x + \varphi(t,x)]\operatorname{sign} x = -\frac{1}{3}d_0|x| + \varphi(t,x)\operatorname{sign} x \geqslant$$
$$\geqslant -\frac{1}{3}d_0|x| + \Psi(t,|x|),$$

and – in view of (20.16) —

$$[(1-\gamma)x + \varphi(t,x)]\operatorname{sign} x = -\frac{1}{3}d_0|x| + \Psi_1(t,|x|). \tag{20.18}$$

On the other hand, based on (20.14) we have

$$[(1-\gamma)x + \varphi(t,x)]\operatorname{sign} x = -\frac{1}{3}d_0|x| + \frac{1}{3}d_0|x| + d_2 = d_2$$

and from (20.17) we find the inequality

$$[(1-\gamma)x + \varphi(t,x)]\operatorname{sign} x \geqslant \frac{1}{3}d_0|x| - \Psi_1(t,|x|), \quad t \in \Omega, \ x \in \mathbb{R}^1. \tag{20.19}$$

From inequalities (20.18) and (20.19) we get the estimate

$$|(1-\gamma)x + \varphi(t,x)| \leqslant \frac{1}{3}d_0|x| - \Psi_1(t,|x|), \quad t \in \Omega, \ x \in \mathbb{R}^1. \tag{20.20}$$

Estimate (20.20) yields condition (8.5) in Theorem 8.1 applied to equation (20.12). The existence of at least one solution of equation (20.12), and, consequently, of equation (20.11), follows from Theorem 8.1.

To complete the proof of the first step, suppose now that condition (20.6) holds, and choose $\gamma = 1 - \frac{1}{3}d_0$. Equality (20.10) and the estimates (20.13) are still true.

The identity

$$[(1 - \gamma)x + \varphi(t, x)]\operatorname{sign} x \equiv \frac{1}{3}d_0|x| + \varphi(t, x)\operatorname{sign} x, \qquad (20.21)$$

together with (20.6) and (20.16), leads to estimate (20.19). From equality (20.14) it follows that

$$\varphi(t, x) \cdot \operatorname{sign} x \geqslant -\frac{1}{3}d_0|x| - d_2, \quad t \in \Omega, \ x \in \mathbb{R}^1$$

whence, based on (20.21), we have

$$[(1 - \gamma)x + \varphi(t, x)] \cdot \operatorname{sign} x \geqslant - d_2.$$

Therefore, estimate (20.18) follows from (20.17) whenever condition (20.6) is fulfilled and $\gamma = 1 - \frac{1}{3}d_0$. It remains to observe, as above, that from (20.18) and (20.19) we get estimate (20.20), which in view of Theorem 8.1 implies the solvability of equations (20.12) and (20.11).

The first step of the proof — the existence of at least one solution of equation (20.11) — is accomplished.

Step 2. We are in a position to consider the general form of equation (20.1). Let $x^*(t) \in L_\infty$ be one of the solutions of equation (20.3); the existence of such a solution was explicitly assumed in the statement of our theorem. We will look for solutions $y(t)$ of equation (20.1) of the form $y(t) = x(t) + x^*(t)$. Let us rewrite equation (20.1) as

$$x(t) + x^*(t) = A\{x(t) + x^*(t) + \varphi[t, x(t) + x^*(t)]\} + b(t),$$

i.e.,

$$x(t) = A\{x(t) + \varphi_1[t, x(t)]\}, \qquad (20.22)$$

where $\varphi_1(t, x) = \varphi[t, x + x^*(t)]$.

In addition, we introduce the notation

$$M = \sup_{t \in \Omega} |x^*(t)|.$$

If the estimate (20.5) is true for $\varphi(t,x)$, then for the function $\varphi_1(t,x)$ we have the estimate

$$\varphi_1(t,x) \cdot \operatorname{sign} x \geqslant \Psi_0(t,|x|), \quad |x| \geqslant u_0 + M,$$

where $\Psi_0(t,u) = \Psi(t, M+u)$. Similarly, in case $\varphi(t,x)$ satisfies (20.6) the function $\varphi_1(t,x)$ is subject to the estimate

$$\varphi_1(t,x) \cdot \operatorname{sign} x \leqslant -\Psi_0(t,|x|), \quad |x| \geqslant u_0 + M.$$

The function $\Psi_0(t,u)$ belongs to the class $\mathfrak{N}(u_* + M)$, and under the conditions in Theorem 20.1 it satisfies an equality analogous to (20.7), namely,

$$\lim_{\delta \to 0} \sup_{e(t) \in E_0,\ \|e\|=1} \frac{\chi(\delta; e)}{\displaystyle\int_\Omega \Phi_0\left[t, u_0 + R\delta^{-1}|e(t)|\right]\,\mathrm{d}\mu} = 0,$$

for all $u_0 \geqslant u_*$ and $R > 0$, where $\Phi_0(t,u) = u\Psi_0(t,u)$. Equation (20.22) is of the form (20.11) and, therefore, its solvability follows from the first step of the proof. Theorem 20.1 is completely proved.

20.3. TWO-POINT BOUNDARY VALUE PROBLEM. We are next concerned with the existence of solutions of the equation

$$Lx \stackrel{\text{def}}{=} x'' + p(t)x' + q(t)x = \varphi(t,x) + b(t), \tag{20.23}$$

satisfying the boundary conditions

$$x(0) = x(T) = 0. \tag{20.24}$$

Specifically, we assume that the functions $p(t)$, $q(t)$ and $b(t)$ act continuously from $[0, T]$ into $\mathbb{R}^1$, and the function $\varphi(x,t) : [0, T] \times \mathbb{R}^1 \to \mathbb{R}^1$ is a jointly continuous weak nonlinearity, i.e., it satisfies condition (19.4), where $\Omega = [0, T]$ (it should be mentioned that condition (19.3) follows from the continuity of $\varphi(t,x)$).

THEOREM 20.2. *Assume the function $b(t)$ is such that the linearized equation*

$$x'' + p(t)x' + q(t)x = b(t) \tag{20.25}$$

has at least one twice continuously differentiable solution subject to the boundary conditions (20.24). Moreover, suppose that the function $\varphi(t,x)$ satisfies one of the conditions

$$\varphi(t,x)\operatorname{sign} x \geqslant \Psi(|x|), \quad 0 \leqslant t \leqslant T,\ x \in \mathbb{R}^1 \tag{20.26}$$

or

$$\varphi(t, x)\operatorname{sign} x \leqslant -\Psi(|x|), \quad 0 \leqslant t \leqslant T, \ x \in \mathbb{R}^1, \tag{20.27}$$

where $\Psi(u)$ is a bounded function such that for sufficiently large values of u the function $u\Phi(u)$ is nonincreasing, continuous, positive and statisfies the condition

$$\int\limits^{\infty} u\Psi(u)\mathrm{d}u = \infty. \tag{20.28}$$

Then the boundary value problem (20.23), (20.24) has at least one twice continuously differentiable solution.

Proof. Theorem 20.2 is a consequence of Theorem 20.1. Condition (20.7) in Theorem 20.1 follows from the simplicity of every eigenvalue of the operator A of the two-point boundary value problem, and from the estimates

$$c_1\delta \leqslant \chi(\delta; e) \leqslant c_2\delta, \quad 0 \leqslant \delta \leqslant \delta_0 \tag{20.29}$$

(see estimate (7.11) in Section 7.2) of the distributions $\chi(\delta; e)$ of the eigenfunctions $e(t)$ of the operator A. The theorem is proved.

If an eigenfunction $e(t)$ of the differential operator L with the boundary value conditions (20.24) is known, then the Fredholm condition has the explicit form

$$\int\limits_0^T e(t)b(t)\mathrm{d}\mu(t) = 0. \tag{20.30}$$

Integral (20.30) above is computed with respect to the measure μ defined by (7.9).

Theorem 20.2 can be generalized to the case when the functions $\varphi(t, x)$ or $b(t)$ are no longer continuous. In this case, of course, we have to use solutions of equation (20.23) without continuous second order derivatives.

Let us next suppose that equation (20.25) has a unique solution. This assumption is equivalent to the invertibility of the differential operator L with the boundary value conditions (20.24), or to the fact that equation $x'' + p(t)x' + q(t)x = 0$ does not have any nontrivial solutions satisfying (20.24). Then all the restrictions imposed upon the weak nonlinearity $\varphi(t, x)$ in Theorem 20.2 are superfluous. This follows easily from Theorem 19.1. The strength of Theorem 20.2 consists exactly in the absence of the requirement mentioned at the begining of this paragraph.

20.4. FORCED OSCILLATIONS IN WEAK NONLINEAR SYSTEMS. We next consider the equation

$$L\left(\frac{\mathrm{d}}{\mathrm{d}t}\right)x = M\left(\frac{\mathrm{d}}{\mathrm{d}t}\right)[\alpha x + \varphi(t, x) + b(t)]. \tag{20.31}$$

In equation (20.31) above, L and M stand for the polynomials (7.20) and (7.21), the weak nonlinearity $\varphi(t, x)$ is periodic in t with period T and satisfies the Caratheodory condition, $b(t)$ is an integrable T-periodic function, and α is a real number.

We state below two theorems on the existence of T-periodic solutions of equation (20.31).

THEOREM 20.3. *Assume that the algebraic equation*

$$L(p) - \alpha M(p) = 0 \tag{20.32}$$

has $n > 0$ distinct solutions of the form $p = 2k\pi T^{-1}\mathrm{i}$, where k is an integer ($k = k_1, \ldots, k_n$). Moreover, suppose that
   i) $b(t)$ satisfies the Fredholm conditions

$$\int_0^T b(t) \sin k_j t\, \mathrm{d}t = \int_0^T b(t) \cos k_j t\, \mathrm{d}t = 0, \quad k_j \geqslant 0; \tag{20.33}$$

   ii) the weak nonlinearity $\varphi(t, x)$ satisfies either the estimate (20.26) or the estimate (20.27), where $\Psi(u)$ is a bounded function such that for sufficiently large values of u the function $u\Psi(u)$ is nonincreasing, continuous and positive;
   iii) if $n = 2$, then the additional condition (20.28) holds, whereas if $n > 2$, then

$$\lim_{u \to \infty} u^{\frac{1}{n-1}} \Psi(u) = \infty. \tag{20.34}$$

Then equation (20.31) has at least one T-periodic solution.

THEOREM 20.4. *Assume that the equation*

$$L\left(\frac{\mathrm{d}}{\mathrm{d}t}\right) x = M\left(\frac{\mathrm{d}}{\mathrm{d}t}\right) [\alpha x + b(t)] \tag{20.35}$$

has at least one T-periodic solution. Moreover, suppose that the weak nonlinearity $\varphi(t, x)$ satisfies either the estimate (20.26), or the estimate (20.27), where the bounded function $\Psi(u)$ is such that for suficiently large values of u the function $u\Psi(u)$ is nonincreasing, continuous and positive, and if $l > 1$, then

$$\lim_{u \to \infty} u^{\frac{1}{l-1}} \Psi(u) = \infty, \tag{20.36}$$

where $l = \deg L(p)$. Then equation (20.31) has at least one T-periodic solution.

Both Theorems 20.3 and 20.4 follow from Theorem 20.1. Their proofs use estimates for the distributions of trigonometric polynomials (see Subsection 11.5). The complete proofs do not display new ideas and therefore are omitted.

The difference between Theorems 20.3 and 20.4 is obvious: the additional information on the polynomials L and M (as it occurs in Theorem 20.3) enables us to weaken the estimate (20.36) of admissible asymptotics of the function $\Psi(u)$ ($l \leqslant n$).

Under the conditions in Theorems 20.3 and 20.4 the harmonic balance method can be successfully used to construct approximate solutions of equation (20.31)

20.5. ADDITIONAL REMARKS.

A. In both the preceding examples (the two-point boundary value problem and the problem of forced oscillations) we indicated for the equation

$$x = A[\alpha x + \varphi(t, x)] + b(t) \tag{20.37}$$

conditions on the admissible asymptotics (equalities (20.28) and (20.36)), independent of the number α. This favorable circumstance explains the existence of a general estimate $\dim E_0 < \text{const} < \infty$ for all the subspaces E_0 consisting of eigenfunctions of the linear operator A corresponding to an arbitrary eigenvalue. A similar estimate for the Laplace operator with zero boundary conditions does not exist. In order to find consequences of Theorem 20.1 applicable to the Dirichlet problem

$$\Delta u = \alpha u + \varphi(t, x) + b(t), \quad u(t)|_{\partial\Omega} = 0,$$

it is necessary to use the multiplicity of the eigenvalue α.

For the Laplace operator even rough estimates of the distributions of normalized functions from invariant subspaces corresponding to a nonleading eigenvalue are unknown.

B. We may consider equation (20.37) for distinct numbers $\alpha \in [\alpha_1, \alpha_2]$ ($\alpha_1 > 0$). Assume that the operator A has a unique eigenvalue α_0 in the interval $[\alpha_2^{-1}, \alpha_1^{-1}]$ and that this eigenvalue is simple and all the conditions in Theorem 20.1 are satisfied for $\alpha = \alpha_0$. Then there exists a number $\rho > 0$ such that every equation (20.37) with $\alpha \in [\alpha_1, \alpha_2]$ has a solution in the ball $\{\|x\| \leqslant \rho\}$.

The just stated result does not follow from Theorem 20.1, therefore we will sketch its proof.

For each $\alpha \in [\alpha_1, \alpha_2]$ we choose and fix an open neighborhood $\mathfrak{D}(\alpha)$ of α and a number $\rho(\alpha)$ such that equation (20.37) has solutions in the ball $\{\|x\| \leqslant \rho(\alpha)\}$ for any α' in $\mathfrak{D}(\alpha)$.

If $\alpha \neq \alpha_0$, then the existence of the neighborhood $\mathfrak{D}(\alpha)$ and the number $\rho(\alpha)$ is obvious.

If $\alpha = \alpha_0$, then the existence of $\mathfrak{D}(\alpha)$ and $\rho(\alpha)$ follows from Theorem 8.1 invoking arguments similar to those used in the proof of Theorem 20.1.

The proof is concluded by choosing a finite subcovering $\{\mathfrak{D}(\alpha^1), \ldots, \mathfrak{D}(\alpha^n)\}$ of the interval $[\alpha_1, \alpha_2]$ from the covering $\{\mathfrak{D}(\alpha)\}$ and by setting $\rho = \max\{\rho(\alpha_1), \ldots \ldots, \rho(\alpha_n)\}$.

Analogous results can be established for equations more general than (20.37), of the form

$$x = A(\alpha)[\alpha x + \varphi(t, x; \alpha)] + b(t; \alpha).$$

§21. Auxiliary results

21.1. SETTING THE PROBLEM. In this section we deal with the following problem. Let $e(t)$ and $b(t)$ be two given functions lying in the space $L_p = L_p(\Omega)$ $(1 \leqslant p \leqslant \infty)$ and satisfying the inequality

$$\|h(t)\|_p \leqslant a_1 \|\Psi[|\xi e(t) + h(t)|]\|_r + a_2, \tag{21.1}$$

where $1 \leqslant r \leqslant p$, a_1 and a_2 are nonnegative numbers, and $\Psi(u)$ $(u \geqslant 0)$ is a positive continuous function such that

$$\lim_{u \to \infty} u^{-1} \Psi(u) = 0. \tag{21.2}$$

The main goal is to point out conditions under which (21.1) leads to the estimate

$$\|h(t)\|_p \leqslant H(|\xi|), \quad |\xi| \geqslant \xi_0 > 0, \tag{21.3}$$

with a specific function $H(u)$ $(u \geqslant 0)$.

We will separately consider two cases, namely, the case when $\Psi(u)$ is a nondecreasing function, and the case when $\Psi(u)$ is nonincreasing and approaches zero as $u \to \infty$. In the second case we suppose that inequality (21.1) holds with $a_2 = 0$.

21.2. THE NONDECREASING CASE.

LEMMA 21.1. *Assume that inequality (21.1) is true and the function $\Psi(u)$ is nondecreasing. In addition, suppose that either $p = \infty$, or $p < \infty$ and the function*

$\Psi_p(u) = \Psi^p(u^{1/p})$ *is concave for sufficiently large values of* u. *Then estimate* (21.1) *implies the estimate* (21.3) *with*

$$H(u) = c\Psi(cu), \quad u \geqslant \xi_0 \tag{21.4}$$

(c, ξ_0 — *some positive numbers*).

Before proceeding with the proof of Lemma 21.1, we notice that the function $\Psi_p(u)$ is concave for sufficiently large values of u if, for instance, $\Psi(u)$ equals u^β ($\beta \in (0, 1)$) or $\ln^\alpha u$ ($\alpha > 0$) for large values of u. If $p \in (1, \infty)$ and $\Psi_p(u)$ is differentiable, a sufficient condition for the concavity of $\Psi_p(u)$ is available. Specifically, $\Psi_p(u)$ is concave whenever the function $\Psi(u)$ is concave and $u^{-1}\Psi(u)$ approaches zero monotonically in equality (21.2) above.

Proof. We first show that under the conditions in Lemma 21.1 there exists a positive constant c_1 for which the inequality

$$\|\Psi(|\xi e(t) + h(t)|)\|_r \leqslant c_1\Psi(c_1|\xi| + c_1\|h(t)\|_p + c_1) \tag{21.5}$$

is true.

If $p = \infty$, then inequality (21.5) follows from the next chain of relations

$$\|\Psi[|\xi e(t) + h(t)|]\|_r \leqslant \|\Psi[|\xi e(t) + h(t)|]\|_\infty \cdot \|1\|_r =$$
$$= \Psi[\|\xi e(t) + h(t)\|_\infty] \cdot \|1\|_r \leqslant \Psi[|\xi| \cdot \|e(t)\|_\infty + \|h(t)\|_\infty] \cdot \|1\|_r$$

with $c_1 = \max\{1, \|1\|_r, \|e(t)\|_\infty\}$. In the previous relations, as well as below, we let $\|1\|_r$ denote the norm in L_r of the function $f(t) \equiv 1$ ($t \in \Omega$); clearly, if $r = \infty$, then $\|1\|_r = 1$, and if $r < \infty$, then $\|1\|_r = (\mu\Omega)^{1/r}$.

The proof of (21.5) when $p < \infty$ uses Jensen integral inequality (see [Zygmund, 1935]). Based on this inequality we have

$$\int_\Omega \Psi_p[u_0 + |z(t)|]d\mu \leqslant \mu\Omega\Psi_p\left\{\frac{1}{\mu\Omega}\int_\Omega [u_0 + |z(t)|]d\mu\right\}, \tag{21.6}$$

where $u_0 \geqslant 0$ is such that $\Psi_p(u)$ is concave for $u \geqslant u_0$, and $z(t)$ is an arbitrary integrable function. Since

$$\|\Psi(|\xi e(t) + h(t)|)\|_r \leqslant d_{pr}\|\Psi(|\xi e(t) + h(t)|)\|_p,$$

where d_{pr} stands for the norm of the inclusion operator from the space L_p into the space L_r, and because of

$$\|\Psi[|\xi e(t) + h(t)|]\|_p = \left\{\int_\Omega \Psi^p[|\xi e(t) + h(t)|]d\mu\right\}^{\frac{1}{p}} =$$

$$= \left\{\int_\Omega \Psi_p[|\xi e(t) + h(t)|^p]d\mu\right\}^{\frac{1}{p}} \leqslant \left\{\int_\Omega \Psi_p[u_0 + |\xi e(t) + h(t)|^p]d\mu\right\}^{\frac{1}{p}},$$

inequality (21.6) with $z(t) = |\xi e(t) + h(t)|^p$ leads to the inequality

$$\|\Psi[|\xi e(t) + h(t)|]\|_r \leqslant d_{pr}\left\{\mu\Omega\Psi_p\left[\frac{1}{\mu\Omega}\int[u_0 + |\xi e(t) + h(t)|^p]d\mu\right]\right\}^{\frac{1}{p}},$$

which in its turn can be rewritten as

$$\|\Psi[|\xi e(t) + h(t)|]\|_r \leqslant d_{pr}\|1\|_p\left[\||\xi e(t) + h(t)\|_p^p \cdot (\mu\Omega)^{\frac{1}{p}} + u_0\right].$$

It remains to observe that (21.5) follows from the last inequality by taking

$$c_1 = \max\left\{d_{pr}\|1\|_p, \|e(t)\|_p(\mu\Omega)^{-\frac{1}{p}}, (\mu\Omega)^{-\frac{1}{p}}, u_0\right\}.$$

Thus inequality (21.5) is completely proved.

From (21.2) it follows that for every $\varepsilon > 0$ there exists a number $c(\varepsilon) > 0$ such that the estimate

$$\Psi(u) \leqslant \varepsilon u + c(\varepsilon), \quad u \geqslant 0$$

is fulfilled. Therefore, in view of (21.5) we get that for every $\varepsilon > 0$ condition (21.1) implies the inequality

$$\|h(t)\|_p \leqslant a_1 c_1\{\varepsilon[c_1|\xi| + c_1\|h(t)\|_p + c_1] + c(\varepsilon)\} + a_2,$$

i.e.,

$$\|h(t)\|_p \leqslant a_1 c_1^2\varepsilon\|h(t)\|_p + a_1 c_1^2\varepsilon|\xi| + a_1 c_1^2\varepsilon + a_1 c_1 c(\varepsilon) + a_2.$$

In particular, if $2\varepsilon a_1 c_1^2 = 1$, then from the last inequality we get the estimate

$$\|h(t)\|_p \leqslant |\xi| + c_2, \quad c_2 = 1 + 2a_1 c_1 c(\varepsilon) + 2a_2.$$

Hence, and by (21.5), it follows that

$$\|h(t)\|_p \leqslant c_1\Psi(2c_1|\xi| + c_1 + c_1 c_2),$$

i.e., the estimate (21.4) is true. Lemma 21.1 is proved.

21.3. THE NONINCREASING CASE. Throughout the remaining part of this section we will assume that the function $\Psi(u)$ is nonincreasing with $\lim\limits_{u\to\infty}\Psi(u)=0$, and $a_2=0$.

LEMMA 21.2. *Besides the just mentioned conditions, suppose there exists $\gamma>0$ such that*

$$\mu\{t:t\in\Omega,\ |e(t)|\leqslant\gamma\}=0. \tag{21.7}$$

Then inequality (21.1) implies the estimate (21.3), where

$$H(u)=c_1\Psi(c_2u),\quad u\geqslant\xi_0 \tag{21.8}$$

(c_1, c_2, ξ_0 — some positive numbers).

Proof. We first observe that (21.1) with $a_2=0$ leads to the estimate

$$\|h(t)\|_p\leqslant b_1\overset{\mathrm{def}}{=}a_1\Psi(0)\|1\|_r. \tag{21.9}$$

Therefore, if $p=\infty$, then Lemma 21.1 follows from the inequality

$$\|h(t)\|_\infty\leqslant a_1\|1\|_r\Psi(\gamma|\xi|-b_1),\quad|\xi|\geqslant b_1\gamma^{-1},$$

which is clearly true in view of (21.1).

Let us now assume that $p<\infty$. Then from the next chain of relations

$$\|\Psi(|\xi e(t)+h(t)|)\|_r\leqslant d_{pr}\|\Psi(|\xi e(t)+h(t)|)\|_p=$$

$$=d_{pr}\left\{\int_\Omega\Psi^p(|\xi e(t)+h(t)|)\mathrm{d}\mu\right\}^{\frac{1}{p}}=$$

$$=d_{pr}\left\{\int_{\{t:h(t)\geqslant\rho\}}\Psi^p(|\xi e(t)+h(t)|)\mathrm{d}\mu+\int_{\{t:h(t)<\rho\}}\Psi^p(|\xi e(t)+h(t)|)\mathrm{d}\mu\right\}^{\frac{1}{p}}\leqslant$$

$$\leqslant d_{pr}\left\{\Psi^p(0)\cdot\mu\{t:h(t)\geqslant\rho\}+\int_{\{t:h(t)<\rho\}}\Psi^p(|\xi e(t)+h(t)|)\mathrm{d}\mu\right\}^{\frac{1}{p}},$$

where $\rho=a_1d_{pr}\Psi(0)2^{\frac{1}{p}}$, and assuming that $\xi\geqslant 2\rho\gamma^{-1}$, we get

$$\|\Psi(|\xi e(t)+h(t)|)\|_r^p\leqslant d_{pr}\left\{\Psi^p(0)\cdot\mu\{t:h(t)\geqslant\rho\}+\int_{\{t:h(t)<\rho\}}\Psi^p\left(\frac{1}{2}\gamma|\xi|\right)\mathrm{d}\mu\right\}.$$

The last relation and Chebyshev inequality

$$\mu\{t : h(t) \geqslant \rho\} \leqslant \frac{\|h(t)\|_p^p}{\rho^p}$$

imply

$$\|\Psi(|\xi e(t) + h(t)|)\|_r^p \leqslant \frac{d_{pr}\Psi^p(0)\|h(t)\|_p^p}{\rho^p} + d_{pr}\Psi^p\left(\frac{1}{2}\gamma|\xi|\right) \cdot \mu\Omega;$$

hence, in view of (21.1), we conclude that

$$\|h(t)\|_p^p \leqslant \frac{1}{2}\|h(t)\|_p^p + a_1^p d_{pr}^p \Psi^p\left(\frac{1}{2}\gamma|\xi|\right) \cdot \mu\Omega.$$

Thus estimate (21.3) holds true for $\xi \geqslant 2\rho\gamma^{-1}$ with $H(u)$ as in (21.8), where $c_1 = a_1 d_{pr}(2\mu\Omega)^{1/p}$ and $c_2 = \frac{1}{2}\gamma$. Lemma 21.2 is proved.

21.4. MORE ABOUT THE NONINCREASING CASE. For the sake of completeness we now investigate the main case when inequality (21.4) fails for every $\gamma > 0$. This means that the distribution

$$\chi(\delta; e) = \mu\{t : t \in \Omega, \ |e(t)| \leqslant \delta\}, \quad \delta \geqslant 0 \tag{21.10}$$

of the function $e(t)$ is positive for $\delta > 0$. We also assume that $\chi(0; e) = 0$, i.e.,

$$\lim_{\delta \to 0} \chi(\delta; e) = 0 \tag{21.11}$$

LEMMA 21.3. *Let* $a_2 = 0$ *and* $\Psi(u)$ *be as in Lemma 21.2 above, with* $\Psi(u)$ *constant on the interval* $[0, \varepsilon_0]$ $(\varepsilon_0 > 0)$. *Then for each* r $(1 \leqslant r < p)$ *there exists* $\xi_0 > 0$ *such that condition (21.1) implies the estimate (21.3), where*

$$H(u) = 2a_1 \left\|\Psi\left(\frac{1}{2}u|e(t)|\right)\right\|_r, \quad u \geqslant 0. \tag{21.12}$$

The proof of Lemma 21.3 is given in the next subsection.

It should be noticed that for $r = p = \infty$ there are no results analogous to Lemma 21.3, whereas for $r = p < \infty$ such analogues are still available.

Under the conditions in Lemma 21.3 the function (21.12) approaches zero as $u \to \infty$. In order to prove this fact it is enough to observe that

$$[H(t)]^r = 2^r a_1^r \cdot \int_\Omega \Psi^r\left(\frac{1}{2}u|e(t)|\right) d\mu =$$

$$= 2^r a_1^r \left\{ \int_{\{t:\frac{1}{2}u|e(t)| \leqslant \sqrt{u}\}} \Psi^r \left(\frac{1}{2}u|e(t)|\right) d\mu + \int_{\{t:\frac{1}{2}u|e(t)| > \sqrt{u}\}} \Psi^r \left(\frac{1}{2}u|e(t)|\right) d\mu \right\} \leqslant$$

$$\leqslant 2^r a_1^r \Psi^r(0) \cdot \chi\left(\frac{2}{\sqrt{u}}; e\right) + 2^r a_1^r \mu \Omega \cdot \Psi^r(\sqrt{u}) \to 0.$$

To illustrate conclusion (21.12) of Lemma 21.3 we mention two specific examples. As a first one, let

$$\Psi(u) = \begin{cases} cu^{-\beta} & \text{if } u \geqslant 1, \\ c & \text{if } 0 \leqslant u \leqslant 1, \end{cases}$$

where $\beta > 0$, and suppose the function (21.10) satisfies the estimates

$$c_1 \delta^\alpha \leqslant \chi(\delta; e) \leqslant c_2 \delta^\alpha, \quad 0 \leqslant \delta \leqslant \delta_0; \ \alpha > 0. \tag{21.13}$$

Then for sufficiently large values of u we get

$$\left\| \Psi\left(\frac{1}{2}u|e(t)|\right) \right\|_r \leqslant \text{const} \cdot \begin{cases} u^{-\frac{\alpha}{r}} & \text{if } \alpha < \beta r, \\ u^{-\beta} \ln u & \text{if } \alpha = \beta r, \\ u^{-\beta} & \text{if } \alpha > \beta r. \end{cases}$$

The second example deals with the case $r = 1$. We clearly have

$$\left\| \Psi\left(\frac{1}{2}u|e(t)|\right) \right\|_1 = \int_\Omega \Psi\left(\frac{1}{2}u|e(t)|\right) d\mu = \int_0^\infty \Psi\left(\frac{1}{2}u\delta\right) d\chi(\delta; e).$$

If the function (21.10) satisfies the estimate

$$\chi(\delta; e) \leqslant \chi_0(\delta), \quad \delta > 0,$$

then

$$\left\| \Psi\left(\frac{1}{2}u|e(t)|\right) \right\|_1 \leqslant \int_0^\infty \Psi\left(\frac{1}{2}u\delta\right) d\chi_0(\delta).$$

In its turn, the last inequality leads to the estimate (21.3), with

$$H(u) = a_1 \int_0^\infty \Psi\left(\frac{1}{2}u\delta\right) d\chi_0(\delta). \tag{21.14}$$

If the inequalities (21.13) are true, then the function (21.3) satisfies the estimate

$$H(u) \leqslant \text{const} \cdot u^{-\alpha} \cdot \int\limits_0^{cu} z^{\alpha-1}\Psi(z)\mathrm{d}z \leqslant \text{const} \cdot \Psi(u).$$

Analogous arguments can be developed in the case $r > 1$ as well.

21.5. PROOF OF LEMMA 21.3. Suppose the number ξ and the function $h(t)$ satisfy inequality (21.1). Since under the conditions in our lemma the number r is finite, we have

$$\|\Psi(|\xi e(t) + h(t)|)\|_r^r = \int\limits_\Omega \Psi^r(|\xi e(t) + h(t)|)\mathrm{d}\mu.$$

Set $\Xi = \{t : |h(t)| < \frac{1}{2}|\xi e(t)|\}$. From the next chain of relations

$$\|\Psi(|\xi e(t) + h(t)|)\|_r^r = \int\limits_\Xi \Psi^r(|\xi e(t) + h(t)|)\mathrm{d}\mu + \int\limits_{\Omega\setminus\Xi} \Psi^r(|\xi e(t) + h(t)|)\mathrm{d}\mu \leqslant$$

$$\leqslant \int\limits_\Xi \Psi^r\left(\frac{1}{2}|\xi e(t)|\right)\mathrm{d}\mu + \int\limits_{\Omega\setminus\Xi} \Psi^r(|\xi e(t) + h(t)|)\mathrm{d}\mu \leqslant$$

$$\leqslant \int\limits_\Omega \Psi^r\left(\frac{1}{2}|\xi e(t)|\right)\mathrm{d}\mu + \int\limits_{\Omega\setminus\Xi} \left[\Psi^r(|\xi e(t) + h(t)|) - \Psi^r\left(\frac{1}{2}|\xi e(t)|\right)\right]\mathrm{d}\mu$$

we get the estimate

$$\|\Psi(|\xi e(t) + h(t)|)\|_r^r \leqslant \frac{1}{2^r a_1^r} \cdot H^r(|\xi|) + J(\xi, h), \tag{21.15}$$

where $H(u)$ is the function defined by (21.12), and

$$J(\xi, h) = \int\limits_{\Omega\setminus\Xi} \left[\Psi^r(|\xi e(t) + h(t)|) - \Psi^r\left(\frac{1}{2}|\xi e(t)|\right)\right]\mathrm{d}\mu. \tag{21.16}$$

On the other hand let us observe that for the integral (21.16) we have the estimate

$$J(\xi, h) \leqslant \Psi^r(0) \cdot \mu\left\{t : |h(t)| \geqslant \frac{1}{2}|\xi e(t)|\right\}. \tag{21.17}$$

The rest of the proof proceeds with a separate examination of the cases $p = \infty$ and $p < 0$.

We will first consider the case $p = \infty$. From (21.1) it follows that $\|h(t)\|_\infty \leqslant b_1$ (see (21.9)), whence, in view of (21.17) above, we obtain the inequality

$$J(\xi, h) \leqslant \Psi^r(0)\chi\left(\frac{2b_1}{|\xi|}; e\right).$$

Based on (21.1) and (21.15), the last inequality leads to the estimate

$$\|h(t)\|_\infty \leqslant H_0(|\xi|) = \left\{\frac{1}{2^r}H^r(|\xi|) + a_1^r\Psi^r(0)\chi\left(\frac{2b_1}{|\xi|}; e\right)\right\}^{\frac{1}{r}}.$$

The function $H_0(u)$ approaches zero as $u \to \infty$. Therefore, we can choose and fix $\xi_0 > 0$ such that $H_0(u) \leqslant \frac{1}{3}\varepsilon_0$ for all $u \geqslant \xi_0$. Consequently, if $|\xi| \geqslant \xi_0$ and $t \in \{t : |h(t)| \geqslant \frac{1}{2}|\xi e(t)|\}$, then

$$|\xi e(t) + h(t)|, \frac{1}{2}|\xi e(t)| \leqslant \varepsilon_0, \tag{21.18}$$

i.e., the equalities

$$\Psi(|\xi e(t) + h(t)|) \equiv \Psi\left(\frac{1}{2}|\xi e(t)|\right) \equiv \Psi(0) \tag{21.19}$$

hold for every $|\xi| \geqslant \xi_0$ and every $t \in \{t : |h(t)| \geqslant \frac{1}{2}|\xi e(t)|\}$. From (21.16) it follows that

$$J(\xi, h) = 0, \quad |\xi| \geqslant \xi_0. \tag{21.20}$$

Finally, by (21.15) we conclude that

$$\|\Psi(|\xi e(t) + h(t)|)\|_r \leqslant \frac{1}{2a_1}H(|\xi|)$$

for all $\xi| \geqslant \xi_0$, and Lemma 21.3 — when $p \doteq \infty$ — is proved. As a matter of fact, instead of $\|h(t)\|_\infty \leqslant H(|\xi|)$, we got a stronger estimate, namely, $\|h(t)\|_\infty \leqslant \frac{1}{2}H(|\xi|)$.

To complete the proof of Lemma 21.3, we will next consider the case $p < \infty$. We first notice that (21.17) and the obvious inclusion

$$\left\{t : \frac{1}{2}|\xi e(t)| \leqslant |h(t)|\right\} \subset \{t : |h(t)| \geqslant \sqrt{|\xi|}\} \cup \{t : |\xi e(t)| \leqslant 2\sqrt{|\xi|}\}$$

imply the inequality

$$J(\xi, h) \leqslant \Psi^r(0)\left\{\frac{\|h(t)\|_p^p}{\sqrt{|\xi|^p}} + \chi\left[\frac{2}{\sqrt{|\xi|}}; e\right]\right\}. \tag{21.21}$$

On the other hand, if $|\xi| \geqslant \xi_1 = 2^{1/p} \cdot a_1^{2r/p} [\Psi(0)]^{2r/p} \cdot b_1^{2(p-r)/p}$, where b_1 is the number defined by (21.9) above, then

$$\Psi^r(0) \frac{\|h(t)\|_p^{p-r}}{\sqrt{|\xi|}} \leqslant \frac{1}{2a_1^r};$$

therefore, from (21.21) it follows that

$$J(\xi, h) \leqslant \frac{1}{2a_1^r} \|h(t)\|_p^r + \Psi^r(0)\chi \left[\frac{2}{\sqrt{|\xi|}}; e \right]$$

for every $|\xi| \geqslant \xi_1$.

The last estimate, together with (21.15) and (21.1), leads to the inequality

$$\|h(t)\|_p^r \leqslant a_1^r \left\{ \frac{1}{2a_1^r} \cdot H^r(|\xi|) + \frac{1}{2a_1^r} \cdot \|h(t)\|_p^r + \Psi^r(0) \cdot \chi \left[\frac{2}{\sqrt{|\xi|}}; e \right] \right\},$$

that can be rewritten as

$$\|h(t)\|_p^r \leqslant H_1(|\xi|) = \left\{ 2^{1-r} H^r(|\xi|) + 2a_1^r \Psi^r(0)\chi \left[\frac{2}{\sqrt{|\xi|}}; e \right] \right\}^{\frac{1}{r}}.$$

Let us now set $\alpha = (p-r)/2p$. The inclusion

$$\left\{ t : |h(t)| \geqslant \frac{1}{2}|\xi e(t)| \right\} \subset$$

$$\subset \{ t : |h(t) \geqslant H_1^\alpha(|\xi|) \} \cup \left\{ t : \frac{1}{2}|\xi e(t)| \leqslant h(t) \leqslant H_1^\alpha(|\xi|) \right\}$$

clearly implies

$$J(\xi, h) \leqslant \Psi^r(0) \cdot \mu \{ t : |h(t)| \geqslant H_1^\alpha(|\xi|) \} +$$

$$\int\limits_{\{t : \frac{1}{2}|\xi e(t)| \leqslant |h(t)| \leqslant H_1^\alpha(|\xi|)\}} \left[\Psi^r(|\xi e(t) + h(t)|) - \Psi^r\left(\frac{1}{2}|\xi e(t)| \right) \right] d\mu. \qquad (21.22)$$

Since the function $H_1(u)$ approaches zero as $u \to \infty$, there exists $\xi_2 \geqslant \xi_1$ such that $H_1^\alpha(u) \leqslant \frac{1}{3}\varepsilon_0$ for all $u \geqslant \xi_2$. Consequently, if $|\xi| \geqslant \xi_2$ and $t \in \{ t : \frac{1}{2}|\xi e(t)| \leqslant |h(t)| \leqslant H_1^\alpha(|\xi|) \}$, then the estimates (21.18) hold, hence the equalities (21.19) are true. In view of (21.22) we obtain

$$J(\xi, h) \leqslant \Psi^r(0) \cdot \mu \{ t : |h(t)| \geqslant H_1^\alpha(|\xi|) \}.$$

This inequality, together with Chebyshev inequality, leads to

$$J(\xi, h) \leqslant \Psi^r(0) \cdot \frac{\|h(t)\|_p^p}{H_1^{\alpha p}(|\xi|)} \leqslant \frac{\Psi^r(0)\|h(t)\|_p^r H_1^{p-r}(|\xi|)}{H_1^{\alpha p}(|\xi|)} =$$

$$= \Psi^r(0) \cdot [H_1(|\xi|)]^{\frac{1}{2}(p-r)}\|h(t)\|_p^r.$$

If we now choose $\xi_3 \geqslant \xi_2$ such that

$$\Psi^r(0)[H_1(u)]^{\frac{1}{2}(p-r)} \leqslant \frac{1}{2}a_1^{-r}$$

whenever $u \geqslant \xi_3$, then for every $|\xi| \geqslant \xi_3$ we have the estimate $J(\xi, h) \leqslant \frac{1}{2}a_1^{-r}\|h(t)\|_p^r$. Finally, from this relation and from (21.1) and (21.15) we get the inequality

$$\|h(t)\|_p^r \leqslant \frac{1}{2^r}H^r(|\xi|) + \frac{1}{2}\|h(t)\|_p^r,$$

that can be rewritten as

$$\|h(t)\|_p \leqslant 2^{\frac{1-r}{r}} H(|\xi|) \leqslant H(|\xi|).$$

Lemma 21.3 is completely proved.

§22. Equations with nonnormal operators

22.1. SETTING THE PROBLEM. In this section as well as later on, we continue our study of the equation (19.1). For sake of convenience, we first recall the general form of that equation:

$$x(t) = A[x + \varphi(t, x)] + b(t). \tag{22.1}$$

As in Section 20, we assume that 1 is an eigenvalue of the operator A and that $\varphi(t, x)$ is a weak nonlinearity. Moreover, we suppose that $\varphi(t, x)$ satisfies a condition analogous to (19.3) and (19.4), namely,

$$|\varphi(t, x)| \leqslant \Psi(|x|), \quad t \in \Omega, \ -\infty < x < \infty. \tag{22.2}$$

The function $\Psi(u)$ $(u \geqslant 0)$ in (22.2) is positive, continuous and monotonous (either nondecreasing or nonincreasing and vanishing at infinity). We also impose the condition

$$\lim_{u \to \infty} u^{-1}\Psi(u) = 0. \tag{22.3}$$

In the sequel the operator A is supposed to be a completely continuous operator acting on a fixed space L_p ($1 \leqslant p \leqslant \infty$). No conditions analogous to normality or self-adjointness are required. Eventually we assume that $b(t) \in L_p$.

From (22.2) and (22.3) above we easily get the estimate

$$|\varphi(t, x)| \leqslant c_1 |x| + c_2, \quad t \in \Omega, \quad -\infty < x < \infty,$$

for some positive numbers c_1 and c_2. It follows that the superposition operator $x(t) \to \lambda x(t) + \varphi(t, x)$ acts continuously on each space L_p for every λ. Therefore the operator

$$B_\lambda x = A[\lambda x(t) + \varphi(t, x)] + b(t) \tag{22.4}$$

is a completely continuous operator on the space L_p. Let us choose $\lambda_0 \neq 0$ such that λ_0^{-1} is not an eigenvalue of the operator A. Then the rotation γ of the vector field $x - B_{\lambda_0} x$ on the sphere $\{\|x\|_p = \rho\}$ in L_p with a sufficiently large radius ρ equals 1 or -1 ($\gamma = (-1)^\beta$, where β is the sum of the multiplicities of the real eigenvalues of the operator $\lambda_0 A x$ that are greater than 1).

If we could prove that the vector field $x - B_\lambda x$ is nonsingular on a sphere $\{\|x\|_p = \rho\}$ (with ρ sufficiently large) for all $\lambda \in [\lambda_0, 1]$ (or $\lambda \in [1, \lambda_0]$, in case $\lambda_0 > 1$), then by general fixed point principles (see, for instance, [M. A. Krasnoselskii, 1956], [Krasnoselskii, Zabreiko, 1975]) we would get the solvability of equation (22.1).

In what follows we are going to formulate nonsingularity criteria for the vector field $x - B_\lambda x$ on spheres with sufficiently large radii and to derive theorems on the solvability of equation (22.1).

22.2. BIFURCATION SYSTEM. Suppose 1 is a simple eigenvalue of the operator A. Then 1 also is a simple eigenvalue of the adjoint operator A^* acting on the space $(L_p)^*$. We denote the corresponding eigenfunctions of the operators A and A^* by $e(t) \in L_p$ and $g(t) \in L_q$, respectively, i.e., $Ae(t) = e(t)$, $A^*g(t) = g(t)$. If $p < \infty$, then $(L_p)^* = L_q$, where $q = p(p-1)^{-1}$; if $p = \infty$, then $(L_p)^* \supset L_1$. In the latter case we assume, in addition, that $g(t) \in L_1$. Let P and Q denote the linear operators defined by

$$Px(t) = e(t) \int_\Omega g(\tau) x(\tau) d\mu, \quad x(t) \in L_p \tag{22.5}$$

and

$$Qx(t) = x(t) - Px(t), \quad x(t) \in L_p. \tag{22.6}$$

Throughout this section we suppose that the functions $e(t)$ and $g(t)$ are normalized such that the equality $Pe(t) = e(t)$ holds. The operators P and Q acting on L_p

are projections. Specifically, the operator (22.5) projects the whole space onto the one-dimensional subspace

$$E_0 = \left\{ x(t) \in L_p, \ x(t) = \xi e(t), \ \xi \in \mathbb{R}^1 \right\},$$

whereas the operator (22.6) is a projection onto an infinite dimensional subspace $E_1 \subset L_p$, which is invariant for A and has codimension 1. As a simple consequence of (22.5) and (22.6) above we notice the identity

$$\int\limits_\Omega g(t) Q x(t) \mathrm{d}\mu = 0, \quad x(t) \in L_p. \tag{22.7}$$

Let us denote by $u(t)$ and $h(t)$ the projections $Px(t)$ and $Qx(t)$ of a solution $x(t)$ of the equation $x - B_\lambda x = 0$ for a given λ. For these projections we have the equalities $u = PB_\lambda(u + h)$ and $h = QB_\lambda(u + h)$, that can be rewritten as

$$u(t) = \lambda u(t) + P\varphi[t, u(t) + h(t)] + Pb(t), \tag{22.8}$$

$$h(t) = \lambda AQh(t) + AQ\varphi[t, u(t) + h(t)] + Qb(t). \tag{22.9}$$

The system (22.8)–(22.9) is referred to as the bifurcation system of the equation $x - B_\lambda x = 0$ (see [Fučik, Kufner, 1980]).

22.3. A LEMMA ON THE SOLUTIONS OF EQUATION $x - B_\lambda x = 0$. We suppose that $\lambda_0 > 0$ and that the Fredholm condition $Pb(t) = 0$ is fulfilled, i.e.,

$$\int\limits_\Omega b(t) g(t) \mathrm{d}\mu = 0. \tag{22.10}$$

Let $u(t) = \xi e(t)$. Then from (22.8) ($\lambda \in [1, \lambda_0]$) it follows that

$$\xi \int\limits_\Omega g(t) \varphi[t, \xi e(t) + h(t)] \mathrm{d}\mu \leqslant 0. \tag{22.11}$$

We next proceed with the analysis of the second equation (22.9) of the bifurcation system.

Assume the number $\lambda^{-1} \in [\lambda_0^{-1}, 1)$ is not an eigenvalue of the operator A. This assumption can be always achieved by choosing λ_0 sufficiently close to 1. Then the operators $(I - \lambda AQ)^{-1}Q$ ($\lambda \in [1, \lambda_0]$) are continuous on L_p and the estimate

$$\|(I - \lambda QA)^{-1}Qx\|_p \leqslant a_0 \|Qx\|_p, \quad x \in L_p, \ \lambda \in [1, \lambda_0] \tag{22.12}$$

holds for a certain $a_0 > 0$. Assume, in addition, that the operator A acts not only on L_p, but it also determines a continuous operator from a space L_r $(1 \leqslant r \leqslant p)$ into L_p. Then by (22.12) it follows that

$$\|(I - \lambda QA)^{-1} QAx\|_p \leqslant a_1 \|x\|_r, \quad x \in L_r, \ \lambda \in [1, \lambda_0], \tag{22.13}$$

where $a_1 = a_0 \|A\|_{L_r \to L_p}$. Hence and from (22.9), rewritten as

$$h(t) = (I - \lambda QA)^{-1} QA\{\varphi[t, \xi e(t) + h(t)]\} + (I - \lambda QA)^{-1} b(t),$$

we get the inequality

$$\|h\|_p \leqslant a_1 \|\varphi[t, \xi e(t) + h(t)]\|_r + a_2, \tag{22.14}$$

where $a_2 = a_0 \|b(t)\|_p$. From (22.14), in view of (22.2), we finally find the estimate

$$\|h\|_p \leqslant a_1 \|\Psi[|\xi e(t) + h(t)|]\|_r + a_2. \tag{22.15}$$

Thus we just proved the following result:

LEMMA 22.1. *Suppose there are no eigenvalues of the operator A in the interval $[\lambda_0^{-1}, 1)$, and A acts continuously from L_r into L_p. Then the components ξ and $h(t)$ of every solution $x(t) = \xi e(t) + h(t)$ of each equation $x - B_\lambda x = 0$ satisfy the inequalities (22.11) and (22.15) for all $\lambda \in [1, \lambda_0]$.*

Based on the results proved in the previous section it follows from (22.15) above that the components ξ and $h(t)$ also satisfy the estimate

$$\|h\|_p \leqslant H(|\xi|). \tag{22.16}$$

The function $H(u)$ an (22.16) is determined by the function $\Psi(u)$ and the distribution $\chi(\delta, e)$ of the function $e(t)$. On the other hand, the results obtained in Section 4 enable us to formulate conditions for the existence of a priori estimates for the solutions $x(t)$ of the equations $x - B_\lambda x = 0$, i.e., conditions for the solvability of equation (22.1).

22.4. EXISTENCE THEOREMS (BOUNDED NONLINEARITIES). Throughout the remaining part of this section we assume that the eigenfunctions $e(t)$ and $g(t)$ of the operators A and A^* are sign-compatible, i.e.,

$$\mu\{t : t \in \Omega, \ e(t) \cdot g(t) \leqslant 0\} = 0. \tag{22.17}$$

By $\Phi(t, u)$ $(t \in \Omega,\ u \geqslant 0)$ we denote a given function in one of the classes $\mathfrak{N}(u_0)$ (see Section 2 for the definition of these classes).

THEOREM 22.1. *Suppose the following five conditions are fulfilled:*

i) *the superpositionally measurable and bounded function $\varphi(t, x)$ satisfies the estimate*

$$\varphi(t, x)\text{sign}\,x \geqslant \Phi(t, |x|), \quad t \in \Omega,\ -\infty < x < \infty; \tag{22.18}$$

ii) *the functions $e(t)$ and $g(t)$ are sign-compatible and the function*

$$\chi(\delta; e, g) \overset{\text{def}}{=} \mu\{t : t \in \Omega,\ |e(t)| \leqslant \delta\} \equiv \int\limits_{\{t:t\in\Omega,\ |e(t)|\leqslant\delta\}} |g(t)|\mathrm{d}\mu \tag{22.19}$$

is positive for $\delta > 0$;

iii) *the operator A acts on L_∞, i.e., $p = \infty$;*

iv) *the function $b(t)$ in (22.1) satisfies the Fredholm condition (22.10);*

v) *the equality*

$$\lim_{u\to\infty} \frac{1}{\chi(Ru^{-1}; e, g)} \int\limits_{\Omega} |g(t)|\Phi[t, u_0 + u|e(t)|]\mathrm{d}\mu = \infty \tag{22.20}$$

is true for every $R > 0$.

Then equation (22.1) has at least one solution $x_*(t) \in L_\infty$.

Proof. As we have already explained at the beginning of this section, in order to prove the existence of a solution $x_*(t)$ of equation (22.1) it is enough to establish an a priori estimate

$$\|x(t)\|_\infty \leqslant \text{const} < \infty \tag{22.21}$$

for all the solutions $x(t)$ of the equations $x - B_\lambda x = 0$, where $\lambda \in [1,\ \lambda_0]$ and $\lambda_0 > 1$ is a sufficiently close to 1 number. To this end, let $x(t)$ be a solution of the equation $x - B_\lambda x = 0$, with $\lambda \in [1,\ \lambda_0]$ an arbitrarily chosen number. Then, in view of Lemma 22.1, the components ξ and $h(t)$ of this solution, i.e., $\xi e(t) = Px(t)$, $h(t) = Qx(t)$, satisfy the inequalities (22.11) and (22.15). Since the function $\Psi(u)$ in (22.15) can be assumed constant $(\Psi(u) \equiv \sup\{|\Psi(t, x)| : t \in \Omega,\ x \in \mathbb{R}^1\})$, the estimate (22.16) is true with $H(u)$ a constant function, too.

Therefore, all the conditions in Theorem 4.1 are fulfilled. The a priori estimate (22.21) follows now from that theorem. Theorem 22.1 is proved.

Before proceeding with a few additional remarks, let us notice that condition (22.20) above coincides with condition (4.15) for $\Psi(u)$, $H(u) = \text{const}$.

The conclusion of Theorem 22.1 remains unaltered if instead of condition (22.18) we assume that

$$\varphi(t,x)\operatorname{sign} x \leqslant -\Phi(t,|x|), \quad t \in \Omega, \infty < x < \infty. \tag{22.22}$$

For the proof of this version of the existence theorem it will suffice to develop the arguments presented in Subsection 22.2 for $\lambda_0 < 1$.

If function (22.19) vanishes for some $\delta > 0$, then condition (22.20) is no longer necessary (see Theorem 4.2).

If $p < \infty$ and the operator A acts continuously from L_r into L_p ($r < p$), then condition (22.20) has to be replaced by

$$\lim_{u \to \infty} \frac{1}{\chi[2\theta(Ru^{-1}); e, g]} \int_\Omega |g(t)||\Phi[t, u_0 + u|e(t)|]|\mathrm{d}\mu = \infty, \tag{22.23}$$

where

$$\theta(z) = \sup \left\{ \delta : \delta[x(\delta; e, g)]^{\frac{r}{p}} \leqslant z \right\}, \quad z \geqslant 0 \tag{22.24}$$

(see Theorem 4.3; condition (22.23) stands for condition (4.23)).

If $p < \infty$ and the operator A acts continuously from L_r into L_p ($r \leqslant p$), and the function (22.19) vanishes for some $\delta > 0$, then the analogue to (22.20) condition has the form

$$\lim_{u \to \infty} u^{\frac{p}{r}} \cdot \int_\Omega |g(t)||\Phi[t, u_0 + u|e(t)|]|\mathrm{d}\mu = \infty \tag{22.25}$$

(see Theorem 4.4).

22.5. EXISTENCE THEOREMS (UNBOUNDED NONLINEARITIES). If $\varphi(t,x)$ is an unbounded function, the proper substitute for condition (22.20) is more intricate. Throughout this subsection we assume that $\varphi(t,x) \in \mathfrak{M}(\Phi, \Psi)$ (see Subsection 4.2), i.e., the inequalities (22.2) and (22.18) hold true.

THEOREM 22.2. *Suppose the following five conditions are fulfilled:*
  i) *the function $\varphi(t,x)$ lies in the class $\mathfrak{M}(\Phi, \Psi)$;*
  ii) *the functions $e(t)$ and $g(t)$ are sign-compatible and the function (22.19) is positive for $\delta > 0$;*
  iii) *$p = \infty$;*
  iv) *the function $b(t)$ in (22.1) satisfies the Fredholm condition (22.10);*

v) *the equality*

$$\lim_{u \to \infty} \frac{\displaystyle\int_\Omega |g(t)| \Phi[t, u_0 + u|e(t)|] \mathrm{d}\mu}{\chi\left(\dfrac{R\Psi(Ru)}{u}; e, g\right) \cdot \Psi[R\Psi(Ru)]} = \infty \qquad (22.26)$$

is true for every $R > 0$.

Then equation (22.1) has at least one solution $x_*(t) \in L_\infty$.

The proof of Theorem 22.2 is completely analogous to the proof of Theorem 22.1. Under the assumptions in Theorem 22.2, the estimate $\|h(t)\|_\infty \leqslant H(|\xi|)$ follows from Theorem 21.1, where the function $H(u)$ is defined by equality (21.4).

The conclusion of Theorem 22.2 remains the same if instead of condition (22.18) we use condition (22.22).

If function (22.19) vanishes for some $\delta > 0$, then condition (22.26) may be dropped.

Two other versions of Theorem 22.2 are still available, with appropriate modifications, if $p < \infty$. More precisely, if $p < \infty$ and the functions $\Psi(u)$ and $\Psi^p(u^{1/p})$ are concave for sufficiently large values of u, the operator A acts from L_r into L_p $(r \leqslant p)$, and the function (22.19) is positive for every $\delta > 0$, then condition (22.26) has to be replaced by

$$\lim_{u \to \infty} \frac{1}{\bar\chi(u)} \int_\Omega |g(t)| \Phi[t, u_0 + 2u|e(t)|] \mathrm{d}\mu = \infty, \qquad (22.27)$$

where $\bar\chi(u)$ denotes the function (see (22.24))

$$\bar\chi(u) = \chi\left\{2\theta\left[\frac{R\Psi(Ru)}{u}; e, g\right]\right\} \cdot \Psi[R\Psi(Ru)].$$

If $p < \infty$ and the functions $\Psi(u)$ and $\Psi^p(u^{1/p})$ are concave for sufficiently large values of u, the operator A acts from L_r into L_p $(r \leqslant p)$, and the function (22.19) vanishes for some $\delta > 0$, then condition (22.26) is replaced by

$$\lim_{u \to \infty} \frac{u^{\frac{p}{r}}}{[\Psi(Ru)]^{1+\frac{p}{r}}} \int_\Omega |g(t)| \Phi[t, u_0 + 2u|e(t)|] \mathrm{d}\mu = \infty. \qquad (22.28)$$

22.6. EXISTENCE THEOREMS (DECREASING NONLINEARITIES). In what follows we let $\varphi(t, x)$ be a function from a fixed class $\mathfrak{M}(\Phi, \Psi)$ (see Subsection 4.2), where $\Psi(u)$ $(u \geqslant 0)$ is nonincreasing, $\Psi(u) \to 0$ as $u \to \infty$, and $\Psi(\varepsilon_0) = \Psi(0)$ $(\varepsilon_0 > 0)$. If equation

(22.1) is considered in L_∞, then the specific behavior of $\varphi(t, x)$ at infinity has no significance, and all the conditions we need in order to get the existence of solutions for equation (22.1) are those from Theorem 22.1.

Let us next assume that $p < \infty$ and that the operator A acts from the space L_r $(r < p)$ into L_p.

We first consider the case when function $b(t)$ in (22.1) is identically equall to zero. Then the components ξ and $h(t)$ of every solution $x(t)$ of each equation $x - B_\lambda x = 0$ for $\lambda \in [1, \lambda_0]$ satisfy the estimate (22.15) with $a_2 = 0$. If function (22.19) is positive for all $\delta > 0$, then in view of Lemma 21.3 the components ξ and $h(t)$ also satisfy the estimate

$$\|h(t)\|_p \leqslant H(|\xi|), \quad |\xi| \leqslant \xi_0, \tag{22.29}$$

where

$$H(u) = 2a_1 \left\| \Psi\left(\frac{1}{2}u|e(t)|\right) \right\|_r. \tag{22.30}$$

Finally, assume that equality (22.27) holds true for every $R > 0$, where

$$\bar{\chi}(u) = \chi\left\{ \frac{u_0}{u} + \theta\left[\frac{H(u)}{u} \right]; e, g \right\}, \quad u \geqslant 0, \tag{22.31}$$

with $H(u)$ defined by formula (22.30). Then equation (22.1) has at least one solution $x_*(t) \in L_p$.

If the function (22.19) vanishes for some $\delta > 0$, then the estimate (22.29) is fulfilled — in view of Theorem 21.2 — for a function $H(u)$ given by

$$H(u) = c_1 \Psi(c_2 u). \tag{22.32}$$

Therefore, condition (22.27) that ensures the existence of solutions for equation (22.1) has to be replaced by

$$\lim_{u \to \infty} \frac{u^{\frac{p}{r}} \int_\Omega |g(t)| \Phi[t, u_o + 2u|e(t)|] \mathrm{d}\mu}{[\Phi(Ru)]^{\frac{p}{r}}} = \infty. \tag{22.33}$$

We now consider equation (22.1) in the general case of an arbitrary function $b(t)$.

THEOREM 22.3. *Suppose that the linearized equation $x(t) = Ax + b(t)$ has nontrivial solutions and at least one of them is bounded. In addition, assume that for every $u_* \geqslant u_0$ one of the following two conditions is fulfilled:*

i) *either* $\chi(\delta; e, g) > 0$ *for each* $\delta > 0$ *and the equality*

$$\lim_{u \to \infty} \frac{1}{\bar{\chi}(u)} \int_{\Omega} |g(t)| \Phi[t, u_* + 2u|e(t)|] \mathrm{d}\mu = \infty \qquad (22.34)$$

holds true, where $\bar{\chi}(u) = \chi \left\{ \dfrac{u^*}{u} + \theta \left[\dfrac{H(u)}{u} \right]; e, g \right\}$ *and* $H(u)$ *is given by formula* (22.30), *or*

ii) $\chi(\delta; e, g) = 0$ *for some* $\delta > 0$ *and*

$$\lim_{u \to \infty} \frac{u^{\frac{p}{r}} \displaystyle\int_{\Omega} |g(t)| \Phi[t, u_* + 2u|e(t)|] \mathrm{d}\mu}{[\Psi(Ru)]^{\frac{p}{r}}} = \infty. \qquad (22.35)$$

Then equation (22.1) *has at least one solution* $x_*(t) \in L_p$.

To prove Theorem 22.3 we may use an approach similar to the second step in the proof of Theorem 20.1.

22.7. THE CASE WHEN 1 IS A MULTIPLE EIGENVALUE. Natural analogues of all the results formulated along the present section are also available in case the operator A has 1 as a multiple eigenvalue. To illustrate the point we only mention the analogue of Theorem 22.1.

Suppose the weak nonlinearity $\varphi(t, x)$ is bounded and satisfies condition (22.18), where $\Phi(t, u) \in \mathfrak{N}(u_0)$. Let A be a completely continuous operator acting on the space L_∞, and let

$$\Xi_0 = \{e(t) : e(t) \in L_\infty, \ Ae(t) = e(t), \ \|e(t)\| = 1\}$$

and

$$\Xi_0^* = \{g(t) : g(t) \in (L_\infty)^*, \ A^*g(t) = g(t), \ \|g(t)\| = 1\}.$$

Assume that $\Xi_0^* \subset L_1$ and that the family Ξ_0 is sign-compatible with the family Ξ_0^* (see Subsection 6.1). Let $\Pi : \Xi_0 \to \Xi_0^*$ be a fixed mapping so that any $e(t) \in \Xi_0$ and $g(t) = \Pi e(t) \in \Xi_0^*$ satisfy condition (22.17). Moreover, suppose the superpositionally measurable and bounded function $\chi(\delta; e) = \chi(\delta; e, \Pi e)$ (see (22.10)) is positive for $\delta > 0$.

THEOREM 22.4. *Let* $b(t) \in L_\infty$ *be a function satifying the Fredholm condition*

$$\int_{\Omega} b(t)g(t)\mathrm{d}\mu = 0, \quad g(t) \in \Xi_0^*,$$

and assume the equality

$$\lim_{u \to \infty} \frac{1}{\chi(Ru^{-1}; e)} \int_\Omega |\Pi e(t)| \Phi[t, u_0 + u|e(t)|] d\mu = \infty$$

is true for every $R > 0$.

Then equation (22.1) has at least one bounded solution.

§23. Integral equations with nonnegative kernels

23.1. PRELIMINARIES. An area where the results of Section 22 naturally evidence their strength is the theory of nonlinear integral equations

$$x(t) = \int_\Omega G(t, s)[\alpha x(s) + \varphi(s, x)] ds + b(t) \tag{23.1}$$

with nonnegative kernels $G(t, s)$:

$$G(t, s) \geqslant 0, \quad t, s \in \Omega. \tag{23.2}$$

The naturalness is explained by the fact that under widely applicable assumptions the spectra of the linear integral operators with nonnegative kernels

$$Ax(t) = \int_\Omega G(t, s)x(s) ds \tag{23.3}$$

and

$$Bx(t) = \int_\Omega G(s, t)x(s) ds \tag{23.4}$$

have a special structure. More precisely, they contain the leading positive eigenvalue

$$\lambda_0 = r(A) = r(B),$$

that corresponds to two nonnegative eigenfunctions $e(t)$ and $g(t)$ of A and B, respectively, and the remaining parts of the spectra of A and B are located inside the circle $|\lambda| = q\lambda_0$, where $q < 1$. A large amount of work has been devoted to the disclosure of conditions that imply the specific structure described above of the spectra of operators (23.3) and (23.4) (see, for instance, [Krasnoselskii, Lifshits, Sobolev, 1985] and the references included there).

We confine ourselves to two simple examples. First, if the operator (23.3) is completely continuous on a space L_p and the kernel $G(t, s)$ is positive, then the spectra of the operators (23.3) and (23.4) have the above mentioned structure. Second, the same conclusion is reached in case there exists a fixed nonnegative and nonzero function $u(t) \in L_p$ such that for every nonnegative and nonzero function $x(t) \in L_p$ there are two numbers $\alpha, \beta > 0$ satisfying the condition

$$\alpha u(t) \leqslant Ax(t) \leqslant \beta u(t), \quad t \in \Omega.$$

For a subsequent use we introduce a new definition. The kernel $G(t, s)$ is called proper with respect to a space $L_p = L_p(\Omega)$ if the operators (23.3) and (23.4) have equal spectra and the operator (23.4) acts on the space $L_q = L_q(\Omega)$, where $q = p(p-1)^{-1}$, as the adjoint operator A^*. Of course, the latter condition has to be imposed only in the case $p = \infty$ and $q = 1$.

23.2. A FIRST EXAMPLE. Suppose the kernel $G(t, s)$ is nonnegative and

$$\alpha_1 v_1(t)v_2(s) \leqslant G(t, s) \leqslant \alpha_2 v_1(t)v_2(s), \quad t, s \in \Omega, \tag{23.5}$$

where $v_1, v_2 : \Omega \to \mathbb{R}^1$ are two bounded and almost everywhere positive measurable functions, and $\alpha_1, \alpha_2 > 0$. In addition, assume that $\mu\{t : v_1(t) \leqslant \delta\} > 0$ for every $\delta > 0$. Under these conditions the kernel $G(t, s)$ is proper with respect to every space L_p $(1 \leqslant p \leqslant \infty)$. Moreover, the operators (23.3) and (23.4) act and are completely continuous on each space L_p. The eigenfunctions $e(t)$ and $g(t)$ of the operators A and A^* satisfy the estimates

$$c_2 v_1(t) \leqslant e(t) \leqslant c_1 v_1(t), \quad c_2 v_2(t) \leqslant g(t) \leqslant c_1 v_2(t), \qquad t \in \Omega, \tag{23.6}$$

for some constants $c_1, c_2 > 0$.

Therefore, the function

$$\chi(\delta; e, g) = \int\limits_{\{t \in \Omega : |e(t)| \leqslant \delta\}} |g(t)| d\mu$$

satisfies the inequalities

$$c_2 \chi\left(\frac{\delta}{c_1}\right) \leqslant \chi(\delta; e, g) \leqslant c_1 \chi\left(\frac{\delta}{c_2}\right), \quad \delta \geqslant 0, \tag{23.7}$$

where $\chi(\delta) = \chi(\delta; v_1, v_2)$. If the function $\chi(\delta)$ satisfies for every $k > 0$ some natural restrictions at zero of the form

$$k_2 \chi(\delta) \leqslant \chi(k\delta) \leqslant k_1 \chi(\delta), \quad \delta \leqslant 0; \ k_1, k_2 > 0,$$

then from (23.7) we get the inequalities

$$b\chi(\delta) \leqslant \chi(\delta; e, g) \leqslant a\chi(\delta).$$

Let $\varphi(t, x)$ be a function subject to the conditions

$$|\varphi(t, x)| \leqslant \text{const}, \quad t \in \Omega, \ x \in \mathbb{R}^1,$$

$$\varphi(t, x)\text{sign}\, x \geqslant \varPhi(|x|), \quad t \in \Omega, \ x \in \mathbb{R}^1,$$

where the function $\varPhi(u)$ is bounded for $u \geqslant 0$, and continuous, positive and nonincreasing for $u \geqslant u_0$, with

$$\lim_{u \to \infty} \frac{1}{\chi(u^{-1})} \int_{u_0}^{\infty} \varPhi(z)\mathrm{d}\chi(z) = \infty.$$

If $b(t) \in L_\infty$ satisfies the Fredholm condition

$$\int_{\Omega} b(t)g(t)\mathrm{d}\mu = 0,$$

then equation (23.1) has at least one bounded solution.

23.3. A SECOND EXAMPLE. For our next example we let the kernel $G(t, s)$ be strictly positive $(G(t, s) \geqslant \varepsilon > 0)$ and bounded. Suppose the function $\varphi(t, x)\text{sign}\, x$ has a constant sign for sufficiently large values of $|x|$, and let $b(t) \in L_\infty$ be such that the linearized equation $x = r^{-1}Ax + b(t)$ has at least one integrable solution (i.e., the Fredholm condition is satisfied). Then equation (23.1) has at least one bounded solution.

§24. Landesman-Lazer type theorems

In this section we continue our study of the operator (19.1) with weak nonlinearities.

24.1. A PROPERTY OF SUPERPOSITION OPERATORS. The first subsection investigates the behavior of the superposition operator

$$\mathfrak{f}x = f[t, x(t)] \tag{24.1}$$

as $\|x(t)\| \to \infty$, in case $x(t) = \xi e(t) + h(t)$, where $\|h(t)\| \leqslant \text{const}$, and $e(t)$ is a fixed function. We assume that the operator (24.1) is defined by a bounded function $f(t,x) : \Omega \times \mathbb{R}^1 \to \mathbb{R}^1$ satisfying the Caratheodory conditions and the next two relations:

$$\lim_{\xi \to +\infty} f(t,\xi) = f_+(t) \tag{24.2}$$

and

$$\lim_{\xi \to -\infty} f(t,\xi) = f_-(t). \tag{24.3}$$

The functions $f_+(t), f_-(t) : \Omega \to \mathbb{R}^1$ are supposed to be measurable and bounded. The limits in equalities (24.2) and (24.3) are computed with respect to the uniform convergence.

LEMMA 24.1. *Let* $e(t), g(t) \in L_1$ *be two given functions such that*

$$\mu\{t : t \in \Omega, \ e(t) = 0\} = 0. \tag{24.4}$$

Then the equality

$$\lim_{\substack{\xi \to \infty \\ \|h(t)\|_{L_1} \leqslant c}} \sup \int_{\Omega} g(t) f[t, \xi e(t) + h(t)] \mathrm{d}\mu = l_+ \tag{24.5}$$

holds true for every $c > 0$, *where* l_+ *stands for the number*

$$l_+ = \int_{\{t:e(t)>0\}} g(t) f_+(t) \mathrm{d}\mu + \int_{\{t:e(t)<0\}} g(t) f_-(t) \mathrm{d}\mu \tag{24.6}$$

This lemma and its analogs were used by many authors (see [Landesman, Lazer, 1970a], [Fučik, 1980], [Mawhin 1974a]).

Before proceeding with the proof of Lemma 24.1, let us make two remarks. First we notice that under the assumptions in Lemma 24.1 we also get the equality

$$\lim_{\substack{\xi \to -\infty \\ \|h(t)\|_{L_1} \leqslant c}} \sup \int_{\Omega} g(t) f[t, \xi e(t) + h(t)] \mathrm{d}\mu = l_-, \tag{24.7}$$

where

$$l_- = \int_{\{t:e(t)>0\}} g(t) f_-(t) \mathrm{d}\mu + \int_{\{t:e(t)<0\}} g(t) f_+(t) \mathrm{d}\mu. \tag{24.8}$$

Both the limits l_+ and l_- are finite and they might or might not be equal.

On the other hand, the uniform convergence used in relations (24.2) and (24.3) can be replaced by other kinds of convergence. For instance, it is enough to suppose that

$$\lim_{\xi \to +\infty} \mu \left[\bigcup_{u \geqslant \xi} \{t : t \in \Omega, |f(t, u) - f_+(t)| \geqslant \varepsilon\} \right] = 0$$

and

$$\lim_{\xi \to -\infty} \mu \left[\bigcup_{u \leqslant \xi} \{t : t \in \Omega, |f(t, u) - f_-(t)| \geqslant \varepsilon\} \right] = 0,$$

for every $\varepsilon > 0$. These equalities are evidently true if (24.2) and (24.3) above are fulfilled. However, it should be mentioned that the usual convergence in measure does not suffice.

Proof. We first split the set Ω into four subsets

$$\Omega_1(\xi) = \left\{t : t \in \Omega, \ |e(t)| \leqslant \frac{2}{\sqrt{\xi}} \right\},$$

$$\Omega_2(\xi) = \left\{t : t \in \Omega, \ |e(t)| > \frac{2}{\sqrt{\xi}}, \ |h(t)| > \sqrt{\xi} \right\},$$

$$\Omega_3(\xi) = \left\{t : t \in \Omega, \ e(t) > \frac{2}{\sqrt{\xi}}, \ |h(t)| \leqslant \sqrt{\xi} \right\},$$

and

$$\Omega_4(\xi) = \left\{t : t \in \Omega, \ e(t) < -\frac{2}{\sqrt{\xi}}, \ |h(t)| \leqslant \sqrt{\xi} \right\}.$$

Next, we estimate each of the four numbers

$$J_i \stackrel{\text{def}}{=} \lim_{\xi \to +\infty} \sup_{\|h(t)\|_{L_1} \leqslant c} \int_{\Omega_i(\xi)} g(t) f[t, \xi e(t) + h(t)] d\mu, \quad i = 1, 2, 3, 4.$$

To start with, observe that (24.4) implies the relation

$$\lim_{\xi \to +\infty} \mu \Omega_1(\xi) = 0,$$

whence, in view of the uniform continuity of the integral, we get

$$J_1 = 0. \tag{24.9}$$

From the Chebyshev inequality

$$\mu\{t : t \in \Omega, \ |h(t)| > \sqrt{\xi}\} \leqslant \frac{\displaystyle\int_{\Omega} |h(t)| d\mu}{\sqrt{\xi}}$$

and from the inclusion $\Omega_2(\xi) \subset \{t : t \in \Omega, |h(t)| > \sqrt{\xi}\}$ it follows that

$$\mu\Omega_2(\xi) \leqslant \frac{c}{\sqrt{\xi}},$$

i.e.,

$$\lim_{\xi \to +\infty} \mu\Omega_2(\xi) = 0$$

and

$$J_2 = 0. \tag{24.10}$$

We claim that

$$J_3 = \int\limits_{\{t:e(t)>0\}} g(t)f_+(t)\mathrm{d}\mu. \tag{24.11}$$

Indeed, since $\xi e(t) + h(t) > \sqrt{\xi}$ for any $\xi \in \Omega_3(\xi)$, relation (24.11) is a consequence of the equality

$$J_3 - \int\limits_{\{t:e(t)>0\}} g(t)f_+(t)\mathrm{d}\mu =$$

$$= \lim_{\xi \to +\infty} \sup_{\|h(t)\|_{L_1} \leqslant c} \left\{ \int\limits_{\Omega_3(\xi)} g(t)[f[t, \xi e(t) + h(t)] - f_-(t)]\mathrm{d}\mu + \right.$$

$$\left. + \int\limits_{\{t:e(t)>0\}\backslash\Omega_3(\xi)} g(t)f_+(t)\mathrm{d}\mu \right\},$$

and the relations

$$\lim_{\xi \to +\infty} \sup_{\|h(t)\|_{L_1} \leqslant c} \int\limits_{\{t:e(t)>0\}\backslash\Omega_3(\xi)} g(t)f_+(t)\mathrm{d}\mu = 0$$

(observe that $\{t : e(t) > 0\} \backslash \Omega_3(\xi) \subset \Omega_1(\xi) \cup \Omega_2(\xi)$) and

$$\lim_{\xi \to +\infty} \sup_{\|h(t)\|_{L_1} \leqslant c} \int\limits_{\Omega_3(\xi)} g(t)[f[t, \xi e(t) + h(t)] - f_-(t)]\mathrm{d}\mu \leqslant$$

$$\leqslant \lim_{\xi \to +\infty} \int\limits_{\Omega_3(\xi)} |g(t)| \cdot \left\{ \sup_{y(t) \geqslant \xi} |f[t, y(t)] - f_+(t)| \right\} \mathrm{d}\mu = 0.$$

A similar proof yields

$$J_4 = \int\limits_{\{t:e(t)<0\}} g(t)f_-(t)\mathrm{d}\mu. \tag{24.12}$$

The assertion in Lemma 24.1 follows from equalities (24.9)–(24.12).

24.2. FIELDS WITH ZERO ROTATION. The present and the next subsections are concerned with the equation

$$x(t) = A\{x(t) + f(t,x)\} - b(t), \tag{24.13}$$

where A is a completely continuous linear operator acting on a fixed space $L_p = L_p(\Omega, \mathbb{R}^1)$ $(1 \leqslant p \leqslant \infty)$, with 1 as a simple eigenvalue. The function $f(t,x)$ is supposed to be bounded and to satisfy conditions (24.2) and (24.3), and $b(t) \in L_p$. We fix an eigenfunction $e(t)$ of the operator A corresponding to the eigenvalue 1, and assume that $e(t)$ satisfies condition (24.4). By $g(t)$ we denote an eigenfunction of the adjoint operator A^* — acting on $(L_p)^*$ —, corresponding to the same eigenvalue (if $p = \infty$, then we suppose that $g(t) \in L_1$). Without any loss of generality we can assume that

$$\int_\Omega g(t)e(t)\mathrm{d}\mu = 1.$$

THEOREM 24.1. *Suppose that either*

$$\beta \overset{\text{def}}{=} \int_\Omega g(t)b(t)\mathrm{d}\mu > \max\{l_+, l_-\}, \tag{24.14}$$

or

$$\beta < \min\{l_+, l_-\}. \tag{24.15}$$

Then the rotation of the completely continuous vector field in L_p

$$x(t) - A\{x(t) + f(t,x)\} + b(t) \tag{24.16}$$

on every sphere

$$S_\rho = \{x(t) : x(t) \in L_p, \ \|x(t)\|_{L_p} = \rho\}$$

with a sufficiently large radius ρ equals 0.

Proof. We will only consider the case corresponding to condition (24.14). The proof in case condition (24.15) is fulfilled goes analogously.

Choose an arbitrary number $\varepsilon < 0$ and consider the field

$$x(t) - Ax(t) - \varepsilon e(t). \tag{24.17}$$

Since the field (24.17) is nonsingular (the projection of this field onto the line through $e(t)$ equals $-\varepsilon e(t) \neq 0$), its rotation on every sphere S_ρ is 0. In order to prove our theorem, it will be enough to check that the linear homotopy

$$\Xi(\lambda, x) = x(t) - Ax(t) - \lambda Af(t, x) + \lambda b(t) - (1 - \lambda)\varepsilon e(t), \quad 0 \leqslant \lambda \leqslant 1$$

is nonsingular on spheres S_ρ with a sufficiently large radius ρ. For $\lambda = 0$ the field $\Xi(\lambda, x)$ coincides with the field (24.17), whereas for $\lambda = 1$ it equals the field (24.16). Let us represent any element $x(t) \in L_p$ as $x(t) = \xi e(t) + h(t)$, where

$$\xi = \int_\Omega x(t) \cdot g(t)\mathrm{d}\mu, \quad h(t) = x(t) - \xi e(t).$$

The set of all such functions $h(t)$ provides an invariant subspace E_1 of the operator A with codimension 1. Clearly 1 is a regular value for the restriction of A to this subspace. The projection $x(t) \mapsto h(t)$ is denoted by Q.

Every singular point of the homotopy $\Xi(\lambda, x)$ $(0 \leqslant \lambda \leqslant 1)$ satisfies the equalities (bifurcation system):

$$-\lambda \int_\Omega g(t)\{f[t, \xi e(t) + h(t)] - b(t)\}\mathrm{d}\mu = (1 - \lambda)\varepsilon \tag{24.18}$$

and

$$h(t) = Ah(t) + \lambda AQf(t, x) - \lambda b_1(t) \quad (b(t) = \beta e(t) + b_1(t)). \tag{24.19}$$

From equality (24.19) we get the estimate

$$\|h(t)\|_{L_p} \leqslant c_1 \stackrel{\text{def}}{=} \|(I - A)^{-1}\|_{E_1 \to E_1} \left\{ \|A\| \sup_{t \in \Omega, \, x \in \mathbb{R}^1} (\mu\Omega)^{-1} + \|b_1(t)\| \right\},$$

that in its turn leads to an a priori estimate

$$\|h(t)\|_{L_1} \leqslant c \tag{24.20}$$

for the components $h(t)$ of any singular point $x(t) = \xi e(t) + h(t)$ of the homotopy $\Xi(\lambda, x)$ for $0 \leqslant \lambda \leqslant 1$.

Since

$$\varlimsup_{\substack{|\xi| \to \infty \\ \|h(t)\|_{L_1} \leqslant c}} \sup \int_\Omega g(t)\{f[t, \xi e(t) + h(t)] - b(t)\}\mathrm{d}\mu = \max\{l_+, l_-\} - \beta < 0,$$

equality (24.18) fails for sufficiently large values of $|\xi|$ and consequently, there exists an a priori norm estimate for all the singular points $x(t)$.

Theorem 24.1 is proved.

It should be mentioned that the just proved result is similar in many respects to a Landesman-Lazer type theorem. More precisely, the similarity refers to that part concerned with the sufficiency of the Landesman-Lazer condition.

The simplest instance of Theorem 24.1 occurs when $f_+(t) \equiv f_-(t) \equiv 0$ ($t \in \Omega$), i.e., when both conditions (24.2) and (24.3) reduce to

$$\lim_{|\xi| \to \infty} \sup_{t \in \Omega} |f(t, \xi)| = 0. \tag{24.21}$$

Under this assumption we clearly have $l_+ = l_- = 0$. For the sake of convenience we state this particular case separately.

THEOREM 24.2. *Suppose condition (24.21) is fulfilled and let $b(t) \in L_p$ be a function that does not satisfy the Fredholm condition*

$$\int_\Omega b(t) g(t) \mathrm{d}\mu = 0. \tag{24.22}$$

Then the rotation of the completely continuous vector field (24.16) on every sphere S_ρ with a sufficiently large radius ρ equals 0.

24.3. FIELDS WITH NONZERO ROTATION.

THEOREM 24.3. *Suppose the estimates*

$$\min\{l_+, l_-\} < \beta \overset{\text{def}}{=} \int_\Omega g(t) b(t) \mathrm{d}\mu < \max\{l_+, l_-\} \tag{24.23}$$

hold true. Then the rotation $\gamma(\infty)$ of the completely continuous vector field in L_p defined by (24.16) on every sphere S_ρ with a sufficiently large radius ρ is different from 0.

Under the conditions in Theorem 24.3 equation (24.13) has at least one solution.

Proof. Throughout this proof we assume that $l_+ > l_-$. In this case the estimates (24.23) have the form

$$l_- < \beta < l_+. \tag{24.24}$$

The proof uses, once again, the representation $x(t) = \xi e(t) + h(t)$ of every function $x(t)$, the subspace E_1, the projection Q, and the function $b_1(t)$ as in the proof of Theorem 24.1. This time we consider the completely continuous homotopy

$$\Xi(\lambda, x) = x(t) - \lambda A x(t) - A f(t, x) + b(t), \quad 1 \leqslant \lambda \leqslant \lambda_0 \qquad (24.25)$$

that relates the vector field (24.16) to the field

$$x(t) - \lambda_0 A x(t) - A f(t, x) + b(t). \qquad (24.26)$$

The parameter λ in (24.25) runs through the interval $[1, \lambda_0]$, where $\lambda_0 > 1$ is a sufficiently close to 1 number. Specifically, we assume that the interval $[\lambda^{-1}, 1)$ does not contain points from the spectrum of A. Then we have the estimate

$$\|(I - \lambda A)^{-1} h(t)\|_{L_p} \leqslant a_1 \|h(t)\|_{L_p}, \quad h(t) \in E_1, \ \lambda \in [1, \lambda_0], \qquad (24.27)$$

where the constant a_1 is independent of λ and $h(t)$.

Since the number λ_0^{-1} is not an eigenvalue of A, the rotation of the vector field (24.26) is different from zero (it equals $(-1)^n$, where n counts the multiplicity of all the real eigenvalues of A greater than 1). We claim that the homotopy (24.25) is nonsingular on spheres S_ρ with sufficiently large radius ρ.

Let $x(t) = \xi e(t) + h(t)$ be a singular point of the homotopy (24.23) for a fixed $\lambda \in [1, \lambda_0]$. Then the equalities

$$\xi(\lambda - 1) + \int_\Omega g(t)\{f[t, \xi e(t) + h(t)] - b(t)\} d\mu = 0, \qquad (24.28)$$

and

$$h(t) - \{\lambda A h(t) + A Q f(t, x) - b_1(t)\} = 0 \qquad (24.29)$$

are true. Equality (24.29) and estimate (24.27) lead to the estimate (24.20) with a positive constant c. On the other hand, in view of Lemma 24.1 we know that

$$\lim_{\xi \to +\infty} \sup_{\|h(t)\|_{L_1} \leqslant c} \int_\Omega g(t)\{f[t, \xi e(t) + h(t)] - b(t)\} d\mu = l_+ - \beta > 0$$

and

$$\lim_{\xi \to -\infty} \sup_{\|h(t)\|_{L_1} \leqslant c} \int_\Omega g(t)\{f[t, \xi e(t) + h(t)] - b(t)\} d\mu = l_- - \beta < 0,$$

i.e., either for $\xi > 0$, or for $\xi < 0$, if $|\xi|$ is sufficiently large, then

$$\xi(\lambda - 1) + \int_{\Omega} g(t)\{f[t, \xi e(t) + h(t)] - b(t)\}\mathrm{d}\mu \neq 0.$$

The last relation, when compared with (24.28), implies the existence of an a priori norm estimate of all the singular points of the homotopy (24.25).

Theorem 24.3 is proved.

24.4. ADDITIONAL REMARKS. **A.** The theorems discussed so far in this section can be generalized to the case when the eigenvalue 1 of the operator A is no longer simple. To illustrate the point, we next indicate a generalization of the most simple, previously proved result, Theorem 24.2.

Suppose that the operator A has the eigenvalue 1, and let E denote the corresponding eigenspace. In addition, assume there are no generalized eigenvectors of A for the eigenvalue 1, i.e., the space L_p splits into the direct sum of two invariant subspaces E_0 and E_1 of A, where $E_0 = \{e(t) : Ae(t) = e(t)\}$ and E_1 is such that 1 is a regular value for the restriction of A to E_1. Since the operator A is completely continuous we clearly have $\dim E_0 < \infty$. Under these assumptions, the adjoint operator A^* acting on the space $(L_p)^*$ has the eigenvalue 1, too, and there are no generalized eigenvectors of A^* for this eigenvalue. Let $E_0^* \subset (L_p)^*$ denote the eigenspace of the operator A^* corresponding to the eigenvalue 1. If $p = \infty$, we additionally suppose that $E_0^* \subset L_1$.

THEOREM 24.4. *Assume that condition (24.21) is fulfilled and that condition (24.4) holds true for every function* $e(t)$. *Let* $b(t)$ *be a function that does not satisfy the Fredholm condition, i.e., equality (24.22) fails for some functions* $g(t) \in E_0^*$. *Then the rotation of the completely continuous vector field in* L_p *defined by (24.16) on every sphere* S_ρ *with a sufficiently large radius* ρ *equals 0.*

B. Different generalizations of the above proved theorems are also available in case equation (24.13) is considered in a space $L_p = L_p(\Omega, \mathbb{R}^n)$ $(n > 1)$ of vector-valued functions. Because their formulation is quite cumbersome, we confine ourselves to give details only in the case when $n = 2$ and $p = 2$, and μ is the planar Lebesgue measure.

Let $\boldsymbol{e}(t) = (e_1(t), e_2(t))$ and $\boldsymbol{g}(t) = (g_1(t), g_2(t))$ be eigenfunctions of the completely continuous operator A and of its adjoint, respectively, corresponding to the simple eigenvalue 1. For the function $\boldsymbol{e}(t)$ we impose condition (24.4), i.e.,

$$\mu\{t : t \in \Omega, \ \boldsymbol{e}(t) = 0\} = 0.$$

Suppose the operator (24.1) is defined by a function $f(t, x) = \{f_1(t, x_1, x_2),$ $f_2(t, x_1, x_2)\} : \Omega \times \mathbb{R}^2 \to \mathbb{R}^2$, such that the uniform limit

$$\lim_{\xi \to +\infty} f(t, \xi x) = F(t, x) = \{(F_1(t, x_1, x_2), F_2(t, x_1, x_2)\}$$

exists for $|x|^2 = x_1^2 + x_2^2 = 1$. The function $F(t, x)$ is defined on the set $\Omega \times \mathbb{S}^1$, the product of Ω and the sphere $\mathbb{S}^1 \subset \mathbb{R}^2$, it takes values in $\mathbb{R}^2$, and we assume that $F(t, x)$ is jointly uniformly continuous in its variables $t \in \Omega$ and $x \in \mathbb{S}^1$.

Using $e(t)$ we introduce the function

$$u(t) = \{u_1(t), u_2(t)\}, \quad u_i(t) = \frac{e_i(t)}{\sqrt{e_1^2(t) + e_2^2(t)}}, \qquad i = 1, 2. \tag{24.30}$$

This function is defined for almost all $t \in \Omega$ and takes values in $\mathbb{S}^1$.

Let us also consider the numbers

$$l_+ = \int_\Omega (g(t), F[t, u(t)]) \mathrm{d}\mu = \sum_{i=1}^2 \int_\Omega g_i(t) F_i[t, u_1(t), u_2(t)] \mathrm{d}\mu,$$

$$l_- = \int_\Omega (g(t), F[t, -u(t)]) \mathrm{d}\mu = \sum_{i=1}^2 \int_\Omega g_i(t) F_i[t, -u_1(t), -u_2(t)] \mathrm{d}\mu,$$

and

$$\beta = \int_\Omega (b_1(t) g_1(t) + b_2(t) g_2(t)) \mathrm{d}\mu \quad (b(t) = \{b_1(t), b_2(t)\}).$$

THEOREM 24.5. *If one of the conditions (24.14) or (24.15) is fulfilled, then the rotation of the completely continuous vector field (24.16) in $L_p(\Omega, \mathbb{R}^2)$ on every sphere S_ρ with a sufficiently large radius ρ equals 0. If condition (24.23) is fulfilled, then the rotation is different from zero.*

The proof of Theorem 24.5 uses the next generalization of Lemma 24.1.

LEMMA 24.2. *Let $f(t, x) : \Omega \times \mathbb{R}^2 \to \mathbb{R}^1$ be a function such that the uniform limit*

$$\lim_{\xi \to +\infty} f(t, \xi x) = F(t, x), \quad |x|^2 = x_1^2 + x_2^2 = 1 \tag{24.31}$$

exists. Then

$$\lim_{\xi \to +\infty} \sup_{\|h(t)\|_{L_1} \leqslant c} \int_\Omega g(t) f[t, \xi e(t) + h(t)] \mathrm{d}\mu = \int_\Omega F[t, u(t)] \mathrm{d}\mu,$$

where $u(t)$ is the function (24.30).

Let us notice that condition (24.31) is completely analogous to conditions (24.2) and (24.3).

24.5. SINGULAR CASES. Theorems 24.1 and 24.3 do not give any information about the rotation of the vector field (24.16) on speres S_ρ with large radii in the case when either $l_+ = \beta$, or $l_- = \beta$, or $l_+ = l_- = \beta$. A possible way of handling these cases is based on the analysis of the asymptotic behavior of the nonlinearity at infinity. The theorems in Section 22 provide conditions under which the rotation is different from zero (in case $l_+ = l_- = 0$). This subsection will focus on conditions leading to zero rotation. These conditions rely on the constructions developed in Subsection 6.3.

To start with, let us, once again, consider the field (24.16). Suppose the bounded nonlinearity $f(t, x)$ is subject to condition (24.21) and

$$g(t) \cdot f(t, x) \geqslant \Phi(t, |x|), \quad t \in \Omega, \ x \in \mathbb{R}^1.$$

We next introduce the function

$$\chi(\delta; e) = \{t : t \in \Omega, \ |e(t)| \leqslant \delta\},$$

and assume that $\chi(\delta; e)$ is positive for $\delta > 0$. In addition, set

$$\theta(z) = \sup \left\{ \delta : \delta \cdot [x(\delta; e)]^{\frac{1}{p}} \leqslant z \right\} \quad (z \geqslant 0).$$

THEOREM 24.5. *Suppose the Fredholm condition (24.22) is fulfilled. Let the function $\Phi(t, u) \in \mathfrak{N}(u_0)$ be such that the equality*

$$\lim_{u \to \infty} \frac{1}{\chi\left[R\theta\left(\dfrac{R}{u}\right); e\right]} \cdot \int_\Omega \Phi[t, u_0^- + u|e(t)|]\mathrm{d}\mu = \infty$$

holds, for any $R > 0$.

Then the rotation of the completely continuous vector field (24.16) in L_p on every sphere S_ρ with a sufficiently large radius ρ equals 0.

For the proof of Theorem 24.3 it suffices to check the nonsingularity of the linear homotopy that joins the field (24.16) and the nonsingular field (24.17) with $\varepsilon < 0$.

24.6. TWO-POINT BOUNDARY VALUE PROBLEMS. There are of course situations when but one of the numbers l_+ or l_- is equal to β. In such a case it is still possible to compute the rotation of the vector field (24.16) on spheres S_ρ with large radii. We will only state the precise result for a specific operator A, the operator of the two-point boundary value problem. The proof can be obtained following some already used arguments and therefore we omit it. However, it should be mentioned that analogous results are true for more general operators.

Let us consider the two-point boundary value problem

$$-x'' = x + f(t,x), \quad x(0) = x(\pi) = 0. \tag{24.32}$$

The additional term $b(t)$ in equation (24.32) is missing; in fact it was included in $f(t,x)$. Let A denote the operator of the two-point boundary value problem, i.e.,

$$Ax(t) = \frac{2}{\pi} \sum_{n=1}^{\infty} \frac{1}{n^2} [\sin n\pi, x(t)] \sin nt.$$

This operator is self-adjoint on L_2 and it also is a completely continuous operator from L_2 into C (see Section 7). The vector field (24.16) has the form

$$x(t) - A\{x(t) + f(t,x)\}. \tag{24.33}$$

Suppose the function $f(t,x)$ satisfies both the conditions (24.2) and (24.3), as well as the conditions

$$l_- \equiv \int_0^\pi f_-(t) \cdot \sin t \, dt > 0,$$

and

$$l_+ \equiv \int_0^\pi f_+(t) \cdot \sin t \, dt = 0.$$

Besides (24.2) we also assume that either

$$f(t,x) - f_+(t) \geqslant \Phi(t,x), \quad 0 \leqslant t \leqslant \pi, \ x \geqslant u_0 \tag{24.34}$$

or

$$f(t,x) - f_+(t) \leqslant -\Phi(t,x), \quad 0 \leqslant t \leqslant \pi, \ x \geqslant u_0, \tag{24.35}$$

where $\Phi(t,u) \in \mathfrak{N}(u_0)$.

THEOREM 24.7. *Suppose the equality*

$$\lim_{u \to \infty} u^2 \int_0^\pi \sin t \cdot \Phi[t, u_* + Ru \sin t] dt = \infty \tag{24.36}$$

holds true, for every $u_* \geqslant u_0$ *and every* $R > 0$.

If condition (24.34) is fulfilled, then the rotation of the vector field (24.33) on spheres with sufficiently large radii equals zero. If condition (24.35) is fulfilled, then the rotation is different from zero.

Condition (24.36) becomes particularly simple if $\Phi(t, u) \equiv \Phi(u)$ $(u \geqslant u_0)$, namely, it reduces to

$$\int_{u_0}^{\infty} u\Phi(u)du = \infty.$$

24.7. NONQUASILINEAR PROBLEMS. Similar arguments can be successfully used in dealing with nonquasilinear problems with bounded nonlinearities.

Let us consider in L_p the completely continuous vector field defined by

$$x(t) - A\{x(t) + f(t, Bx, Cx)\} + b(t), \tag{24.37}$$

where B and C are bounded operators on L_p, and A is a completely continuous operator on L_p having 1 as a simple eigenvalue. As before, we let $e(t)$ and $g(t)$ denote two fixed eigenfunctions of the operators A and A^*, respectively, corresponding to the eigenvalue 1. The function $f(t, \boldsymbol{x}) = f(t, x_1, x_2) : \Omega \times \mathbb{R}^1 \times \mathbb{R}^1 \to \mathbb{R}^1$ is bounded and jointly continuous, and it also satisfies condition (24.31). We assume that

$$\mu\{t : t \in \Omega, \; [Be(t)]^2 + [Ce(t)]^2 = 0\} = 0,$$

and introduce the vector-function

$$\boldsymbol{u}(t) = \left\{ \frac{Be(t)}{\sqrt{[Be(t)]^2 + [Ce(t)]^2}}, \frac{Ce(t)}{\sqrt{[Be(t)]^2 + [Ce(t)]^2}} \right\}$$

as a substitute of (24.30).

Finally, set

$$l_+ = \int_{\Omega} g(t)F[t, \boldsymbol{u}(t)]d\mu, \quad l_- = \int_{\Omega} g(t)F[t, -\boldsymbol{u}(t)]d\mu.$$

THEOREM 24.8. *If one of the conditions (24.14) or (24.15) is fulfilled, then the rotation of the completely continuous vector field (24.37) on spheres S_ρ with a sufficiently large radius ρ equals 0. If condition (24.23) is fulfilled, then the rotation is different from zero.*

The proof of Theorem 24.8 uses Lemma 24.2 and some arguments similar to those already carried out in the proofs of Theorems 24.1 and 24.3.

§25. Asymptotic bifurcation points

25.1. DEFINITIONS. To start with, let us consider the equation

$$\mathcal{F}(x;\lambda) = 0, \tag{25.1}$$

where x is a vector in a Banach space X, and $\lambda \in \Lambda = [a, b]$ is a real parameter. Our specific goal is to figure out conditions that imply the existence of solutions $x \in X$ for equation (25.1) with arbitrarily large norms. The problem is related to the notion of *asymptotic bifurcation points*, first introduced by M. A. Krasnoselskii (see [M. A. Krasnoselskii, 1956]).

DEFINITION. A number λ_0 is called an asymptotic bifurcation point of equation (25.1) if for every $\varepsilon > 0$ there exists a number $\lambda = \lambda_\varepsilon \in \Lambda \cap (\lambda_0 - \varepsilon, \lambda_0 + \varepsilon)$ such that the corresponding equation (25.1) has a solution $x = x_\varepsilon$ with a norm greater than ε^{-1}.

The theorems on asymptotic bifurcation points due to M. A. Krasnoselskii were obtained assuming that the operator $\mathcal{F}(x, \lambda)$ is linearizable at infinity, and they are formulated in terms of the linearized operator. The results proved in the preceding sections enable us to establish the existence of asymptotic bifurcation points relying on properties of weak nonlinearities.

Specifically, we will deal with operator equations of the form

$$x(t) = A(\lambda)\{x(t) + f[t, x(t); \lambda]\} - b(t; \lambda), \tag{25.2}$$

where $\lambda \in \Lambda$ is a real parameter. Equation (25.2) is considered in a space $L_p = L_p(\Omega, \mathbb{R}^1)$ $(1 \leqslant p \leqslant \infty)$ of scalar-valued functions $x(t) : \Omega \to \mathbb{R}^1$. The symbol $A(\lambda)$ $(\lambda \in \Lambda)$ stands for a completely continuous linear operator on L_p, and the operator-valued function $\lambda \mapsto A(\lambda)$ is continuous in the norm topology. The functions

$$f(t, x, \lambda) : \Omega \times \mathbb{R}^1 \times \Lambda \to \mathbb{R}^1 \quad \text{and} \quad b(t; \lambda) : \Omega \times \Lambda \to \mathbb{R}^1$$

are supposed to be jointly continuous. We will study equation (25.2) for values of λ close to a given point $\lambda_0 \in \Lambda$.

The definition of an asymptotic bifurcation point of equation (25.2) is, of course, a particular instance of the previous general definition.

Throughout this section we assume that $f(t, x; \lambda)$ is a weak nonlinearity for every value of the parameter λ, in the sense that

$$\lim_{|x|\to\infty} \sup_{t\in\Omega} x^{-1}|f(t, x; \lambda)| = 0, \quad \lambda \in \Lambda. \tag{25.3}$$

25.2. THE CHANGING INDEX PRINCIPLE. Our next purpose is to formulate general conditions under which a given value λ of the parameter is an asymptotic bifurcation point. The main result of this subsection is stated following [Krasnoselskii, Zabreiko, 1975].

We consider the completely continuous vector field

$$x(t) - A(\lambda)\{x(t) + f[t, x(t); \lambda]\} + b(t; \lambda). \tag{25.4}$$

Suppose that for a given value $\lambda = \lambda^*$ the vector field (25.4) does not have singular points (i.e., solutions of equation (25.2)) outside a ball $\mathfrak{B}_r = \{\|x(t)\| < r\}$ (the radius r of this ball may depend on λ^*). Then (see, for instance, [Krasnoslskii, Zabreiko, 1975], [Krasnoselskii, M. A., 1956a]) the rotation of the field (25.4) on the boundary of any domain including the ball $\mathfrak{B}_r$ is well-defined, and it does not depend on the choice of the domain. This rotation is called the *index (at infinity)* of the field (25.4) for $\lambda = \lambda^*$, and is denoted by $\mathrm{ind}(\lambda^*)$.

THEOREM 25.1. *Suppose there exist two sequences $\lambda_{n,1}$ and $\lambda_{n,2}$ of values of the parameter λ such that*

$$\lim_{n\to\infty} \lambda_{n,1} = \lim_{n\to\infty} \lambda_{n,2} = \lambda_0 \tag{25.5}$$

and

$$\mathrm{ind}(\lambda_{n,1}) \neq \mathrm{ind}(\lambda_{n,2}), \quad n = 1, 2, \dots. \tag{25.6}$$

Then λ_0 is an asymptotic bifurcation point of equation (25.2).

If $\mathrm{ind}(\lambda_0)$ exists, then, without any loss of generality, we can reformulate Theorem 25.1 assuming that one of the sequences $\lambda_{n,i}$ $(i = 1, 2)$ is constant.

We also mention that under the conditions in Theorem 25.1 the solutions approaching infinity of equation (25.2) yield continuous branches.

Suppose next that 1 is an eigenvalue of multiplicity k of the operator $A(\lambda_0)$. Then (see, for instance, [Kato, 1966]) for values of λ close to λ_0 the operator $A(\lambda)$ has exactly k eigenvalues (counting, of course, the multiplicities), which are close to 1. Let $\pi(\lambda)$ denote the sum of the multiplicities of all real eigenvalues of $A(\lambda)$ that are close to 1 and greater than 1. If there exist two sequences $\lambda_{n,1}$ and $\lambda_{n,2}$ of values of the parameter λ, such that equality (25.5) holds true and, in addition, the number $\pi(\lambda_{n,1}) + \pi(\lambda_{n,2})$ is odd, then λ_0 is an asymptotic bifurcation point. This assertion is a straightforward consequence of Theorem 25.1.

The simplest case that illustrates the previous result occurs when $A(\lambda) = \lambda A$, where A is a given completely continuous linear operator. If $\lambda_0 \neq 0$ is such that

λ_0^{-1} is an eigenvalue of the operator A of an odd multiplicity (for instance, a simple eigenvalue), then λ_0 is an asymptotic bifurcation point of equation (25.2).

Theorem 25.1, together with the theorems on equations with weak nonlinearities already proved in the present chapter, can be used to find out new criteria for asymptotic bifurcation points. The reason is that the theorems in Sections 20, 22, and 24 above display conditions that lead to the exact computation of the index of the vector fields under consideration.

25.3. EQUATIONS WITH CONSTANT LINEAR PART. This subsection is concerned with equation (25.2) in case the operator $A(\lambda)$ does not depend on λ, i.e., $A(\lambda) \equiv A$ ($\lambda \in \Lambda$). We will state below two theorems on asymptotic bifurcation points of equation (25.2) corresponding to this particular case.

We first recall some notations and assumptions. Let $e(t)$ and $g(t)$ denote two fixed eigenfunctions of the operators A and A^*, respectively, associated with the simple eigenvalue 1. Suppose the nonlinearity $f(t, x; \lambda)$ is independent of λ, i.e., $f(t, x; \lambda) \equiv f(t, x)$ ($t \in \Omega$, $x \in \mathbb{R}^1$, $\lambda \in \Lambda$), where $f(t, x)$ is subject to conditions (24.2) and (24.3). Let l_+ and l_- be the numbers determined by the functions $f_+(t)$, $f_-(t)$, $e(t)$, and $g(t)$ according to formulas (24.6) and (24.8), and assume that $l_+ \neq l_-$.

THEOREM 25.2. *Suppose the function*

$$l_+ - \int_\Omega b(t; \lambda) g(t) \mathrm{d}\mu$$

changes its sign on every neighborhood of λ_0. Then λ_0 is an asymptotic bifurcation point of equation (25.2).

Proof. In view of Theorem 25.1, it is enough to observe that under the assumptions in our theorem, any neighborhood of λ_0 contains a point λ_1 at which the index of the field (25.4) equals zero (see Theorem 24.1), as well as a point λ_2 at which the index of the same field is different from zero (see again Theorem 24.1).

The proof is complete.

An analogous proof shows that λ_0 is an asymptotic bifurcation point if the function

$$l_- - \int_\Omega b(t; \lambda) g(t) \mathrm{d}\mu$$

takes different signs on any neighborhood of λ_0

We also notice that the hypothesis in Theorem 25.2 implies the equality

$$l_+ = \int_\Omega b(t; \lambda_0) g(t) \mathrm{d}\mu.$$

Our second theorem on asymptotic bifurcation points of equation (25.2) with $A(\lambda) \equiv A$ is based on the results in Section 20.

This time equation (25.2) is considered in the Hilbert space L_2. The operator A is supposed to be normal, and 1 is an eigenvalue of A of an odd multiplicity. In addition, we assume that the linearized equation $x = Ax - b(t; \lambda)$ has at least one bounded solution for each $\lambda \in \Lambda$

THEOREM 25.3. *Suppose that in every neighborhood of λ_0 there exist at least one point λ_1 such that*

$$f(t, x; \lambda_1) \mathrm{sign}\, x \leqslant -\Phi(t, |x|; \lambda_1), \quad t \in \Omega, \ x \in \mathbb{R}^1, \tag{25.7}$$

and at least one point λ_2 such that

$$f(t, x; \lambda_2) \mathrm{sign}\, x \geqslant \Phi(t, |x|; \lambda_2), \quad t \in \Omega, \ x \in \mathbb{R}^1, \tag{25.8}$$

where the function $\Phi(t, u) = \Phi(t, u; \lambda)$ belongs to a class $\mathfrak{N}[u_*(\lambda)]$ for each λ and satisfies condition (20.7) for every $R > 0$ and $u_0 \geqslant u_*(\lambda)$. Then λ_0 is an asymptotic bifurcation point for equation (25.2).

It should be noticed that under the assumptions in Theorem 25.3 the index at infinity of the field (25.4) with $\lambda = \lambda_0$ may not exist. Theorem 25.3 is also a consequence of the change of the index principle.

25.4. EQUATIONS WITH LINEAR PART DEPENDING ON THE PARAMETER. We assume that for $\lambda = \lambda_0$ the operator $A(\lambda)$ has the eigenvalue 1, and there are no generalized eigenvectors of $A(\lambda_0)$ associated to this eigenvalue (this happens, for instance, if $A(\lambda_0)$ is a normal operator on L_2). Let E_0 denote the corresponding eigenspace of the operator $A(\lambda_0)$. The space L_p splits into the direct sum of two invariant subspaces E_0 and E_1 of $A(\lambda_0)$, where $E_0 = \{e(t) : Ae(t) = e(t)\}$, and E_1 is such that 1 is a regular value for the restriction of the operator $A(\lambda_0)$ to this subspace. Since the operator $A(\lambda_0)$ is completely continuous, we have $\dim E_0 < \infty$. The adjoint operator $A(\lambda_0)^*$ acts on the space $(L_p)^*$, and it also has 1 as an eigenvalue. There are no corresponding generalized eigenvectors of $A(\lambda_0)^*$. The set of all eigenvectors forms a subspace $E^* \subset (L_p)^*$. If $p < \infty$, then $(L_p)^* = L_q$, where $q = p(p-1)^{-1}$; if $p = \infty$, then $(L_p)^* \supset L_1$. In the latter case we assume, in addition, that $E^* \subset L_1$.

THEOREM 25.4. *Suppose that*

$$\lim_{|x|\to\infty} \sup_{t\in\Omega} |f(t,x;\lambda_0)| = 0 \tag{25.9}$$

and

$$\mu\{t : t \in \Omega, \ e(t) = 0\} = 0, \tag{25.10}$$

and let $g(t) \in E^*$ *be such that*

$$\int_\Omega g(t)b(t;\lambda_0)\mathrm{d}\mu \stackrel{\text{def}}{=} \beta \neq 0. \tag{25.11}$$

Let $\lambda_n \in \Lambda$ $(n = 1, 2, \ldots)$ *be a sequence with* $\lambda_n \to \lambda_0$ *and such that 1 is a regular value of any linear operator* $A(\lambda_n)$. *Then* λ_0 *is an asymptotic bifurcation point for equation* (25.2).

Condition (25.11), in view of Fredholm Theorem, means that the linearized equation

$$x(t) = A(\lambda_0)x(t) + b(t;\lambda_0)$$

does not have solutions.

For the proof of Theorem 25.4 it suffices to observe that the index at infinity of the field (25.4) equals zero for $\lambda = \lambda_0$, and is different from zero for $\lambda = \lambda_n$.

This chapter consists mainly of results obtained in [Krasnoselskii, A. M., 1991]. Theorem 25.4 was jointly proved by the author and Yu. Appell (see [Appell, Krasnoselskii, A. M., 1992]). Since the publication of the Russian edition of this book, the author has published some new results close to the ones in Section 24 (see [Krasnoselskii, A. M., 1992], [Krasnoselskii, A. M., Mawhin, 1992]).

Chapter 5

One-sided estimates for nonlinearities

§26. Positive linear operators

26.1. PRELIMINARY DEFINITIONS. Let H be a fixed real Hilbert space with the inner product $[\cdot, \cdot]$. A linear operator A acting on H is called *positive* (see [Krasnoselskii, Zabreiko, 1975]) if there exists $\mu > 0$ such that

$$[Ax, Ax] \leqslant \mu[x, Ax], \tag{26.1}$$

for all $x \in H$. The infimum of all such values of μ is called the *positivity coefficient* of A and is denoted by $\mu(A)$ or $\mu(A; H)$.

The simplest example of a positive operator is provided by any bounded self-adjoint and positive definite operator A on H. The positivity coefficient of such an operator A is given by the equality

$$\mu(A; H) = \|A\|_{H \to H}.$$

Let us now consider a Banach space E consisting of functions $x(t) : \Omega \to \mathbb{R}^n$. We denote, as usual, the dual space of E by E^* and let $E^\square = E \cap E^*$. If $E = L_p = L_p(\Omega, \mathbb{R}^n)$ $(1 < p < \infty)$ and $\mu\Omega < 0$, then the set $E^\square$ coincides with one of the sets E or E^*. The meaning of the notation $E^\square$ can be equally well made clear for some other specific function spaces. Suppose A is an operator that takes E into $E^\square$. If there exists $\mu > 0$ such that inequality (26.1) is true for every $x \in E$, then A is called a positive operator on the Banach space E.

The next subsection presents two simple positivity conditions for operators acting on Hilbert spaces.

26.2. POSITIVITY CONDITIONS. Let A be a continuous operator on a Hilbert space H. We let A^* denote the adjoint of A and define the self-adjoint and the skew-symmetric parts of A by the formulas $A_+ = \frac{1}{2}(A + A^*)$ and $A_- = \frac{1}{2}(A - A^*)$, respectively. The operators A^*, A_+ and A_- are continuous, too. Moreover, we clearly have $[Ax, x] = [A_+x, x]$, $[A_-x, x] = 0$ $(x \in H)$, $A = A_+ + A_-$, $A_-^* = -A_-$.

LEMMA 26.1 ([Krasnoselskii, Zabreiko, 1975]). *Suppose there exists $\varepsilon > 0$ such that the self-adjoint operator $A_+ + \varepsilon A_-^2$ is positive semidefinite. Then A is a positive operator on H.*

Proof. We start with two remarks. First, the operator $-A_-^2$ is self-adjoint and positive semidefinite. This fact easily follows from the relations $[-A_-^2 x, x] = [A_-x, A_-x] \geqslant 0$. Second, under our assumptions the operator A_+ is positive semidefinite, too, and $[A_+x, x] \geqslant \varepsilon[-A_-^2 x, x]$.

Since

$$[Ax, Ax] = \|A_+x + A_-x\|^2 \leqslant 2\left(\|A_+x\|^2 + \|A_-x\|^2\right) = 2\left(\|A_+x\|^2 - [A_-^2 x, x]\right),$$

and because the operator A_+ is positive with the positivity coefficient $\|A_+\|$, we get

$$[Ax, Ax] \leqslant 2\|A_+\|[A_+x, x] + \frac{2}{\varepsilon}[A_+x, x] = 2\left(\|A_+\| + \frac{2}{\varepsilon}\right)[Ax, x].$$

Therefore, the operator A is positive and its positivity coefficient does not exceed $2\left(\|A_+\| + \frac{2}{\varepsilon}\right)$.

The lemma is proved.

Let us now suppose that the bounded operator A is the inverse of an unbounded operator M. In many specific situations the role of M is played by a differential operator corresponding to a boundary value problem. Assume M is of the form $M = M_+ + M_-$, where M_+ and M_- are two unbounded operators sharing the same domain of definition $D \subset H$ and such that M_+ is self-adjoint and M_- is skew-symmetric.

LEMMA 26.2 ([Krasnoselskii, Zabreiko, 1975]). *Suppose the operator M_+ is positive definite and there exists $a > 0$ such that the estimate*

$$[M_+x, x] \geqslant a[x, x], \quad x \in D \tag{26.2}$$

holds. Then $A = M^{-1}$ is a positive operator on the Hilbert space H.

For the proof it suffices to take account of the relations

$$[Ax, Ax] \leqslant \frac{1}{a}[M_+ Ax, x] = \frac{1}{a}[MAx, Ax] = \frac{1}{a}[x, Ax].$$

We notice that the positive semidefiniteness of the self-adjoint part M_+ of the operator $M = A^{-1}$ is a necessary condition for the positivity of A. Under same natural assumptions, the estimate (26.2) is a consequence of inequality (26.1).

26.3. POSITIVE NORMAL OPERATORS. The main theorem in this subsection singles out a necessary and sufficient condition for a completely continuous and normal operator A on a Hilbert space H to be positive. To start with, suppose that the operator A is fixed, and let $E_n \subset H$ $(n \geqslant 0)$ be a full sequence of mutually orthogonal invariant subspaces of A of dimensions 1 or 2 (for details, see Section 7). Each subspace E_n corresponds to an eigenvalue λ_n of A, which in its turn determines a rotation U_n on E_n. If we let P_n denote the orthogonal projection onto E_n, then the operator A has the spectral representation

$$Ax = \sum_{n=0}^{\infty} |\lambda_n| U_n P_n x, \quad x \in H, \tag{26.3}$$

where $\lambda_n \to 0$ as $n \to \infty$.

THEOREM 26.1. *The completely continuous and normal operator A is positive on H if and only if its spectrum $\sigma(A)$ is included in a disk*

$$K_\mu = \left\{ \lambda : \left| \lambda - \frac{1}{2}\mu \right| \leqslant \frac{1}{2}\mu \right\}, \quad \mu > 0, \tag{26.4}$$

in the complex plane. Moreover, if the operator A is positive, then its positivity coefficient $\mu(A; H)$ is given by the equality

$$\mu(A; H) = \left[\inf_{\lambda \in \sigma(A), \; \lambda \neq 0} \mathrm{Re}(\lambda^{-1}) \right]^{-1}, \tag{26.5}$$

and coincides with the smallest value of μ for which $\sigma(A) \subset K_\mu$.

Proof. Let us consider the self-adjoint operators $A^*A = AA^*$ and $A_+ = \frac{1}{2}(A + A^*)$. The decomposition (26.3) of A yields the relations

$$A^*Ax = \sum_{n=0}^{\infty} |\lambda_n|^2 P_n x, \quad x \in H \tag{26.6}$$

and

$$A_+ x = \sum_{n=0}^{\infty} \operatorname{Re} \lambda_n P_n x, \quad x \in H. \tag{26.7}$$

Since the operators (26.3), (26.6) and (26.7) share the same system of invariant subspaces E_n, they commute with each other.

From (26.6) and (26.7) above we get

$$\mu[Ax, x] - [Ax, Ax] = [(\mu_+ - AA^*)x, x] = \sum_{n-0}^{\infty} \left(\mu \operatorname{Re} \lambda_n - |\lambda_n|^2\right) \|P_n x\|^2,$$

hence condition (26.1) for the operator A is equivalent to

$$\mu \operatorname{Re} \lambda \geqslant |\lambda|^2, \quad \lambda \in \sigma(A). \tag{26.8}$$

It remains to observe that (26.8) is an explicit reformulation of condition $\sigma(A) \subset K_\mu$.

The proof of equality (26.5) follows from another simple remark. More precisely, condition $\lambda \in K_\mu$ ($\lambda \neq 0$) is equivalent to (26.8), where both the numbers μ and $\operatorname{Re} \lambda$ are positive, and inequality (26.8) clearly can be rewritten as $\mu \geqslant [\operatorname{Re}(\lambda^{-1})]^{-1}$. Theorem 26.1 is completely proved.

The next result — a positivity test — is a consequence of Theorem 26.1.

COROLLARY 26.1. *Suppose A is a completely continuous normal operator on a Hilbert space H. Then A is positive on H if and only if the number (26.5) is well-defined and positive.*

In case the completely continuous normal operator A on H is the inverse of an unbounded operator M with spectrum $\sigma(M)$, the positivity of A reduces to the uniform positivity of the set of real numbers $\{\operatorname{Re} \lambda : \lambda \in \sigma(M)\}$. Explicitly, all we need in such a case is to observe that equality (26.5) can be rewritten as

$$\mu(A; H) \inf_{\lambda \in \sigma(M)} \operatorname{Re} \lambda = 1.$$

26.4. POTENTIALLY POSITIVE OPERATORS. We first introduce a new definition. An operator A acting on a Hilbert space H is called *potentially positive*, or, more accurately, *potentially positive from below*, if there exists $\gamma \in \mathbb{R}$ such that the operator $I - \gamma A$ is continuously invertible and the operator

$$A_\gamma = (I - \gamma A)^{-1} A, \quad x \in H \tag{26.9}$$

is positive on H.

The concept of a potentially positive from above operator is defined analogously. More precisely, an operator A is called potentially positive from above, if the operator $-A$ is potentially positive from below.

The simplest examples of potentially positive operators are provided by any positive operator on H.

We also notice that a completely continuous self-adjoint operator A on H is potentially positive if and only if it has a finite number of negative eigenvalues.

LEMMA 26.3. *A completely continuous operator A is potentially positive if and only if there exists $a \in \mathbb{R}$ such that*

$$a[Ax, Ax] \leqslant [x, Ax], \quad x \in H. \tag{26.10}$$

Before proceeding with the proof of Lemma 26.3, let us notice the analogy between condition (26.10) above and the positivity condition (26.1). The main difference is that the constant a is not necessarily a positive number.

Proof. We will separately prove the necessity and the sufficiency of condition (26.10).

Accordingly, suppose first that the operator A is potentially positive. Then there exists $\gamma \in \mathbb{R}$ such that

$$[(I - \gamma A)^{-1} Ay, (I - \gamma A)^{-1} Ay] \leqslant \mu(A_\gamma; H)[y, (I - \gamma A)^{-1} Ay], \quad y \in H. \tag{26.11}$$

Set $x = (I - \gamma A)^{-1}y$. From (26.11) we get

$$[Ax, Ax] \leqslant \mu(A_\gamma; H)[(I - \gamma A)x, Ax], \quad x \in H, \tag{26.12}$$

an inequality that clearly implies (26.10) with $a = \gamma + [\mu(A_\gamma; H)]^{-1}$.

Let us next assume that A satisfies inequality (26.10). Without any loss of generality we may suppose that $a \neq 1$. Under these two assumptions, we claim that the operator $(I - \gamma A)^{-1}$ is bounded for $\gamma = a - 1$. The proof goes as follows. If $(I - \gamma A)^{-1}$ is unbounded, then γ^{-1} is a point in the spectrum $\sigma(A)$ of the operator A. Since A is completely continuous, γ^{-1} must be an eigenvalue of A, hence there exists a vector $x \in H$ with $\|x\| = 1$, such that $x = \gamma Ax$. It remains to observe that inequality (26.10) fails for this specific x. The just reached contradiction proves our claim.

To conclude the proof of Lemma 26.3, we take an arbitrary vector $y \in H$ and use (26.10) for $x = (1 - \gamma A)^{-1}y$. We get $[A_\gamma x, A_\gamma y] \leqslant [y, A_\gamma y]$, an inequality that clearly proves the positivity of A_γ. Lemma 26.3 is completely proved.

For a subsequent use we let $\gamma_H(A)$ denote the largest number a such that inequality (26.10) holds true. If the operator A is positive, then $\gamma_H(A) \cdot \mu(A; H) = 1$.

LEMMA 26.4. *Suppose A is a completely continuous potentially positive operator. Then for every $\gamma < \gamma_H(A)$ the operator A_γ is positive and $\gamma + [\mu(A_\gamma; H)]^{-1} = \gamma_H(A)$.*

Proof. We first observe that the operator $I - \gamma A$ is continuously invertible for each $\gamma < \gamma_H(A)$. If not, then there exists a nonzero vector $x \in H$ such that $x = \gamma A x$, an equality which apparently contradicts the inequalities

$$\gamma_H(A)[Ax, Ax] \leqslant [x, Ax], \quad x \in H \tag{26.13}$$

and $\gamma < \gamma_H(A)$. The positivity of the operator A_γ and the equality $\mu(A_\gamma; H) = (\gamma_H(A) - \gamma)^{-1}$ follow from the inequality

$$[A_\gamma x, A_\gamma x] \leqslant (\gamma_H(A) - \gamma)^{-1}[x, A_\gamma x], \quad x \in H$$

and from the fact that this inequality is equivalent to (26.13). The last assertion may be easily proved by simple computations.

The proof of Lemma 26.4 is complete.

Analogues of both Lemmas 26.3 and 26.4 can be formulated and proved for operators that are potentially positive from above.

The next result provides a criterion for the potential positivity of normal operators in terms of their spectra.

THEOREM 26.2. *A normal completely continuous operator A on H is potentially positive if and only if there exists a positive number μ such that the nonzero spectrum $\sigma(A) \setminus \{0\}$ of A lies outside the disk*

$$K_{-\mu} = \left\{ \lambda : \left| \lambda + \frac{1}{2}\mu \right| \leqslant \frac{1}{2}\mu \right\}. \tag{26.14}$$

Proof. The spectrum $\sigma(A_\gamma)$ of the operator (26.9) consists of all the numbers

$$\nu = \frac{\lambda}{1 - \gamma\lambda},$$

where $\lambda \in \sigma(A)$. Since every operator A_γ is also normal and completely continuous, its positivity (see Corollary 26.1) is equivalent to the positivity of the number

$$\inf_{\lambda \in \sigma(A), \ \lambda \neq 0} \text{Re}\left(\frac{1 - \gamma\lambda}{\lambda}\right) = -\gamma + \inf_{\lambda \in \sigma(A), \ \lambda \neq 0} \text{Re}(\lambda^{-1}) \tag{26.15}$$

Suppose now that the operator A is potentially positive. Then there exists $\gamma \in \mathbb{R}$ such that the operator (26.9) is positive. Therefore,

$$\inf_{\lambda \in \sigma(A),\ \lambda \neq 0} \operatorname{Re}(\lambda^{-1}) > \gamma,$$

i.e., $\operatorname{Re}\lambda > (\gamma+\varepsilon)|\lambda|^2$ ($\lambda \in \sigma(A)$, $\lambda \neq 0$) for some $\varepsilon > 0$. Consequently (since without any loss of generality we may assume $\gamma + \varepsilon < 0$) we get

$$\left|\lambda + \frac{1}{2|\gamma+\varepsilon|}\right|^2 > \frac{1}{4|\gamma+\varepsilon|^2}, \quad \lambda \in \sigma(A),\ \lambda \neq 0$$

hence $\lambda \notin K_{-\mu}$ ($\lambda \in \sigma(A)$, $\lambda \neq 0$) for $\mu = |\gamma+\varepsilon|$.

Suppose next that the nonzero spectrum of the operator A lies outside the disk (26.14). This means that $\mu \operatorname{Re}\lambda + |\lambda|^2 \geqslant 0$ ($\lambda \in \sigma(A)$, $\lambda \neq 0$), hence

$$\mu \operatorname{Re}(\lambda^{-1}) + 1 \geqslant 0, \quad \lambda \in \sigma(A),\ \lambda \neq 0. \tag{26.16}$$

Choose $\lambda < \mu^{-1}$. Then (26.16) implies the positivity of the number (26.15). Theorem 26.2 is proved.

From Theorem 26.2 it follows that a normal completely continuous operator A is potentially positive from above if and only if its nonzero spectrum lies outside a disk K_μ as in (26.4).

Some operators are simultaneously potentially positive from above and from below. For instance, every skew-symmetric completely continuous operators enjoys this property. Other examples will be discussed in Subsection 29.3.

For a normal completely continuous potentially positive operator A the constant $\gamma_H(A)$ is defined by

$$\gamma_H(A) = \inf_{\lambda \in \sigma(A),\ \lambda \neq 0} \operatorname{Re}(\lambda^{-1}). \tag{26.17}$$

In case the operator A is potentially positive from above, a special attention has to be paid to the constant

$$\gamma_B(A) = \sup_{\lambda \in \sigma(A),\ \lambda \neq 0} \operatorname{Re}(\lambda^{-1}). \tag{26.18}$$

The numbers (26.17) and (26.18) will be often used in the present chapter.

26.5. THE AUXILIARY OPERATOR. Suppose that the operator A is acting on a Hilbert space and is completely continuous, normal, and positive. Let (26.3) be its spectral representation. We associate to A the auxiliary operator $A^\square$ defined by

$$A^\square x = \sum_{n=0}^{\infty} \frac{|\lambda|}{\sqrt{\operatorname{Re}\lambda_n}} U_n P_n x, \quad x \in H. \tag{26.19}$$

The operator (26.19) can be equally well defined by the formula

$$A^\square x = \sum_{n=0}^{\infty} [\mathrm{Re}(\lambda_n^{-1})]^{-1} U_n P_n x, \quad x \in H.$$

Clearly $A^\square$ is a normal operator on H.

LEMMA 26.5. *The operator (26.19) is continuous on H and*

$$\|A^\square\|_{H \to H} = \sqrt{\mu(A; H)}. \tag{26.20}$$

Proof. We first observe that the positivity of A implies the inclusion $\sigma(A) \subset K_\mu$, where $\mu = \mu(A; H)$, whence $|\lambda|^2 \leqslant \mu(A; H)\mathrm{Re}\,\lambda$ ($\lambda \in \sigma(A)$). On the other hand, there exists $\lambda_0 \in \sigma(A)$, $\lambda_0 \neq 0$, such that $|\lambda_0|^2 = \mu(A; H)\mathrm{Re}\,\lambda_0$. Consequently

$$\|A^\square\|_{H \to H} = \sup_{\lambda \in \sigma(A),\ \lambda \neq 0} \frac{|\lambda|}{\sqrt{\mathrm{Re}\,\lambda}} = \sqrt{\mu(A; H)},$$

and the proof is complete.

If the operator A is positive, then the self-adjoint operator A_+ is positive semidefinite. Then it makes sense to consider the square root $(A_+)^{1/2}$ of the operator A_+, which also is self-adjoint and positive semidefinite. The inverse of $(A_+)^{1/2}$, denoted by $(A_+)^{-1/2}$, is an unbounded operator defined on a dense subspace $D[(A_+)^{-1/2}]$ of $H/\mathrm{Ker}\,A$. The auxiliary operator (26.19) can be written as

$$A^\square = (A_+)^{-1/2}A = A(A_+)^{-1/2}. \tag{26.21}$$

If the operator A is normal and potentially positive, then for every $\gamma < \gamma_H(A)$ we may define the operator $A_\gamma^\square$. Its spectral representation is explicitely given by

$$A_\gamma^\square x = \sum_{n=0}^{\infty} [\mathrm{Re}(\lambda_n^{-1}) - \gamma]^{-1} U_n P_n x, \quad x \in H. \tag{26.22}$$

It should be mentioned that the operators (26.21) can be likewise introduced for operators A that are not necessarily normal.

26.6. STRICTLY POSITIVE NORMAL OPERATORS. A positive operator A acting on H is called *strictly positive* if the space H splits into a direct sum $E_0 \oplus E_1$ of two orthogonal invariant subspaces of A, such that E_0 is finite-dimensional, $\mathrm{Ker}\,A \subset E_1$, and

$$[Ax, Ax] = \mu(A; H)[x, Ax], \quad x \in E_0 \tag{26.23}$$

and

$$[Ax, Ax] \leqslant \mu_1(A; H)[x, Ax], \quad x \in E_1 \tag{26.24}$$

where

$$\mu_1(A; H) < \mu(A; H). \tag{26.25}$$

THEOREM 26.3. *A normal, positive, and completely continuous operator A is strictly positive if and only if but a finite number of its nonzero eigenvalues lie on the circle*

$$\partial K_{\mu(A;H)} = \left\{ \lambda : \left| \lambda - \frac{1}{2}\mu(A; H) \right| = \frac{1}{2}\mu(A; H) \right\}, \tag{26.26}$$

and all the other eigenvalues lie on a disk K_μ as in (26.4) with a radius $\frac{1}{2}\mu$ less than $\frac{1}{2}\mu(A; H)$.

Theorem 26.3 is a straightforward consequence of Theorem 26.1 and of its proof. The fact that the subspace E_0 is finite-dimensional clearly follows from the complete continuity of the operator A.

In view of Theorem 26.3 we get that every self-adjoint, completely continuous, and positive definite operator is strictly positive.

A normal, completely continuous and positive operator A is strictly positive if and only if there exists a finite set $\Lambda \subset \sigma(A)$ $(0 \notin \Lambda)$ such that

$$\gamma_H(A) = \text{Re}(\lambda^{-1}) \quad (\lambda \in \Lambda), \qquad \inf_{\lambda \in \sigma(A),\ \lambda \neq 0,\ \lambda \notin \Lambda} \text{Re}(\lambda^{-1}) > \gamma_H(A). \tag{26.27}$$

§27. Solvability of nonlinear operator equations with positive linear part

27.1. THE SIMPLEST CONDITIONS. This section will focus on the quasilinear operator equation

$$x = A\mathfrak{f}x, \tag{27.1}$$

where the completely continuous linear operator A and the nonlinear superposition operator

$$\mathfrak{f}x(t) = f[t, x(t)], \tag{27.2}$$

are acting on spaces of functions $x(t) : \Omega \to \mathbb{R}^n$. We will constantly assume that the function $f(t, x) : \Omega \times \mathbb{R}^n \to \mathbb{R}^n$ satisfies the Caratheodory condition. Our main objective is to find conditions for the solvability of equation (27.1). These conditions are formulated here as one-sided constraints of the form

$$(x, f(t, x)) \leqslant k|x|^2 + b(t), \quad t \in \Omega, \ x \in \mathbb{R}^n, \ b(t) \in L_1(\Omega, \mathbb{R}^1), \quad (27.3)$$

satisfied by the nonlinearity $f(x, t)$. A few slightly different versions of (27.3) above will be also considered. Another basic assumption repeatedly used in what follows is the positivity or semiboundedness of the operator A.

Recall that by $|\cdot|$ and $(\cdot, \cdot)$ we denote the norm and the inner product in the space $\mathbb{R}^n$.

ASSERTION. *Suppose that the operator A acting on the space $L_2 = L_2(\Omega, \mathbb{R}^n)$ is completely continuous and positive, and that the function $f(t, x)$ satisfies the estimate*

$$|f(t, x)| \leqslant c|x| + c(t), \quad t \in \Omega, \ x \in \mathbb{R}^n, \ c(t) \in L_2(\Omega, \mathbb{R}^1). \quad (27.4)$$

In addition, assume that $f(t, x)$ is subject to the one-sided constraint (27.3), where

$$k\mu(A; L_2) < 1. \quad (27.5)$$

Then equation (27.1) has at least one solution $x(t) \in L_2$.

For the proof of our assertion (some forthcoming comments will explain why it was not formulated as a theorem) we may proceed in the following way. The estimate (27.4) implies that the operator (27.2) acts continuously on the space L_2, hence the operator $A\mathfrak{f}$ is completely continuous on L_2. Let $y(t) \in L_2$ be a solution of an equation $y = \lambda A\mathfrak{f}y$, where $0 \leqslant \lambda \leqslant 1$. Then the function $x(t) = \lambda \mathfrak{f}y \in L_1(\Omega, \mathbb{R}^n)$ is a solution of the equation $x = \lambda \mathfrak{f}Ax$. We clearly have $y(t) = Ax(t)$. Observe next that, on one hand

$$[x, Ax] = [\lambda \mathfrak{f}Ax, Ax] \leqslant |[\mathfrak{f}Ax, Ax]| \leqslant k\|Ax\|^2 + \|b(t)\|_{L_1(\Omega, \mathbb{R})},$$

and, on the other hand,

$$[x, Ax] \geqslant \frac{1}{\mu(A; L_2)} \|Ax\|^2.$$

Therefore

$$\|y\|^2 = \|Ax\|^2 \leqslant [1 - k\mu(A, L_2)]^{-1} \mu(A, L_2)\|b(t)\|_{L_1(\Omega, \mathbb{R})},$$

and the solvability of equation (27.1) follows from the Leray-Schauder Principle.

Let us next analyze the previous assertion in case of scalar-valued functions $x(t)$. The estimate (27.3) reduces to

$$xf(t, x) \leqslant kx^2 + b(t),$$

and from this inequality, together with (27.4), we get the estimate

$$\left| f(t,x) - \frac{k-c}{2}x \right| \leqslant \frac{k+c}{2}|x| + b_1(t), \quad b_1(t) \in L_2(\Omega, \mathbb{R}^1). \tag{27.6}$$

Further, suppose that $k > 0$ and $c > k$. The operator equation (27.1) can be rewritten in the equivalent form

$$x = A_\gamma \mathfrak{f}_\gamma x, \tag{27.7}$$

where A_γ is the operator defined by (26.6),

$$\mathfrak{f}_\gamma x = \mathfrak{f}x - \gamma x(t), \tag{27.8}$$

and

$$\gamma = \frac{k-c}{2}. \tag{27.9}$$

Since the operator A is positive and $\gamma < 0$, it turns out that the operator (26.6) is continuously invertible, hence — in view of (27.6) — from Theorem 7.1 we get that equation (27.7) is solvable in case

$$\|A_\gamma\| \frac{c+k}{2} < 1. \tag{27.10}$$

Assume that the positive operator A is also normal. Then the solvability condition (27.5) has the form

$$k < \inf_{\lambda \in \sigma(A), \ \lambda \neq 0} \operatorname{Re}(\lambda^{-1}). \tag{27.11}$$

Condition (27.10) — after some simple transformations — may be rewritten as

$$\left(\frac{c+k}{2} \right)^2 \leqslant \inf_{\lambda \in \sigma(A), \ \lambda \neq 0} \left\{ \left[\operatorname{Re}(\lambda^{-1}) + \frac{c+k}{2} \right]^2 + (\operatorname{Im}\lambda)^2 \right\}. \tag{27.12}$$

A straightforward check shows that inequality (27.12) follows from inequality (27.11). In this respect, the above formulated assertion is weaker than Theorem 7.1.

It should be mentioned that conditions (27.11) and (27.12) are equivalent if and only if the boundary of the disk (26.4) — i.e., the circle ∂K_μ — with $\mu = \mu(A; L_2)$ contains only real eigenvalues of A.

In the following subsections we will state solvability conditions that do not require two-sided linear estimates of the type (27.4).

27.2. EQUATIONS WITH NORMAL OPERATORS.

THEOREM 27.1. *Suppose A is a positive operator on L_2 acting as a completely continuous operator from L_1 into L_2. Assume the nonlinearity $f(t,x)$ satisfies the one-sided constraint (27.3) with a constant k subject to condition (27.5), as well as the inequality*

$$|f(t,x)| \leqslant c|x|^2 + c_1(t), \quad t \in \Omega, \ x \in \mathbb{R}^n, \ c_1(t) \in L_1(\Omega, \mathbb{R}^1). \tag{27.13}$$

Then equation (27.1) has at least one solution $x(t) \in L_2$

In the particular case of scalar-valued functions, conditions (27.3) and (27.13) above lead to the inequalities

$$-cx^2 - \cdots \leqslant f(x,t) \leqslant kx + \cdots \tag{27.14}$$

whenever $x \geqslant 0$, and

$$-k|x| - \cdots \leqslant f(x,t) \leqslant cx^2 + \cdots \tag{27.15}$$

if $x \leqslant 0$.

Condition (27.5) can be rewritten as

$$k < \gamma_H(A). \tag{27.16}$$

Under the assumptions in Theorem 27.1, the operator $\mathfrak{f}$ acts continuously from L_2 into L_1; hence $A\mathfrak{f}$ is a completely continuous operator on L_2. The proof of Theorem 27.1 can be easily completed following the arguments already used in the proof of the assertion made in Subsection 27.1. The details are left to the reader.

If the operator A satisfies additional constraints, then condition (27.13) can be considerably weakened.

We next suppose that the nonlinearity $f(t,x)$ is "locally square integrable", i.e.,

$$\alpha(t,r) \stackrel{\text{def}}{=} \sup_{|x| \leqslant r} |f(t,x)| \in L_2(\Omega, \mathbb{R}^1). \tag{27.17}$$

Condition (27.17) means that the superposition operator (27.2) acts continuously from the space $L_\infty = L_\infty(\Omega, \mathbb{R}^n)$ into L_2. As simple examples of functions $f(t,x)$ satisfying condition (27.17) we mention any function of the form $a(t) + f_1(x)$ $(a(t) \in L_2)$, where $f_1(x)$ is a locally bounded function that may have a quite fast growth as $|x| \to \infty$.

THEOREM 27.2. *Let A be a completely continuous operator acting from L_2 into L_∞, and suppose that it is normal and positive as an operator on L_2, with the positivity* coefficient $\mu(A; L_2)$. *Assume that the auxiliary operator* (26.19) *acts continuously from* L_2 *into* L_∞. *Let* $f(t, x)$ *be a nonlinearity subject to the one-sided constraint* (27.3) *with a constant* k *satisfying inequality* (27.5), *and such that condition* (27.17) *is fulfilled, too. Then equation* (27.1) *has at least one bounded solution.*

Proof. We first replace equation (27.1) by an equivalent equation. As a matter of fact, it is enough to observe that whenever $y(t) \in L_2$ is a solution of the equation

$$y = fAy, \tag{27.18}$$

the function

$$x = Ay \tag{27.19}$$

is a bounded solution of equation (27.1). Consequently, our specific goal consists of proving the solvability of equation (27.18) in L_2.

We next notice that under the conditions in our theorem, the superposition operator f acts continuously from L_∞ into L_2. Therefore, the operator fA is completely continuous on the space L_2. In view of the Leray-Schauder Principle, all we need in order to prove the solvability of equation (27.18) in L_2 is to check that there exists an a priori norm estimate of all the solutions $y(t) \in L_2$ of the equations

$$y = \xi fAy, \quad 0 \leqslant \xi \leqslant 1. \tag{27.20}$$

To this end, let $y(t) \in L_2$ be an arbitrary solution of the equation (27.20) for a fixed $\xi \in [0, 1]$. Taking now the inner products in L_2 with the function $Ay \in L_2$, and using the positivity of the operator A, from (27.20) we get the estimates

$$0 \leqslant [y, Ay] = \xi[fAy, Ay] \leqslant [fAy, Ay].$$

The last inequality combined with (27.3) leads to the following relations

$$[y, Ay] = \int_\Omega (f[t, Ay(t)], Ay(t)) \mathrm{d}\mu \leqslant \int_\Omega \{k|Ay(t)|^2 + b(t)\} \mathrm{d}\mu =$$
$$= k\|Ay\|^2 + \|b(t)\|_{L_1(\Omega, \mathbb{R})}.$$

Based once again on the positivity of the operator A, we obtain the inequality

$$[y, Ay] \leqslant k\mu(A; L_2)[y, Ay] + \|b(t)\|_{L_1(\Omega, \mathbb{R})}.$$

Thus we just proved the estimate

$$[y, Ay] \leqslant c_2^2 \stackrel{\text{def}}{=} \|b(t)\|_{L_1(\Omega, \mathbb{R})}[1 - k\mu(A; L_2)]^{-1}. \tag{27.21}$$

Let $A_+ = \frac{1}{2}(A + A^*)$ denote the self-adjoint part of the normal operator A (see Section 26). Since A is a positive operator, it follows that A_+ is a positive semidefinite operator, hence its square root

$$A_+^{1/2}x = \sum_{n=0}^{\infty} \sqrt{\operatorname{Re}\lambda_n}\,P_n x$$

is a positive semidefinite operator, too. On the other hand,

$$[z, Az] = [z, A_+z] = [A_+^{1/2}z, A_+^{1/2}z], \quad z \in L_2,$$

therefore estimate (27.21) can be rewritten as

$$\|A_+^{1/2}y\| \leqslant c_2. \tag{27.22}$$

From (26.21) we know that

$$Az = A^{\square}A_+^{1/2}z, \quad z \in L_2;$$

consequently, (27.22) yields the estimate

$$\|Ay\|_{L_\infty} \leqslant \|A^{\square}\|_{L_2 \to L_\infty} \cdot c_2 \overset{\text{def}}{=} c_3.$$

In its turn, this estimate together with (27.17) implies

$$|y(t)| = \xi|f[t, Ay(t)]| \leqslant \alpha(t, c_3) \in L_2$$

hence

$$\|y(t)\|_{L_2} \leqslant \|\alpha(t, c_3)\|_{L_2} < \infty.$$

Thus the existence of an a priori norm estimate of all the solutions $y(t)$ in L_2 of equations (27.20) is proved.

The proof of Theorem 27.2 is now complete.

27.3. THE SCHAEFER-M. A. KRASNOSELSKII METHOD. We now assume that instead of conditions (27.13) and (27.17) above, we have the "intermediate" estimate

$$|f(t, x)| \leqslant c_1\|x\|^{p/2} + b(t), \quad t \in \Omega, \ x \in \mathbb{R}^n, \tag{27.23}$$

where $b(t) \in L_2(\Omega, \mathbb{R}^1)$ and $p > 4$. In this case the operator (27.2) acts continuously from the space L_p into the space L_2.

THEOREM 27.3. *Let A be a completely continuous operator acting from L_2 into L_p, and suppose that A is normal and positive as an operator on L_2, with the positivity coefficient $\mu(A; L_2)$. Assume that the auxiliary operator (26.19) acts continuously from L_2 into L_p. Let $f(t, x)$ be a nonlinearity subject to the one-sided constraint (27.3) with a constant k satisfying inequality (27.5), and such that condition (27.23) is fulfilled, too. Then equation (27.1) has at least one solution $x(t) \in L_p$.*

The proof of Theorem 27.3 is completely analogous to the proof of Theorem 27.2, and for this reason we will omit it.

However, it should be mentioned that more sutle solvability criteria for the equation (27.1) with nonlinearities satisfying growth conditions of type (27.23) are known (see [Schaefer, 1955], [Krasnoselskii, Zabreiko, 1975]). The next theorem, aimed to illustrate the point, is stated following [Krasnoselskii, Zabreiko, 1975]. It can be successfully used in case the linear operator A is not necessarily normal. In order to formulate the result, we will first recall the concept of convergence in measure. Specifically, a sequence $x_N(t)$ of functions converges in measure to a function $x_*(t)$ if

$$\lim_{N \to \infty} \mu\{t : t \in \Omega, \ |x_N(t) - x_*(t)| \geqslant \varepsilon\} = 0, \qquad (27.24)$$

for every $\varepsilon > 0$. Some other equivalent definitions are possible (see, for instance, [Dunford, Schwartz, 1958], [Yosida, 1965]). The convergence in measure follows from the usual convergence in L_p, as well as from the pointwise convergence. Let us also recall that every number $p \in (1, \infty)$ is related to the number $q = p/(p-1)$, such that the Lebesgue spaces L_p and L_q are conjugate to each other.

Throughout the rest of this subsection, equation (27.1) is considered in spaces of scalar-valued functions, i.e., $n = 1$, and A is supposed to be an integral operator, i.e.,

$$Ax(t) = \int_\Omega G(t, s)x(s)\mathrm{d}\mu(s). \qquad (27.25)$$

The operator (27.25) is defined not only for functions $x(t)$ in a Lebesgue space L_p, but also for arbitrary integrable functions $x(t)$.

THEOREM 27.4. *Suppose that the operator A satisfies the following conditions:*

a) *A is completely continuous and positive on L_2, with the positivity coefficient $\mu(A; L_2)$;*

b) *the boundedness of the numerical sequence $[Ax_N, x_N]$ implies the compactness of the sequence Ax_N relatively to the convergence in measure;*

c) *if a sequence x_N converges in measure to x_* and is bounded in the norm of L_p $(1 < p < 2)$, and if the sequence Ax_N converges in measure to y_*, then $Ax_* = y_*$.*

Let $f(t, x)$ be a nonlinearity subject to the one-sided constraint (27.3) with a constant k satisfying inequality (27.5), and such that the corresponding superposition operator (27.2) acts continuously from L_q into L_p, i.e.,

$$|f(t, x)| \leqslant c|x|^{\frac{1}{p-1}} + c(t), \quad t \in \Omega, \ x \in \mathbb{R}^1, \ c(t) \in L_p. \tag{27.26}$$

Then equation (27.1) has at least one solution.

Before proceeding with the proof of Theorem 27.4 we make two remarks.

First, the conditions in this theorem are satisfied, for instance, by operators that are inverses of elliptic differential operators.

The second remark is related to condition c) in Theorem 27.4. Let $\Omega = [0, 1]$. The operator

$$Ax(t) = \int\limits_0^T x(s)ds$$

takes the sequence

$$x_N(t) = \begin{cases} N & \text{if } 0 \leqslant t \leqslant \dfrac{1}{N} \\[2mm] 0 & \text{if } \dfrac{1}{N} < t \leqslant 1, \end{cases} \tag{27.27}$$

that converges in measure to the function $x_*(t) \equiv 0$, into the constant sequence $Ax_N(t) \equiv 1$. The norm of every function $x_N(t)$ in L_1 equals 1, whereas the norm in L_p $(1 < p < \infty)$ is equal to $N^{1-(1/p)}$. This example shows that even a "nice" operator A (of finite rank!) fails to satisfy condition c).

Returning now to the general setting, we notice that if the operator A is completely continuous as an operator from L_p into a space L_r $(r > 1)$, then condition c) is fulfilled.

Proof of Theorem 27.4. For the sake of convenience we split the proof into four steps.

Step 1. Set, for each natural number N,

$$f_N(t, x) = \begin{cases} f(t, x) & \text{if } |f(t, x)| \leqslant N \\[3mm] \dfrac{N \cdot f(t, x)}{|f(t, x)|} & \text{if } |f(t, x)| > N, \end{cases} \tag{27.28}$$

and denote by $\mathfrak{f}_N$ the superposition operator corresponding to the function (27.28). Since every function $f_N(t, x)$, as well as the function $f(t, x)$, satisfies the Caratheodory condition and is bounded ($\sup |f_N(t, x)| \leqslant N$), all the operators $\mathfrak{f}_N$ act continuously on the space L_2. Therefore — based on condition a) in our theorem — each operator

$\mathfrak{f}_N A$ acts and is completely continuous on the space L_2. Moreover, the operator $\mathfrak{f}_N A$ sends the ball $\{x : \|x\|_{L_2} \leqslant N\sqrt{\mu\Omega}\} \subset L_2$ into itself. From the Schauder Principle we conclude that every operator $\mathfrak{f}_N A$ has a fixed point $x_N \in L_2$.

Step 2. We next observe that $xf_N(t,x) = xf(t,x)$ whenever $|f(t,x)| \leqslant N$, and $xf_N(t,x) = \dfrac{N}{|f_N(t,x)|}f(t,x)$ whenever $|f(t,x)| > N$. At the same time, without any loss of generality, we may assume that the constant k and the function $b(t)$ in (27.3) are positive. Consequently, for each function (27.28) we get the estimate

$$x_N f(t,x) \leqslant kx^2 + b(t), \quad t \in \Omega, \; x \in \mathbb{R}^1, \; b(t) \in L_1, \; N = 1, 2, \ldots,$$

which is analogous to (27.3). Let us now consider the inner products $[x_N, Ax_N]$. From the next chain of relations

$$[x_N, Ax_N] = [\mathfrak{f}_N Ax_N, Ax_N] = \int_\Omega \mathfrak{f}_N[t, Ax_N(t)] \cdot Ax_N(t)\mathrm{d}\mu \leqslant$$

$$\leqslant \int_\Omega \{k|Ax_N(t)|^2 + b(t)\}\,\mathrm{d}\mu = k\|Ax_N\|^2 + \|b(t)\|_{L_1} \leqslant$$

$$\leqslant k\mu(A; L_2)[x_N, Ax_N] + \|b(t)\|_{L_1}$$

it follows that

$$[x_N, Ax_N] \leqslant \|b(t)\|_{L_1}[1 - k\mu(A; L_2)]^{-1} \overset{\text{def}}{=} b_1.$$

Based on condition b) in our theorem, the sequence Ax_N is compact with respect to the convergence in measure. This enables us to choose a subsequence $Ax_{N(j)}$ $(j = 1, 2, \ldots)$ that converges in measure to a function $z(t)$. The subsequence $x_{N(j)} = \mathfrak{f}_{N(j)}Ax_{N(j)}$, converges in measure to the function $\mathfrak{f}z(t)$ (see [Caratheodory, 1918], [Nemyckii, 1934]).

Suppose for a while that the norms in L_p of the functions x_N are uniformly bounded. Then from condition c) we conclude that $z = A\mathfrak{f}z$. Thus, all we need in order to complete the proof of Theorem 27.4 is to prove the uniform estimate

$$\|x_N\|_{L_p} \leqslant \text{const} < \infty. \tag{27.29}$$

Step 3. Our subsequent goal is to estimate the integrals

$$J_N \overset{\text{def}}{=} \int_\Omega |\mathfrak{f}_N[t, Ax_N(t)]Ax_N(t)|\mathrm{d}\mu = \int_\Omega |x_N(t)Ax_N(t)|\mathrm{d}\mu. \tag{27.30}$$

Set

$$\Omega(N) = \{t : t \in \Omega, \; Ax_N(t)x_N(t) \geqslant 0\}.$$

From (27.3) we get the estimate

$$\int\limits_{\Omega(N)} |Ax_N(t)x_N(t)\mathrm{d}\mu \leqslant \int\limits_{\Omega(N)} \left\{ k[Ax_N(t)]^2 + b(t) \right\} \mathrm{d}\mu \leqslant$$

$$\leqslant k\|Ax_N\|^2 + \|b(t)\|_{L_1} \leqslant k\mu(A; L_2)b_1 + \|b(t)\|_{L_1} \overset{\text{def}}{=} b_2.$$

Since all the functions under consideration are scalar-valued, we conclude that

$$J_N = \int\limits_{\Omega(N)} Ax_N(t)x_N(t)\mathrm{d}\mu - \int\limits_{\Omega\setminus\Omega(N)} Ax_N(t)x_N(t)\mathrm{d}\mu =$$

$$= 2 \int\limits_{\Omega(N)} Ax_N(t)x_N(t)\mathrm{d}\mu - \int\limits_{\Omega} Ax_N(t)x_N(t)\mathrm{d}\mu \leqslant 2b_2 + b_1.$$

Step 4. Since the operator $\mathfrak{f}$ acts continuously from L_q into L_p, then (see [Krasnoselskii, *et al.*, 1966])

$$\sup_{\|v(t)\|_{L_q} \leqslant 1} \|\mathfrak{f}v(t)\|_{L_p} = M < \infty;$$

therefore $(|\mathfrak{f}_N(t,x)| \leqslant |f(t,x)|)$ we have the estimates

$$\sup_{\|v(t)\|_{L_q} \leqslant 1} \|\mathfrak{f}_N v(t)\|_{L_p} \leqslant M.$$

Consequently, for every set $\Omega_1 \subset \Omega$ it follows that

$$\left[\int\limits_{\Omega_1} |\mathfrak{f}_N v(t)|^p \mathrm{d}\mu \right]^{\frac{1}{p}} \leqslant M, \quad v(t) \in L_q, \ \|v(t)\|_{L_q} \leqslant 1. \qquad (27.31)$$

Let $v(t) \in L_q$ with $\|v(t)\|_{L_q} \leqslant 1$ be fixed. Set

$$G(N,v) = \{t : t \in \Omega, \ |v(t)| \geqslant |Ax_N(t)|\},$$

and

$$y_N = \begin{cases} Ax_N(t) & \text{if } t \in G(N,v), \\ 0 & \text{if } t \notin G(N,v). \end{cases}$$

From the next chain of relations

$$\|y_N(t)\|_{L_q} = \left[\int\limits_{\Omega} |y_n(t)|^q \mathrm{d}\mu \right]^{\frac{1}{q}} = \left[\int\limits_{G(N,v)} |Ax_n(t)|^q \mathrm{d}\mu \right]^{\frac{1}{q}} \leqslant$$

$$\leqslant \left[\int\limits_{G(N,v)} |v(t)|^q \mathrm{d}\mu \right]^{\frac{1}{q}} \leqslant \|v(t)\|_{L_q} \leqslant 1$$

and from (27.31), where $\Omega_1 = G(N, v)$, we get the inequality

$$\left[\int\limits_{G(N,v)} |\mathfrak{f}_N y_N(t)|^p \mathrm{d}\mu \right]^{\frac{1}{p}} \leqslant M.$$

Therefore

$$\left[\int\limits_{G(N,v)} |x_N(t)|^p \mathrm{d}\mu \right]^{\frac{1}{p}} = \left[\int\limits_{G(N,v)} |\mathfrak{f}_N A x_N(t)|^p \mathrm{d}\mu \right]^{\frac{1}{p}} \leqslant M,$$

and, in view of Hölder inequality, we obtain

$$\int\limits_{G(N,v)} |v(t) x_N(t)| \mathrm{d}\mu \leqslant \left[\int\limits_{G(N,v)} |v(t)|^p \mathrm{d}\mu \right]^{\frac{1}{p}} \cdot M \leqslant M.$$

It remains now to observe that the estimate $J_N \leqslant 2b_2 + b_1$ proved at the previous step, together with the relations

$$\|x_N\|_{L_p} = \sup_{\|v(t)\|_{L_q}} \int\limits_\Omega v(t) x_N(t) \mathrm{d}\mu \leqslant$$

$$\leqslant \sup_{\|v(t)\|_{L_q}} \left\{ \int\limits_{\Omega \backslash G(N,v)} |x_N(t) A x_N(t)| \mathrm{d}\mu + \int\limits_{G(N,v)} |v(t) x_N(t)| \mathrm{d}\mu \right\} \leqslant J_N + M,$$

leads to the estimate (27.28), with

$$\mathrm{const} = 2b_2 + b_1 + M.$$

Theorem 27.4 is completely proved.

27.4. FRACTIONAL POWERS OF SELF-ADOINT OPERATORS. In case the linear operator A is self-adjoint and positive on L_2, we may follow a procedure based on the use of fractional powers of operator A.

To be more specific, let E be a Banach space of functions $x(t) : \Omega \to \mathbb{R}^n$, such that $E^* \subset L_2 \subset E$. Assume that the operator A acts as a completely continuous operator from E into E^* and that as an operator on L_2 it is self-adjoint and positive semidefinite. Then ([Krasnoselskii, *et al.*, 1966]) the operator A can be represented as

$$A = KK^*,$$

where K is a completely continuous operator acting from L_2 into E^*, and K^* is a completely continuous operator acting from E into L_2. The operator K restricted to the space L_2 coincides with the self-adjoint positive semidefinite square root of the operator A. We have the equality

$$\|K\|_{L_2 \to L_2} = \sqrt{\lambda},$$

where λ is the leading eigenvalue of the operator A.

Suppose, in addition, that the superposition operator (27.2) acts continuously from E^* into E. If $E = L_p$ $(1 < p < 2)$, then the last assumption means exactly that the estimate (27.26) holds true. We deduce that the operator

$$Bx = K^* \mathfrak{f} K x \qquad (27.32)$$

is completely continuous on L_2. Let us also assume that condition (27.3) is fulfilled. Then every solution $x(t)$ of any equation

$$x = \xi K^* \mathfrak{f} K x, \quad 0 \leqslant \xi \leqslant 1$$

satisfies the relations

$$\|x\|^2 = \xi[x, K^* \mathfrak{f} K x] \leqslant [x, K^* \mathfrak{f} K x] \leqslant [Kx, \mathfrak{f} K x] \leqslant k\|Kx\|^2 + \|b\|_{L_1(\Omega, \mathbb{R}^n)} \leqslant$$
$$\leqslant k\lambda\|x\|^2 + \|b\|_{L_1(\Omega, \mathbb{R}^n)}.$$

If $k\lambda < 1$, then the Leray-Schauder Principle implies the existence of at least one fixed point $x(t)$ of the operator (27.32) lying in the ball $\{x : \|x\|_{L_2} \leqslant \|b\|_{L_1(\Omega_1, \mathbb{R}^n)}(1 - k\lambda)^{-1}\}$.

Every fixed point $x(t)$ of the operator (27.32) corresponds to a solution $Kx \in L_q = L_p^*$ of equation (27.1).

27.5. UNIQUENESS CONDITIONS. We next assume that $f(t, x) : \Omega \times \mathbb{R}^n \to \mathbb{R}^n$ is a function satisfying the "one-sided Lipshitz condition"

$$(x - y, f(t, x) - f(t, y)) \leqslant k(x - y, x - y), \quad t \in \Omega, \ x, y \in \mathbb{R}^n. \qquad (27.33)$$

THEOREM 27.5. *Let A be a continuous positive operator on L_2 with the positivity constant $\mu(A; L_2)$, and such that $[Ax, x] > 0$ for $x \neq 0$. Suppose that the constant k in (27.33) satisfies condition (27.5). Then equation (27.1) has no two different solutions x_1 and x_2 with $\mathfrak{f}x_1 - \mathfrak{f}x_2 \in L_2$.*

Proof. Let x_1 and x_2 be two solutions of equation (27.1), i.e.,

$$x_1 = A\mathfrak{f}x_1, \quad x_2 = A\mathfrak{f}x_2.$$

If $\mathfrak{f}x_1 - \mathfrak{f}x_2 \in L_2$, then the relations

$$\int_\Omega |x_1(t) - x_2(t)|^2 \mathrm{d}\mu = \int_\Omega |A[\mathfrak{f}x_1(t) - \mathfrak{f}x_2(t)]|^2 \mathrm{d}\mu \leqslant$$

$$\leqslant \mu(A; L_2) \cdot \int_\Omega (\mathfrak{f}x_1(t) - \mathfrak{f}x_2(t), A[\mathfrak{f}x_1(t) - \mathfrak{f}x_2(t)])\mathrm{d}\mu =$$

$$= \mu(A; L_2) \cdot \int_\Omega (\mathfrak{f}x_1(t) - \mathfrak{f}x_2(t), x_1(t) - x_2(t))\mathrm{d}\mu \leqslant$$

$$\leqslant k\mu(A; L_2) \cdot \int_\Omega |x_1(t) - x_2(t)|^2 \mathrm{d}\mu$$

imply the equality $\|x_1 - x_2\| = 0$. The theorem is prooved.

In case the function $f(t, x) : [a, b] \times \mathbb{R}^1 \to \mathbb{R}^1$ is sufficiently smooth, condition (27.33) above is equivalent to the estimate $f'_x(t, x) \leqslant k$.

27.6. EQUATIONS WITH POTENTIALLY POSITIVE OPERATORS. The so far developed arguments may be carried over, without major changes, to equations of the form (27.1) with potentially positive (from below or from above) operators A. All we need in order to handle properly this more general situation is to replace equation (27.1) by the equivalent equation

$$x = A_\gamma \mathfrak{f}_\gamma x, \tag{27.34}$$

where A_γ stands for the operator (26.9), and the superposition operator

$$\mathfrak{f}_\gamma x = \mathfrak{f}_\gamma[t, x(t)]$$

corresponds to the function

$$f_\gamma(t, x) = f(t, x) - \gamma x. \tag{27.35}$$

If the given operator A is potentially positive, then the operator A_γ is positive for every $\gamma < \gamma_H(A)$, and, consequently, all the previous results can be applied to equation (27.34). The new superposition operator $\mathfrak{f}_\gamma$ acts from and takes values into the same spaces of functions as the initial superposition operator $\mathfrak{f}$. We are interested in the particular form of conditions (27.3) and (27.5) when these conditions are imposed on equation (27.34).

Suppose that the constraint (27.3) is fulfilled. Then the function (27.35) satisfies the inequality

$$(x, f_\gamma(t, x)) \leqslant (k - \gamma)|x|^2 + b(t), \quad t \in \Omega, \ x \in \mathbb{R}^n, \ b(t) \in L_1(\Omega, \mathbb{R}^1). \qquad (27.36)$$

Therefore, condition (27.5) imposed on equation (27.34) becomes

$$(k - \gamma)\mu(A_\gamma; L_2) < 1,$$

and, since in view of Lemma 26.4 we have

$$\mu(A_\gamma : L_2) = \frac{1}{\gamma_H(A) - \gamma},$$

condition (27.5) reduces to

$$k < \gamma_H(A). \qquad (27.37)$$

As a matter of fact, condition (27.37) has been already introduced and used in Subsection 27.1.

To summarize the preceding discussion, the transition from equations with positive operators to equations with potentially positive operators requires the replacement of condition (27.5) by (27.37). At the same time, it is irrelevant what specific value of γ is used as long as $\gamma < \gamma_H(A)$.

Of course, if $\gamma_H(A) \leqslant 0$, then the admissible values of k in (27.3) must be negative.

We state below, without proofs, two solvability criteria for the equation (27.1) in case A is a potentially positive operator.

THEOREM 27.6. *Suppose A is a potentially positive operator on L_2 acting as a completely continuous operator from L_1 into L_2. Let $f(t, x)$ be a nonlinear function subject to the constraint (27.3) with a constant k satisfying inequality (27.37), and such that condition (27.1) is fulfilled. Then equation (27.1) has at least one solution $x(t) \in L_2$.*

In order to formulate the second criterion — analogous to Theorem 27.2 — we need a preliminary result.

LEMMA 27.1. *Let A be a normal and potentially positive operator on L_2. Suppose there exists $\gamma_0 < \gamma_H(A)$ such that the operator $A_{\gamma_0}^\square$ acts continuously from L_2 into L_∞. Then the operator $A_\gamma^\square$ acts continuously from L_2 into L_∞ for every $\gamma < \gamma_H(A)$.*

For the proof of Lemma 27.1 it suffices to represent the operator (26.30) as

$$A_\gamma^\square x = A_{\gamma_0}^\square Bx,$$

where the operator

$$Bx = \sum_{n=0}^{\infty} \left(\frac{\text{Re}(\lambda_n^{-1}) - \gamma}{\text{Re}(\lambda_n^{-1}) - \gamma_0} \right)^{\frac{1}{2}} P_n x$$

acts continuously on the space L_2.

THEOREM 27.7. *Suppose A is a normal and potentially positive operator on L_2 acting as a completely continuous operator from L_2 into L_∞. Assume there exists $\gamma < \gamma_H(A)$ such that the auxiliary operator $A_\gamma^\square$ acts continuously from L_2 into L_∞. Let $f(t,x)$ be a nonlinear function subject to the constraint (27.3) with a constant k satisfying inequality (27.37), and such that condition (27.17) is fulfilled. Then equation (27.1) has at least one bounded solution.*

The uniqueness condition (27.33) can be also used for equations with potentialy positive operators. Once again, all we need is to assume that the constant k occurring there is less than $\gamma_H(A)$.

Before concluding this section, we notice that all the preceding arguments and theorems have natural counterparts in case of equations with semibounded from above operators A. The one-sided constraint (27.3) has to be rewritten as

$$(x, f(t,x)) \geqslant k|x|^2 + b(t), \quad t \in \Omega, \ x \in \mathbb{R}^n, \ b(t) \in L_1(\Omega, \mathbb{R}^1),$$

where the constant k is supposed to satisfy an analogous to (27.37) condition, namely

$$k > \gamma_B(A).$$

§28. Equations with strictly positive operators

28.1. THE MAIN THEOREM. In this section we will be concerned with the equation

$$x = A\mathfrak{f}x, \tag{28.1}$$

where A is a linear strictly positive operator and $\mathfrak{f}$ is the nonlinear superposition operator (27.2). The particular feature of equation (28.1) above, that is, the strict positivity of the operator A, will enable us to weaken significantly condition (27.5) imposed on the constant k in the the one-sided estimate (27.3), in all the subsequent solvability criteria for equation (28.1).

Throughout this section we assume that A is a fixed completely continuous, normal, and strictly positive operator on the space L_2. From time to time, as the need arises, other additional conditions on A will be explicitly stated.

Let

$$\lambda_1, \cdots, \lambda_r \tag{28.2}$$

be all the nonzero eigenvalues of the operator A lying on the circle $\partial K_{\mu(A;L_2)}$ (see (26.4)). The collection (28.2) contains either no real eigenvalues at all, or it contains $\mu(A; L_2)$ as the only real eigenvalue. The nonreal eigenvalues listed in (28.2) occur in pairs, each pair consisting of an eigenvalue and its complex conjugate. The eigenvalues in (28.2) are all supposed distinct.

We let E_0 denote the finite-dimensional subspace

$$E_0 = E_0(A) = \bigoplus \Pi_j \tag{28.3}$$

where the direct sum (28.3) includes all the subspaces Π_j (see Section 7) corresponding to the eigenvalues (28.2). More precisely, if an eigenvalue λ_j from the collection (28.2) corresponds to more than one subspace Π_j, then all that subspaces are considered in the direct sum (28.3). Accordingly, the dimension of the subspace (28.3) is given by

$$\dim E_0(A) = k_1 + k_2 + \cdots + k_r,$$

where k_j $(j = 1, 2, \ldots, r)$ is the multiplicity of the eigenvalue λ_j from (28.2). The subspace $E_0(A)$ is exactly the subspace E_0 used in the definition of strict positivity (see Subsection 26.6).

In all the forthcoming theorems we assume that the nonlinearity $f(t, x)$ generating the superposition operator $\mathfrak{f}$ in equation (28.1) is subject to the one-sided constraint

$$(x, f(t, x)) \leqslant k(x, x) - \Phi(t, |x|), \quad t \in \Omega, \ x \in \mathbb{R}^n, \tag{28.4}$$

where the function $\Phi(t, u)$ $(t \in \Omega, \ u \geqslant 0)$ lies in one of the classes $\mathfrak{N}(u_0)$ (for the definition of these classes see Section 2).

THEOREM 28.1. *Suppose that the operator A acts as a completely continuous operator from the space L_1 into the space L_2. Let $f(t, x)$ be a nonlinear function satisfying condition (27.13). In addition, assume that the equality*

$$\lim_{\delta \to 0} \sup_{e(t) \in E_0; \ \|e\|=1} \frac{\chi(\delta; e)}{\displaystyle\int_\Omega \Phi[t, u_* + R\delta^{-1}|e(t)|]\mathrm{d}\mu} = 0 \tag{28.5}$$

holds true, for every $R > 0$ and every $u_* \leqslant u_0$. Then there exists $\varepsilon > 0$ such that the one-sided constraint (28.4) with a constant k satisfying the inequality

$$k\mu(A; L_2) < 1 + \varepsilon, \qquad (28.6)$$

implies the existence of at least one solution $x(t) \in L_2$ of equation (28.1).

THEOREM 28.2. *Suppose that the operator A acts as a completely continuous operator from the space L_2 into the space L_p ($4 < p \leqslant \infty$), and the auxiliary operator (26.19) acts continuously from L_2 into L_p, and that the superposition operator $\mathfrak{f}$ acts from L_p into L_2. In addition, assume that equality (28.5) holds true for every $R > 0$ and every $u_* \geqslant u_0$. Then there exists $\varepsilon > 0$ such that the one-sided constraint (28.4) with a constant k satisfying inequality (28.6) implies the existence of at least one solution $x(t) \in L_p$ of equation (28.1).*

Condition (28.6) is obviously weaker than condition (27.5) in the previous section.

The proofs of both Theorems 28.1 and 28.2 will be omitted.

THEOREM 28.3. *Suppose that the operator A acts as a completely continuous operator from L_2 into L_p ($4 < p \leqslant \infty$), and that it also acts continuously from L_1 into L_p. Assume one of the following two conditions is satisfied:*
   a) the nonlinearity is subject to the estimate

$$|f(t, x)| \leqslant c|x|^2 + b(t), \quad t \in \Omega, \ x \in \mathbb{R}^n, \ b(t) \in L_2(\Omega, \mathbb{R}^1); \qquad (28.7)$$

   b) the auxiliary operator (26.19) acts continuously from the space L_2 into the space L_p, and the superposition operator $\mathfrak{f}$ acts from L_p into L_2.
   In addition, suppose that equality (28.5) holds true for every $R > 0$ and every $u_* \geqslant u_0$. Then there exists $\varepsilon > 0$ such that the one-sided constraint (28.4) with a constant k satisfying inequality (28.6) implies the existence of at least one solution $x(t) \in L_p$ of equation (28.1).

Although Theorem 28.3 is weaker than both Theorems 28.1 and 28.2 above, it is still quite close to them. Theorem 28.3 will be proved in Subsections 28.2 and 28.3 below.

Condition (28.7) means, in particular, that the operator $\mathfrak{f}$ acts from L_4 into L_2, a property that does not follow from condition (27.13). Nevertheless, both these conditions share a common feature: they provide quadratic estimates for the growth at infinity of the nonlinearity.

Under the conditions in Theorem 28.3 the operator A is completely continuous as an operator from L_p into L_2.

The assumption that the superposition operator $\mathfrak{f}$ acts continuously from L_p into L_2 is equivalent to the estimate (27.17) if $p = \infty$, and to the estimate (27.23) if $p < \infty$. In case $p \leqslant 4$, the estimate (27.23) follows from (28.7); this is the reason why we required $p > 4$.

The basic condition (28.5), used in all three theorems stated above, is analogous to condition (2.7) in Theorem 2.2, as well as to condition (8.7) in Theorem 8.1.

28.2. AUXILIARY RESULTS. Based on Theorem 2.2, under the conditions in Theorem 28.3 the function $\Phi(t, u)$ is compatible with the family

$$\mathfrak{F} = \mathfrak{F}(A) = \{e(t) : e(t) \in L_2, \ e(t) \in E_0(A), \ \|e\| = 1\}. \tag{28.8}$$

Consequently, for any given $\beta > 0$ there exist a positive nonincreasing function $\alpha(u)$ $(u \geqslant 0)$ and a number $c = c(\beta)$ such that for all the solutions $x(t) = \xi e(t) + h(t) \in L_2$ $(e(t) \in \mathfrak{F})$ of every inequality

$$\|h(t)\|_{L_2}^2 \leqslant -\beta \int\limits_{\Omega} \Phi[t, |x(t)|] \mathrm{d}\mu + \beta \cdot \alpha(\|x\|_{L_2}), \tag{28.9}$$

we have the estimate

$$\|x\|_{L_2} \leqslant c(\beta). \tag{28.10}$$

Inequality (28.9) is used below for

$$\beta = \frac{\mu(A; L_2)[\mu_1(A; L_2) + \mu(A; L_2)]}{\mu(A; L_2) - \mu_1(A; L_2)} > 0. \tag{28.11}$$

We also introduce the number

$$\rho = \max \left\{ c(\beta), \ \sqrt{\mu\Omega \cdot \mu(A; L_2) \sup_{t \in \Omega, \ u \geqslant 0} |\Phi(t, u)|} \right\}. \tag{28.12}$$

Using this number we define the nonlinear operator

$$\mathcal{F}x(t) = \begin{cases} f[t, x(t)] & \text{if } \|x\|_{L_2} \leqslant \rho, \\ [(\rho + 1) - \|x\|_{L_2}] f[t, x(t)] & \text{if } \rho < \|x\|_{L_2} < \rho + 1, \\ 0 & \text{if } \rho + 1 \leqslant \|x\|_{L_2}, \end{cases} \tag{28.13}$$

and the positive constant

$$\varepsilon = \frac{\mu(A; L_2)\alpha(\rho + 1)}{(\rho + 1)^2}. \tag{28.14}$$

Under the conditions in Theorem 28.3, the operator (28.13) acts continuously from the space L_p into L_2. The function $\alpha(u)$ occurring in formula (28.14) was defined at the beginning of this subsection.

We will assume that the one-sided constraint (28.4) is fulfilled for a constant k satisfying the inequality

$$k < \frac{1}{\mu(A; L_2)} + \frac{\alpha(\rho + 1)}{(\rho + 1)^2}, \tag{28.15}$$

i.e., the number ε, whose existence is stated in Theorem 28.3, is defined by formula (28.14).

LEMMA 28.1. *If $x(t) \in L_p$, then the value $\mathcal{F}x(t)$ of the operator (28.13) satisfies the estimate*

$$[x, \mathcal{F}x] \leqslant \frac{1}{\mu(A; L_2)} \|x\|_{L_2}^2 - \int_\Omega \Phi[t, |x(t)|] d\mu + \alpha(\|x\|_{L_2}). \tag{28.16}$$

Proof. We first assume that $\|x\|_{L_2} \leqslant \rho$. Then

$$\mathcal{F}x(t) = \mathfrak{f}x(t)$$

and from inequalities (28.4) and (28.15) we get

$$[x, \mathcal{F}x] \leqslant \frac{1}{\mu(A; L_2)} \|x\|_{L_2}^2 - \int_\Omega \Phi[t, |x(t)|] d\mu + \frac{\alpha(\rho + 1)}{(\rho + 1)^2} \|x\|_{L_2}.$$

The estimate (28.16) follows from this inequality combined with the obvious relations

$$\frac{\alpha(\rho + 1)}{(\rho + 1)^2} \|x\|_{L_2} \leqslant \alpha(\rho + 1) \frac{\rho^2}{(\rho + 1)^2} < \alpha(\rho + 1) \leqslant \alpha(\|x\|_{L_2})$$

Suppose now that $\|x\|_{L_2} > \rho$. Since definition (28.11) yields the estimate

$$\rho^2 \geqslant \mu\Omega \cdot \mu(A; L_2) \sup_{t \in \Omega, \, u \geqslant 0} |\Phi(t, u)|,$$

it follows that the right hand side of inequality (28.16) is positive. Therefore, it will be enough to prove inequality (28.16) under the additional assumption that $[x, \mathcal{F}x] > 0$.

If this is the case, then $\rho < \|x\|_{L_2} < \rho + 1$, and the estimate (28.16) follows from the next chain of relations:

$$[x, \mathcal{F}x] = (\rho + 1 - \|x\|_{L_2})[x, \mathcal{F}x] < [x, \mathcal{F}x] \leqslant$$

$$\leqslant \frac{1}{\mu(A; L_2)}\|x\|_{L_2}^2 - \int_{\Omega} \Phi[t, |x(t)|]d\mu + \frac{\alpha(\rho + 1)}{(\rho + 1)^2}\|x\|_{L_2}^2 \leqslant$$

$$\leqslant \frac{1}{\mu(A; L_2)}\|x\|_{L_2}^2 - \int_{\Omega} \Phi[t, |x(t)|]d\mu + \alpha(\|x\|_{L_2}).$$

Lemma 28.1 is proved.

The conditions in Theorem 28.3 imply that the operator A acts from the space L_2 into the space L_p and is completely continuous. Therefore the operator $\mathcal{F}A$ is completely continuous on L_2.

LEMMA 28.2. *For all the solutions $y(t) \in L_2$ of every equation*

$$y = \xi \mathcal{F} A y, \quad 0 \leqslant \xi \leqslant 1, \tag{28.17}$$

we have the estimates

$$\|Ay\|_{L_2} \leqslant c(\beta), \tag{28.18}$$

where $c(\beta)$ is the same constant as the one appearing in estimate (28.10) as well as the estimates

$$[Ay, y] \leqslant c_1 \stackrel{\text{def}}{=} \frac{c(\beta)}{\mu(A; L_2)} + \alpha(0) + \mu\Omega \sup_{t \in \Omega, \, u \geqslant 0} |\Phi(t, u)|. \tag{28.19}$$

Proof. Let $y \in L_2$ be a solution of equation (28.17) for a fixed $\xi \in [0, 1]$. If $\xi = 0$, then the estimates (28.18) and (28.19) are obvious. Therefore, we will assume that $\xi > 0$.

We next consider the orthogonal projections P and Q onto the subspace $E_0(A)$ and onto its orthogonal complement, respectively. The operators P and Q commute with all the operators A, A^*, A_+, A_- and $A^\square$ (see Section 27). From the strict positivity of the operator A and from the definition of the subspace $E_0(A)$, we get the relations

$$\|A^\square Qx\|_{L_2} \leqslant \sqrt{\mu_1(A; L_2)}\|Qx\|_{L_2},$$

$$\|A^\square Px\|_{L_2} \leqslant \sqrt{\mu_1(A; L_2)}\|Px\|_{L_2}, \quad x \in L_2. \tag{28.20}$$

Using them, as well as the equalities

$$\|A\|_{L_2}^2 = \left\|A^\Box A_+^{\frac{1}{2}} y\right\|_{L_2}^2 = \left\|A^\Box A_+^{\frac{1}{2}} Py\right\|_{L_2}^2 + \left\|A^\Box A_+^{\frac{1}{2}} Qy\right\|_{L_2}^2,$$

we find the estimate

$$\|Ay\|_{L_2}^2 \leqslant \mu(A; L_2)\left\|A_+^{\frac{1}{2}} Py\right\|_{L_2}^2 + \mu_1(A; L_2)\left\|A_+^{\frac{1}{2}} Qy\right\|_{L_2}^2,$$

that can be rewritten as

$$\|Ay\|_{L_2}^2 \leqslant \mu(A; L_2)\left\|A_+^{\frac{1}{2}} y\right\|_{L_2}^2 + [\mu(A; L_2) - \mu_1(A; L_2)]\left\|A_+^{\frac{1}{2}} Qy\right\|_{L_2}^2, \qquad (28.21)$$

Since

$$\left\|A_+^{\frac{1}{2}} y\right\|_{L_2}^2 = [A_+ y, y] = [Ay, y],$$

relation (28.21) implies the inequalities

$$\|Ay\|_{L_2}^2 \leqslant \mu(A; L_2)[Ay, y] + [\mu(A; L_2) - \mu_1(A; L_2)] \cdot \left\|A_+^{\frac{1}{2}} Qy\right\|_{L_2}^2 \leqslant$$

$$\leqslant \mu(A; L_2)[Ay, \xi \mathcal{F} Ay] + [\mu(A; L_2) - \mu_1(A; L_2)] \cdot \left\|A_+^{\frac{1}{2}} Qy\right\|_{L_2}^2 \leqslant$$

$$\leqslant \mu(A; L_2)[Ay, \mathcal{F} Ay] + [\mu(A; L_2) - \mu_1(A; L_2)] \cdot \left\|A_+^{\frac{1}{2}} Qy\right\|_{L_2}^2.$$

In their turn, the last inequalities together with Lemma 28.1 show that for every solution $y(t)$ of equation (28.17) we have

$$[\mu(A; L_2) - \mu_1(A; L_2)] \cdot \left\|A_+^{\frac{1}{2}} Qy\right\|_{L_2}^2 \leqslant \mu(A; L_2) \cdot \left[-\int_\Omega \Phi[t, |Ay(t)|]\mathrm{d}\mu + \alpha(\|Ay\|_{L_2})\right],$$

an inequality that can be more conveniently rewritten as

$$\left\|A_+^{\frac{1}{2}} Qy\right\|_{L_2}^2 \leqslant \frac{\mu(A; L_2)}{\mu(A; L_2) - \mu_1(A; L_2)} \cdot \left[-\int_\Omega \Phi[t, |Ay(t)|]\mathrm{d}\mu + \alpha(\|Ay\|_{L_2})\right]. \qquad (28.22)$$

Let us suppose, for a while, that $\mu_1(A; L_2) > 0$. Then the estimate (28.20) leads to

$$\|AQy\|_{L_2}^2 \leqslant \left\|A^\Box A_+^{\frac{1}{2}} Qy\right\|_{L_2}^2 \leqslant \mu_1(A; L_2)\left\|A_+^{\frac{1}{2}} Qy\right\|_{L_2}^2,$$

and therefore, from (28.22) we get the estimate

$$\|AQy\|_{L_2}^2 \leqslant \frac{\mu(A; L_2) \cdot \mu_1(A; L_2)}{\mu(A; L_2) - \mu_1(A; L_2)} \cdot \left[-\int_\Omega \Phi[t, |Ay(t)|] \mathrm{d}\mu + \alpha(\|Ay\|_{L_2}) \right].$$

The last inequality clearly implies the estimate

$$\|AQy\|_{L_2}^2 \leqslant \beta \left[-\int_\Omega \Phi[t, |Ay(t)|] \mathrm{d}\mu + \alpha(\|Ay\|_{L_2}) \right], \tag{28.23}$$

where β is the number defined by (28.23). The estimate (28.23) can be equally well derived from (28.22) if $\mu_1(A; L_2) = 0$, since in this case we have $QAy = QA_+^{1/2}y = 0$.

Inequality (28.23) may be considered as inequality (28.9) for the function $x(t) = Ay(t)$ ($h(t) = Qx(t)$, $e(t) = Px(t)/\|Px\|$, $\xi = \|Px\|_{L_2}$). Therefore, $x(t)$ satisfies the estimate (28.10), which in fact coincides with (28.18).

In order to prove (28.19) we use the relations

$$\left\| A_+^{\frac{1}{2}} y \right\|_{L_2}^2 = [Ay, y] = \xi[Ay, \mathcal{F}Ay] \leqslant [Ay, \mathcal{F}Ay],$$

from which, in view of Lemma 28.1, we conclude with the inequality

$$\left\| A_+^{\frac{1}{2}} y \right\|_{L_2}^2 \leqslant \frac{1}{\mu(A; L_2)} \|Ay\|_{L_2}^2 - \int_\Omega \Phi[t, |Ay(t)|] \mathrm{d}\mu + \alpha(\|Ay\|_{L_2}).$$

The estimate (28.19) follows from this inequality and from the already established estimate (28.18).

Lemma 28.2 is proved.

28.3. PROOF OF THEOREM 28.3. In order to complete the proof of Theorem 28.3 we will develop different arguments depending on whether condition a) or condition b) is satisfied.

To start with, assume condition a) is fulfilled. Then Lemma 28.2 implies that for every solution $y(t)$ of any equation (28.17) we have the estimate

$$\|y\|_{L_1(\Omega, \mathbb{R}^n)} \leqslant c[c(\beta)]^2 + \|b(t)\|_{L_1(\Omega, \mathbb{R}^1)} \stackrel{\text{def}}{=} c_2,$$

whence

$$\|Ay\|_{L_p} \leqslant c_2 \|A\|_{L_1 \to L_p} \stackrel{\text{def}}{=} c_3.$$

Therefore

$$\|y\|_{L_2} \leqslant \|\mathcal{F}Ay\|_{L_2} = \|f[t, Ay(t)]\|_{L_2} \leqslant \|c|Ay(t)|^2 + b(t)\|_{L_2(\Omega,\mathbb{R}^1)} \leqslant$$

$$\leqslant c\|Ay\|_{L_4} + \|b(t)\|_{L_2(\Omega,\mathbb{R}^1)} \leqslant cc_3c_4 + \|b(t)\|_{L_2(\Omega,\mathbb{R}^1)} \stackrel{\text{def}}{=} c_5,$$

where c_4 denotes the norm of the inclusion operator from the space L_p into the space L_4, an operator that clearly is continuous for $p > 4$.

Consequently, condition a) leads to the estimate

$$\|y\|_{L_2} \leqslant c_5. \tag{28.24}$$

Suppose next that condition b) is fulfilled. Then estimate (28.19) implies that for every solution $y(t)$ of any equation (28.17) we have

$$\|Ay\|_{L_p} \leqslant \|A^{\square}\|_{L_2 \to L_p} \left\|A_+^{\frac{1}{2}}\right\|_{L_2} \leqslant c_1\|A^{\square}\|_{L_2 \to L_p} \stackrel{\text{def}}{=} c_6.$$

Since the superposition operator $\mathfrak{f}$ acts from L_p into L_2, then (see [Krasnoselskii, M. A., 1951])

$$\|y\|_{L_2} \leqslant \|\mathcal{F}Ay\|_{L_2} \leqslant \|\mathfrak{f}Ay\|_{L_2} \leqslant \sup_{x(t) \in L_p, \, \|x\|_{L_p} \leqslant c_6} \|\mathfrak{f}x\|_{L_2} \stackrel{\text{def}}{=} c_7 < \infty.$$

Thus, whatever condition in Theorem 28.3 is satisfied, we get the general estimate

$$\|y\|_{L_2} \leqslant \max\{c_5, c_7\},$$

for all solutions of equations (28.17). Based on the Leray-Schauder Principle, it follows that the equation $y = \mathcal{F}Ay$ has at least one solution $y_* \in L_2$. Moreover, every such solution satisfies the estimate $\|Ay_*\|_{L_2} \leqslant c(\beta)$ and, consequently, the estimate $\|Ay_*\|_{L_2} \leqslant \rho$. Therefore, $\mathcal{F}Ay_* = \mathfrak{f}Ay_*$, that is, the function y_* is a solution of the equation $y = f(t, Ay)$. It remains to observe that the function $x_* = Ay_* \in L_p$ is a solution of equation (28.1).

Theorem 28.3 is proved.

28.4. REMARKS.

A. The theorems stated so far in this section are in many respects analogous to Theorem 8.1. Under the conditions in Theorems 28.1–28.3, likewise under the conditions in Theorem 8.1, there is no way to get an a priori norm estimate for the solutions of equation (28.1). Suitable examples can be easily constructed following the examples discussed in Subsection 8.3.

At the same time it is possible to formulate an analogue of Theorem 8.2 asserting that condition (28.5) is "almost necessary".

B. Using an already described procedure, from Theorems 28.1–28.3 we may derive similar results for equations with potentially positive operators. To be more specific, a potentially positive operator $A : H \to H$ is called *potentially strictly positive* if the space H splits into a direct sum of two orthogonal invariant subspaces of A, denoted by E_0 and E_1, such that

$$\gamma_H(A)[Ax, Ax] = [x, Ax], \quad x \in E_0, \tag{28.25}$$

$$\gamma'_H(A)[Ax, Ax] = [x, Ax], \quad x \in E_1, \tag{28.26}$$

where $\dim E_0 < \infty$, $\gamma_H(A) > \gamma'_H(A)$, and $\operatorname{Ker} A \subset E_1$.

The simplest examples of potentially strictly positive operators are provided by any self-adjoint completely continuous operator with a finite number of negative eigenvalues.

THEOREM 28.4. *Let A be a potentially strictly positive operator on the space L_2 that acts as a completely continuous operator from L_1 into L_2. Suppose the nonlinearity $f(t, x)$ satisfies condition (27.13). In addition, assume that equality (28.5) holds true for every $R > 0$ and every $u_* \geqslant u_0$. Then there exists $\varepsilon > 0$ such that the one-sided constraint (28.4) with a constant k satisfying the inequality*

$$k < \gamma_H(A) + \varepsilon, \tag{28.27}$$

implies the existence of at least one solution $x(t) \in L_2$ of equation (28.1).

Assertions analogous to Theorem 28.4 on the solvability of equations with potentially strictly positive operators are also true under conditions similar to the ones in Theorems 28.2 and 28.3.

In the study of equations with strictly positive or potentially strictly positive operators we may successfully use the Schaefer-M. A. Krasnoselskii method.

§29. Two-point boundary value problems (the quasilinear case)

29.1. THE MAIN THEOREM. The objective of this section is to illustrate the theory of equations with positive operators by applying it to a few concrete nonlinear problems with self-adjoint linear part. To begin with, let us once more consider the two-point boundary value problem

$$\begin{cases} Lx(t) \overset{\text{def}}{=} x'' + p(t)x' + q(t)x = f(t,x), \\ x(0) = x(T) = 0. \end{cases} \tag{29.1}$$

We assume that the coefficients $p(t)$ and $q(t)$ are continuous and that the function $f(t,x)$ is jointly continuous in its variables $t \in [0, T]$ and $x \in \mathbb{R}^1$. As it has been already noticed (see Sections 7 and 10), under natural conditions problem (29.1) is equivalent to the operator equation

$$x(t) = A\mathfrak{f}x(t). \tag{29.2}$$

Equation (29.2) is considered in the space L_2 of all square integrable functions with respect to the measure

$$\mu G = \int\limits_G \exp\left\{ \int\limits_0^s p(\tau)\mathrm{d}\tau \right\} \mathrm{d}s \tag{29.3}$$

The usual norm and inner product on this space are denoted by $\| \cdot \|_\mu$ and $[\cdot, \cdot]_\mu$, respectively. The operator A — called the two-point boundary value problem operator — acts on L_2 and is completely continuous and self-adjoint. Moreover, all its eigenvalues are simple. If the interval $[0, T]$ is a nonoscillation interval for the differential operator $Lx(t)$, then the operator A is negative definite.

We let Λ denote the quantity

$$\left[\sup_{\lambda \neq 0, \ \lambda \in \sigma(A)} (\lambda^{-1}) \right]^{-1}.$$

If the operator A is negative semidefinite, then $\Lambda = \inf\{\lambda : \lambda \in \sigma(A)\}$; if the operator A has positive eigenvalues, too (their number is always finite), then we obviously get $\Lambda = \sup\{\lambda : \lambda > 0, \ \lambda \in \sigma(A)\}$.

The basic solvability condition for equation (29.1) is

$$xf(t,x) \geqslant kx^2 + \Phi(|x|), \quad t \in \Omega, \ x \in \mathbb{R}^1, \tag{29.4}$$

where $\Phi(u)$ is a function in one of the classes $\mathfrak{N}(u_0)$.

THEOREM 29.1. *Suppose that*

$$\int\limits_{u_0}^\infty \Phi(u)\mathrm{d}u = \infty. \tag{29.5}$$

Then there exists $\varepsilon_0 > 0$ such that the estimate (29.4), with a constant k satisfying the inequality

$$k \geqslant \frac{1}{\Lambda} - \varepsilon_0, \tag{29.6}$$

implies the existence of at least one twice continuously differentiable solution of equation (29.1).

Theorem 29.1, as well as all the other theorems stated along this section, follows from the results of Section 28.

The just formulated theorem can be extended in many ways: for instance, we may assume that the coefficients of the differential operator and the nonlinearity are no longer continuous, or that the function Φ in estimate (29.4) depends on t. We can also consider the problem (29.1) in spaces of vector-valued functions.

Inequality (29.4) as a solvability condition for the two-point boundary value problem (29.1) was used in [Krasnoselskii, A. M., 1980a] in the case when $k\Lambda = 1$.

If problem (29.1) has the form

$$x'' = f(t, x), \quad x(0) = x(\pi) = 0,$$

then $\Lambda = -1$.

29.2. THE DIRICHLET PROBLEM. In Section 10 we dealt with two examples of Dirichlet problems for the equation

$$\Delta x = f(t, x) \tag{29.7}$$

(Δ — the Laplace operator). Equation (29.7) was considered on the square

$$\Omega = \{t = (t_1, t_2) : 0 \leqslant t_1, t_2 \leqslant \pi\} \tag{29.8}$$

and on domains $\Omega \subset \mathbb{R}^2$ with a smooth boundary. For our next purposes, the basic constraint on the nonlinearity is of the form (29.4).

THEOREM 29.2. *Suppose that*

$$\lim_{u \to \infty} \frac{u\Phi(u)}{\ln u} = \infty.$$

Then there exists $\varepsilon_0 > 0$ such that the estimate (29.4) with a constant k satisfying the inequality $k \geqslant -1 - \varepsilon_0$ implies the existence of at least one solution of the problem (29.7) vanishing on the boundary of the square (29.8).

For the Dirichlet problem on domains with a smooth boundary the condition imposed on the function $\Phi(u)$ is given by (29.5).

29.3. FOURTH ORDER EQUATIONS. We next consider the problem

$$x^{\mathrm{IV}} = f(t, x), \tag{29.9}$$

$$x(0) = x(1) = x'(0) = x'(1) = 0, \tag{29.10}$$

where $f(t, x)$ is a continuous function. The eigenvalues of the differential operator $Lx(t) = x^{\mathrm{IV}}(t)$ with the boundary conditions (29.10) are numbers of the form λ^4, where λ is a nonzero root of the transcendental equation

$$\cos \lambda \cdot \mathrm{ch}\, \lambda = 1. \tag{29.11}$$

In (29.11) above $\mathrm{ch}\, \lambda$ denotes the hyperbolic cosine,

$$\mathrm{ch}\, \lambda = \frac{1}{2}(e^{\lambda} + e^{-\lambda}).$$

The smallest positive root λ_0 of equation (29.11) equals ≈ 4.730. The eigenfunctions of the differential operator L are linear combinations of the functions $\sin \lambda t$, $\cos \lambda t$, $\exp(\lambda t)$, $\exp(-\lambda t)$; every root λ of equation (29.11) corresponds to exactly one such function (up to a constant multiplier). For the function $e_0(t)$ corresponding to the eigenvalue λ_0^4 of the operator L we have the estimates

$$c_2 t^2 (1 - t)^2 \leqslant e(t) \leqslant c_1 t^2 (1 - t)^2 \quad (t \in \Omega = [0,\, 1]);$$

from these estimates we get the following estimates of the distribution $\lambda(\delta; e_0)$ of the function e_0:

$$c_3 \sqrt{\delta} \leqslant \chi(\delta; e_0) \leqslant c_4 \sqrt{\delta}. \tag{29.12}$$

THEOREM 29.3. *Suppose that*

$$\int\limits^{\infty} \frac{\Phi(u)}{\sqrt{u}}\, du = \infty. \tag{29.13}$$

Then there exists $\varepsilon > 0$ such that the estimate

$$xf(t, x) \leqslant \left(\lambda_0^4 + \varepsilon_0\right) x^2 - \Phi(|x|), \quad t \in \Omega,\ x \in \mathbb{R}^1$$

implies the existence of at least one solution of the problem (29.9), (29.10).

§30. Potential positivity
of the periodic problem operator

30.1. POSITIVITY CONDITIONS.   The main goal of this section is to answer some specific questions about the operator

$$Au(t) = \int_0^T G(t - s; T)u(s)\mathrm{d}s \qquad (30.1)$$

of the T-periodic problem (see Section 7) for the linear link with the transfer function

$$W(p) = \frac{M(p)}{L(p)}. \qquad (30.2)$$

In the last formula as well as further on we let $M(p)$ and $L(p)$ denote the polynomials (7.20) and (7.21), respectively, and we assume that $l = \deg L > m = \deg M$. The operator (30.1) is well-defined in case none of the numbers

$$\omega_k \mathrm{i} = 2k\pi T^{-1}\mathrm{i}, \quad k = 0, \pm 1, \pm 2, \ldots \qquad (30.3)$$

is a root of the polynomial $L(p)$, and it takes each integrable function $u(t)$ into a T-periodic solution $x(t) = Au(t)$ of the equation

$$L\left(\frac{\mathrm{d}}{\mathrm{d}t}\right) x(t) = M\left(\frac{\mathrm{d}}{\mathrm{d}t}\right) u(t). \qquad (30.4)$$

The spectral decomposition of the operator (30.1) is explicitly given by formula (7.29) in Subsection 7.4. In what follows we will freely use all the notations introduced in that subsection. For the sake of convenience we next recall a few basic properties of the operator (30.1).

To start with, we recall that the periodic problem operator (30.1) sends every integrable on $[0, T]$ function $u(t)$ into a function $x(t)$ whose derivative of order $l - m - 1$ is absolutely continuous and whose derivative of order $l - m$ exists almost everywhere and is integrable on $[0, T]$. The operator (30.1) is completely continuous and normal on the space L_2. It also acts continuously from L_1 into C. If $l = m + 1$, then the operator $A : L_1 \to C$ is not completely continuous; if $l > m + 1$, then the operator $A : L_1 \to C$ is always completely continuous.

The spectrum $\sigma(A)$ of the operator (30.1) consists of the numbers $W(\omega_k \mathrm{i})$ $(k = 0, \pm 1, \pm 2, \ldots)$ and zero. Its kernel $G(t - s; T)$ is defined by the unit impulse response $G(\tau; T)$ (see [Rosenvasser, 1969]) of the linear link with the transfer function (30.2).

Let us introduce the following notations:

$$M_1(\omega) \overset{\text{def}}{=} \operatorname{Re} M(\omega i), \quad M_2(\omega) \overset{\text{def}}{=} \operatorname{Im} M(\omega i),$$

$$L_1(\omega) \overset{\text{def}}{=} \operatorname{Re} L(\omega i), \quad L_2(\omega) \overset{\text{def}}{=} \operatorname{Im} L(\omega i),$$

(30.5)

$$\Pi(\omega) \overset{\text{def}}{=} \operatorname{Re}[M(\omega i)\overline{L(\omega i)}] = M_1(\omega)L_1(\omega) + M_2(\omega)L_2(\omega). \tag{30.6}$$

All the polynomials (30.5) and (30.6) have real coefficients. Moreover, the polynomials $M_1(\omega)$, $L_1(\omega)$ and $\Pi(\omega)$ contain only even powers of ω, whereas the polynomials $M_2(\omega)$ and $L_2(\omega)$ include only odd powers of ω. The degree of the polynomial (30.6) is denoted by $2\pi(M, L)$.

In view of our next purposes, we also need the rational function

$$R(\omega, W) = \frac{\Pi(\omega)}{M_1^2(\omega) + M_2^2(\omega)} \equiv \frac{\Pi(\omega)}{|M(\omega i)|^2}, \quad -\infty < \omega < \infty. \tag{30.7}$$

The function (30.7) is even, and its behavior at infinity depends upon the numbers m and $\pi(M, L)$. If the difference $l - m$ is an even number then always $\pi(M, L) > m$; if $l = m + 1$, then $\pi(M, L) = m$; if $l - m$ is an odd number distinct from 1, then "almost always" $\pi(M, L) > m$ (however, there are cases when $\pi(M, L) < m!$). We conclude the previous discussion with the obvious remarks that if $\pi(M, L) \leqslant m$, then the function (30.7) is bounded at infinity, and if $\pi(M, L) > m$, then $R(\omega, W) \to \infty$ as $\omega \to \infty$.

THEOREM 30.1. *The periodic problem operator* (30.1) *is positive if and only if*

$$\inf_{k=0,1,\ldots;\ M(\omega_k i)\neq 0} R(\omega_k, W) > 0. \tag{30.8}$$

If condition (30.8) *is fulfilled, the positivity coefficient* $\mu(A; L_2)$ *of the operator* (30.1) *is given by the equality*

$$\mu(A; L_2) = \sup_{k=0,1,\ldots;\ M(\omega_k i)\neq 0} [R(\omega_k, W)]^{-1}. \tag{30.9}$$

For the proof of Theorem 30.1 it is enough to observe that

$$\inf_{\lambda\in\sigma(A),\ \lambda\neq 0} \operatorname{Re}(\lambda^{-1}) = \inf_{k=0,1,\ldots;\ M(\omega_k i)\neq 0} R(\omega_k, W)$$

and then to use Corollary 26.1 to Theorem 26.1.

Condition (30.8) can be fulfilled but in case

$$\pi(M, L) \geqslant m. \tag{30.10}$$

If inequality (30.10) is true, then (30.8) is equivalent to the condition

$$\Pi(\omega_k) > 0, \quad k = 0, 1, \ldots; \quad M(\omega_k i) \neq 0. \tag{30.11}$$

Thus, the operator (30.1) is positive if and only if both the conditions (30.10) and (30.11) hold true.

30.2. STRICT POSITIVITY CONDITIONS. We proceed by defining the number

$$\bar{\mu}(A) = \lim_{\omega \to \infty} [R(\omega, W)]^{-1}. \tag{30.12}$$

If $\pi(M, L) > m$, then $\bar{\mu}(A) = 0$. We also notice that the number (30.12) is finite in case the operator A is positive.

THEOREM 30.2. *Suppose the operator A is positive. Then it is strictly positive if and only if*

$$\bar{\mu}(A) < \mu(A; L_2). \tag{30.13}$$

The simplest case in which condition (30.13) is fulfilled occurs when $\pi(M, L) > m$; actually in this case the positivity of the periodic problem operator implies its strict positivity.

Proof. We first reformulate the definition of strict positivity in the specific case of operator (30.1). Since under our assumptions the operator A on L_2 is completely continuous, normal, and positive, we can use Theorem 26.3.

We get the following criterion. The positive operator (30.1) is strictly positive if and only if there exists a finite subset

$$K(A) = \{k_1, \ldots, k_r\} \tag{30.14}$$

of the set $\{k = 0, 1, \ldots : M(\omega_k i) \neq 0\}$, such that

$$\mu(A; L_2) = [R(\omega_k, W)]^{-1}, \quad k \in K(A), \tag{30.15}$$

and

$$\mu_1(A; L_2) = \sup_{\substack{k=0,1,\ldots; \ M(\omega_k i) \neq 0 \\ k \notin K(A)}} [R(\omega_k, W)]^{-1}. \tag{30.16}$$

Suppose now that the positive operator A satisfies condition (30.13). Then the supremum in equality (30.9) is attained only for a finite number of values of k in the set $\{k = 0, 1, \ldots : M(\omega_k i) \neq 0\}$, and these values yield the subset $K(A)$. Thus equality (30.15) is true. Assume for a while that $\mu_1(A; L_2) = \mu(A; L_2)$. Then there exists a sequence $k(n) \to \infty$ of natural numbers such that $M(\omega_{k(n)} i) \neq 0$, $k(n) \notin K(A)$, and

$$\lim_{n \to \infty} [R(\omega_{k(n)}, W)]^{-1} = \mu(A; L_2).$$

The last equality clearly contradicts condition (30.13). Consequently the equality $\mu_1(A; L_2) = \mu(A; L_2)$ is false, hence relation (30.15) is true.

To complete the proof of Theorem 30.2, let us suppose that the operator (30.1) is strictly positive, i.e., there exists a finite collection of integers (30.14) for which both the relations (30.15) and (30.16) hold true. Then for sufficiently large values of ω (more precisely for those values for which the rational function (30.7) is monotonous) we have the inequality

$$R(\omega, E) \leqslant \mu_1(A; L_2),$$

that clearly implies condition (30.13).

The theorem is proved.

30.3. EXAMPLES.

A. Let $l = 1$ and $m = 0$, and consider the link with the transfer function

$$W(p) = \frac{b_0}{p + a_1},$$

where $b_0 \neq 0$ and $a_1 \neq 0$. For this link we have $R(\omega, W) \equiv a_1 b_0^{-1}$. Therefore the operator (30.1) is positive for every $T > 0$ if and only if $a_1 b_0 > 0$. If this is the case, then $\mu(A; L_2) = a_1^{-1} b_0$. In view of Theorem 30.2 we easily conclude that the operator (30.1) is never strictly positive.

B. Suppose next that $l = 2$ and $m = 0$, and consider the link with the transfer function

$$W(p) = \frac{b_0}{p^2 + a_1 p + a_2}.$$

For this link we have $R(\omega, W) = (a_2 - \omega^2) b_0^{-1}$. Therefore the periodic problem operator is positive for every $T > 0$ such that $\omega_k^2 + i a_1 \omega_k + a_2 \neq 0$ $(k = 0, 1, \ldots)$, if and only if $a_2 < 0$ and $b_0 < 0$. Of course, if $a_2 < 0$, then $-\omega_k^2 + i a_1 \omega_k + a_2 \neq 0$ $(k = 0, 1, \ldots)$. In case the operator (30.1) is positive, its positivity coefficient is given by $\mu(A; L_2) = a_2^{-1}b_0$. Under the same assumptions we get $\bar{\mu}(A) = 0 \neq \mu(A; L_2)$. Consequently, the inequalities $a_2 < 0$ and $b_0 < 0$ provide necessary and sufficient conditions not only for the positivity of the periodic problem operator, but also for its strict positivity.

C. Suppose now that $l = 2$ and $m = 1$, and consider the link with the transfer function

$$W(p) = \frac{b_0 p + b_1}{p^2 + a_1 p + a_2},$$

where $b_0 \neq 0$. For this link we have

$$R(\omega, W) = \frac{(b_0 a_1 - b_1)\omega^2 + b_1 a_2}{b_0^2 \omega^2 + b_1^2}. \tag{30.17}$$

Let us first assume that $b_1 \neq 0$. Then the operator (30.1) is positive if and only if the following two inequalities

$$b_0 a_1 - b_1 > 0, \quad b_1 a_2 > 0 \tag{30.18}$$

are fulfilled. Suppose both the inequalities (30.18) are true. In this case the function $R(\omega, W)$ is continuous for $\omega \geqslant 0$ and it is either strictly increasing, or strictly decreasing, or constant, according to the sign of the number

$$\vartheta(W) = b_0^2 b_1 a_2 - b_0 b_1^2 a_1 + b_1^3.$$

Specifically, if besides inequalities (30.18) we have $\vartheta(W) < 0$ (for instance, if $b_0 = a_2 = -1$, $b_1 = -2$, $a_1 = 0$), then the operator (30.1) is not only positive, but also strictly positive, as a consequence of the following relations

$$\mu(A; L_2) = [R(0, W)]^{-1} = b_1 a_2^{-1},$$

$$\mu_1(A; L_2) = [R(\omega_1, W)]^{-1} = \frac{b_0^2 \omega_1^2 + b_1^2}{(b_0 a_1 - b_1)\omega_1^2 + b_1 a_2},$$

$$\bar{\mu}(A) = [R(\infty, W)]^{-1} = \frac{b_0^2}{b_0 a_1 - b_1} < \mu(A; L_2).$$

If $\vartheta(W) = 0$ (for instance, if $b_0 = -1$, $a_2 = -4$, $b_1 = -2$, $a_1 = 0$), then $\mu(A; L_2) = b_1 a_2^{-1}$ and all the eigenvalues of the operator (30.1) lie on the circle

$$\left| \lambda - \frac{1}{2}\mu(A; L_2) \right| = \frac{1}{2}\mu(A; L_2). \tag{30.19}$$

Consequently, the periodic problem operator does not satisfy the strict positivity condition in this case.

Finally, if $\vartheta(W) > 0$ (for instance, if $b_0 = a_2 = -1$, $b_1 = -\frac{1}{2}$, $a_1 = 0$), then

$$\mu(A; L_2) = \bar{\mu}(A) = [R(\infty, W)]^{-1} = \frac{b_0^2}{b_0 a_1 - b_1},$$

but the circle (30.19) no longer contains nonzero eigenvalues of the operator (30.1). The cases $\vartheta(W) = 0$ and $\vartheta(W) > 0$ were mentioned separately to point out two different situations related to the spectrum of a positive and normal operator when the strict positivity fails.

So far we dealt with the case $b_1 \neq 0$. If $b_1 = 0$, then everything is immediately at hand. We have $M(0) = 0$ and $R(\omega, W) \equiv a_1 b_0^{-1}$ ($\omega \neq 0$). Therefore the operator (30.1) is positive if and only if $a_1 b_0 > 0$. In addition, $\mu(A; L_2) = b_0 a_1^{-1}$, and the strict positivity condition is not fulfilled.

D. In some of our previous examples we considered linear links with transfer functions for which the periodic problem operator was either simultaneously positive and strictly positive for every $T > 0$, or both these properties were missing for each $T > 0$. In order to get a better insight into the general situation, let us consider the link with the transfer function

$$W_\varepsilon(p) = \frac{p^3 + 1}{p^2(p^2 + 4) + \varepsilon(p^3 + 1)}, \tag{30.20}$$

where ε is a positive number. The denominator of $W_\varepsilon(p)$ is different from zero for all $p = \omega i$ ($-\infty < \omega < \infty$), hence the periodic problem operator for the linear link with the transfer function (30.20) exists for every $T > 0$. If $\varepsilon > 0$ is sufficiently large, then the periodic problem operator is strictly positive for all $T > 0$. In case $\varepsilon > 0$ is small, it turns out that the periodic problem operator is strictly positive for small values of $T > 0$, whereas for large values of T it is not even positive.

30.4. POTENTIAL POSITIVITY FOR THE PERIODIC PROBLEM OPERATOR. We return to the study of the periodic problem operator (30.1) for links with the transfer functions of the general form (30.2). It turns out that this operator is always potentially positive on L_2 (either from below, or from above, or both from below and above).

The numbers (26.17) and (26.18) corresponding to the operator (30.1) are given by

$$\gamma_H = \inf_{\substack{k=0,1,\ldots;\ M(\omega_k i) \neq 0}} R(\omega_k, W),$$

$$\gamma_B = \sup_{\substack{k=0,1,\ldots;\ M(\omega_k i) \neq 0}} R(\omega_k, W). \tag{30.21}$$

If

$$\pi(M, L) > m, \tag{30.22}$$

then but one of the numbers (30.21) is finite; if $\pi(M, L) \leqslant m$, then both those numbers are finite. The arguments developed in Subsection 26.4 lead to the next result.

THEOREM 30.3. *The operator* (30.1) *is potentially positive if and only if the number* γ_H *is finite. If this is the case, then the operator* (30.1) *satisfies the inequality*

$$\gamma_H[Ax, Ax] \leqslant [x, Ax], \quad x \in L_2. \tag{30.23}$$

From Theorem 30.3 it follows that the operator (30.1) is potentially positive from above if and only if the number γ_B is finite. In this case we get

$$\gamma_B[Ax, Ax] \geqslant [x, Ax], \quad x \in L_2. \tag{30.24}$$

If $\pi(M, L) \leqslant m$, then both the numbers (30.21) are finite and consequently

$$\gamma_H[Ax, Ax] \leqslant [x, Ax] \leqslant \gamma_B[Ax, Ax], \quad x \in L_2. \tag{30.25}$$

30.5. THE AUXILIARY OPERATOR. In our previous investigations of equations with positive or strictly positive operators we used the auxiliary operator (26.19). In the particular case of the operator (30.1), the operator (26.19) is given by

$$A^\square u(t) = \sum_{k=0,1,\dots;\ M(\omega_k i) \neq 0} [R(\omega_k, W)]^{-\frac{1}{2}} U_k P_k u(t), \quad u(t) \in L_2, \tag{30.26}$$

where U_k are the operators (7.29) and P_k are the projections (7.27) corresponding to the subspaces Π_k.

THEOREM 30.4. *Suppose the positivity condition* (30.8) *for the operator* (30.1) *as well as* (30.22) *hold true. Then the operator* (30.26) *acts from* L_2 *into* L_∞ *and is completely continuous.*

Proof. Set

$$K(t, s) = \frac{1}{T} + \frac{2}{T} \sum_{k=1}^{\infty} \frac{\cos \omega_k(t - s)}{(k + 1)^2}. \tag{30.27}$$

Since the series in the right hand side of formula (30.27) is uniformly convergent, the function $K(t, s)$ is jointly continuous in its variables. Therefore, the linear integral operator

$$Ku(t) = \int_0^T K(t, s)u(s)\mathrm{d}s$$

acts from the space L_1 into the space C and is completely continuous. Its values are given by

$$Ku(t) = \sum_{k=0}^{\infty}(k+1)^{-2}P_k u(t), \quad u(t) \in L_1, \tag{30.28}$$

where $P_k u$ $(u(t) \in L_1)$ stand for the operators

$$P_0 u(t) \equiv \frac{1}{T}\int_0^T u(s)\mathrm{d}s, \quad P_k u(t) \equiv \frac{2}{T}\int_0^T \cos\omega_k(t-s)u(s)\mathrm{d}s$$

(i.e., formulas (7.27) make sense for $u(t) \in L_1$). The representation (30.28) clearly implies that the operator K is self-adjoint, positive definite, and completely continuous on the space L_2. The self-adjoint and positive definite square root of the operator (30.28) is given by

$$K^{\frac{1}{2}}u(t) = \sum_{k=0}^{\infty}(k+1)^{-1}P_k u(t), \quad u(t) \in L_2. \tag{30.29}$$

Since the operator (30.27) acts from L_1 into C and is completely continuous, a well-known theorem due to M. A. Krasnoselskii (see, for instance, [Krasnoselskii *et al.*, 1966]) implies that the operator (30.29) acts from L_2 into C and is completely continuous.

Let us represent the operator (30.26) as

$$A^{\square} = K^{\frac{1}{2}}B,$$

where

$$Bu(t) = \sum_{k=0,1,\ldots;\ M(\omega_k i)\neq 0}(k+1)[R(\omega_k,W)]^{-\frac{1}{2}}U_k P_k u(t), \quad u(t) \in_2. \tag{30.30}$$

Since for $\pi(M,L) > m$ the sequence

$$\alpha_k = (k+1)[R(\omega_k,W)]^{-\frac{1}{2}} =$$

$$= \sqrt{\frac{(k+1)^2\left[M_1^2(\omega_k) + M_2^2(\omega_k)\right]}{\Pi(\omega_k)}} \simeq \mathrm{const}\sqrt{\frac{k^2(k^{2m}+\cdots)}{k^{2\pi(M,L)}+\cdots}}.$$

is bounded, it follows that the operator (30.30) acts on the space L_2 and is bounded, too. Thus, the operator $A^{\square}$ is represented as the composition of the bounded operator B on L_2 and the completely continuous operator $K^{1/2}: L_2 \to C$.

Theorem 30.4 is proved.

In case the number γ_H is finite, we may define the operator $A_\gamma^\square$ for every $\gamma < \gamma_H$. If condition (30.22) is satisfied, then each such operator acts from L_2 into C and is completely continuous.

30.6. EXISTENCE THEOREMS. This subsection continues the study of forced periodic oscillations for systems whose dynamics is described by the equation

$$L\left(\frac{d}{dt}\right)x = M\left(\frac{d}{dt}\right)f(t,x), \tag{30.31}$$

i.e., the study of T-periodic solutions of such equations. We assume that the function $f(t,x)$ satisfies Caratheodory condition and is periodic in t with period $T > 0$, i.e.,

$$f(t,x) \equiv f(t+T,x), \quad -\infty < t, x < \infty. \tag{30.32}$$

In all the theorem that follow we will use one-sided estimates of the form

$$xf(t,x) \leqslant kx^2 + b(t), \quad 0 \leqslant t \leqslant T, \ x \in \mathbb{R}, \ b(t) \in L_1, \tag{30.33}$$

or

$$xf(t,x) \geqslant kx^2 - b(t), \quad 0 \leqslant t \leqslant T, \ x \in \mathbb{R}, \ b(t) \in L_1. \tag{30.34}$$

The coefficient k in (30.33) and (30.34) may have an arbitrary sign; the function $b(t)$ is nonnegative.

The next two results on the existence of T-periodic solutions of equation (30.31) deal separately with the cases when inequality (30.22) holds true or fails.

THEOREM 30.5. *Assume that inequality (30.32) fails and that*

$$|f(t,x)| \leqslant cx^2 + c(t), \quad 0 \leqslant t \leqslant T, \ x \in \mathbb{R}, \ c(t) \in L_1. \tag{30.35}$$

In addition, suppose one of the following condition is fulfilled:
  a) the constraint (30.33) holds true for a constant k satisfying $k < \gamma_H$;
  b) the constraint (30.34) holds true for a constant k satisfying $k > \gamma_B$.
Then there exists at least one T-periodic solution of equation (30.31).

Theorem 30.5 is a straightforward consequence of Theorem 27.6. The T-periodic solution whose existence is quaranteed by this theorem is continuous together with all its derivatives up to the order $l - m - 1$.

THEOREM 30.6. *Assume that inequality* (30.22) *is true and that the function*

$$\alpha(t,r) \overset{\text{def}}{=} \sup_{|x| \leqslant r} |f(t,x)|, \quad 0 \leqslant t \leqslant T, \ r \geqslant 0, \tag{30.36}$$

is square-integrable on the interval $[0, T]$ *for every* $r > 0$. *In addition, suppose one of the following two conditions is fulfilled:*

a) γ_H *is finite and the constraint* (30.33) *holds true for a constant* k *satisying* $k < \gamma_H$;

b) γ_B *is finite and the constraint* (30.34) *holds true for a constant* k *satisfying* $k > \gamma_B$.

Then there exists at least one T*-periodic solution of equation* (30.31).

Theorem 30.6 is a direct corollary to Theorems 27.7 and 30.4.

In case the periodic problem operator is strictly positive (or potentially strictly positive), equation (30.31) can be successfully studied using the results established in Section 28.

To start with, let $\Phi(t, u) \in \mathfrak{N}(u_0)$ be a given function that does not depend on t for $u \geqslant u_0$ and $t \in \Omega_0 \subset [0, T]$, where $\mu\Omega_0 > 0$, i.e.,

$$\Phi(t, u) \equiv \Phi(u), \quad t \in \Omega_0, \ u \geqslant u_0.$$

In the sequel we will impose conditions analogous to the already considered constraints (28.4), (30.33), and (30.34), namely,

$$xf(t, x) \leqslant kx^2 - \Phi(t, |x|), \quad 0 \leqslant t \leqslant T, \ x \in \mathbb{R}, \tag{30.37}$$

and

$$xf(t, x) \geqslant kx^2 + \Phi(t, |x|), \quad 0 \leqslant t \leqslant T, \ x \in \mathbb{R}. \tag{30.38}$$

We also introduce a number κ defined by

$$\kappa = \begin{cases} [2\pi(M, L) - 1]^{-1}, & \text{if } \pi(M, L) > 0 \text{ is even}, \\ [2\pi(M, L) - 2]^{-1}, & \text{if } \pi(M, L) > 1 \text{ is odd}. \end{cases} \tag{30.39}$$

This number (30.39) is of no use for $\pi(M, L) = 0$ or $\pi(M, L) = 1$.

THEOREM 30.7. *Suppose condition* (30.22) *is fulfilled, and in case* $\pi(M, L) > 1$ *assume that*

$$\lim_{u \to \infty} u^\kappa \Phi(u) = \infty, \tag{30.40}$$

where κ is defined as in (30.39) above. In addition, suppose that the function (30.36) is square-integrable on $[0, T]$ for every $r > 0$. Then there exists $\varepsilon > 0$ such that each of the following conditions

a) γ_H is finite and the constraint (30.37) holds true for a constant k satisfying $k < \gamma_H + \varepsilon$;

b) γ_B is finite and the constraint (30.38) holds true for a constant k satisfying $k > \gamma_B - \varepsilon$;

implies the existence of at least one T-periodic solution of equation (30.31).

Theorem 30.7. will be proved in the next subsection.

Recall that by $\bar{\mu}(A)$ we denoted the number (30.12). If $\pi(M, L) \leqslant m$, then $\bar{\mu}(A) \neq 0$. In this case we let $\bar{\gamma}$ denote the number $[\bar{\mu}(A)]^{-1}$, i.e.,

$$\bar{\gamma} = \lim_{\omega \to \infty} R(\omega, W). \tag{30.41}$$

THEOREM 30.8. *Suppose condition (30.22) fails, i.e., $\pi(M, L) \leqslant m$, and in case $m > 0$ assume that condition (30.40) is fulfilled, where*

$$\kappa = \begin{cases} [2m - 1]^{-1}, & \text{if } m \text{ is even,} \\ [2m - 2]^{-1}, & \text{if } m > 1 \text{ is odd.} \end{cases} \tag{30.42}$$

In addition, suppose that the function $f(t, x)$ satisfies the estimate (30.35). Then there exists $\varepsilon > 0$ such that each of the following conditions

a) $\gamma_H \neq \bar{\gamma}$ and the constraint (30.38) holds true for a constant k satisfying $k < \gamma_H + \varepsilon$;

b) $\gamma_B \neq \bar{\gamma}$ and the constraint (30.38) holds true for a constant k satisfying $k > \gamma_B - \varepsilon$;

implies the existence of at least one T-periodic solution of equation (30.31).

Theorem 30.8 will be also proved in the next subsection.

Let us observe that if $\gamma_H \neq \bar{\gamma}$, then clearly $\gamma_H < \bar{\gamma}$; analogously, if $\gamma_B \neq \bar{\gamma}$, then $\gamma_B > \bar{\gamma}$.

It is important to notice that in many concrete cases condition (30.40) can be weakened. To illustrate the point we discuss below the case when estimate (30.37) is used. Let β denote the number of all distinct roots ω of the form (30.30) (without counting the multiplicities) of the equation

$$\gamma_H = R(\omega, W). \tag{30.43}$$

The following four inequalities are proved in Subsection 30.7:

i) $\beta \leqslant 2\pi(M,L)$ whenever $\pi(M,L)$ is even and $\pi(M,L) > m$;

ii) $\beta \leqslant 2\pi(M,L) - 1$ whenever $\pi(M,L)$ is odd and $\pi(M,L) > m$;

iii) $\beta \leqslant 2m$ whenever m is even and $m \geqslant \pi(M,L)$;

iv) $\beta \leqslant 2m - 1$ whenever m is odd and $m \geqslant \max\{\pi(M,L), 1\}$.

If $\beta = 1$ (i.e., 0 is the only root of equation (30.43) of the form (30.31)), then the additional condition (30.40) may be omitted.

If $\beta = 2$ (i.e., equation (30.43) has a pair of roots of the form (30.3) that differ by their sign only) and $\Omega_0 = [0, T]$, then condition (30.40) may be replaced by

$$\int\limits_{0}^{\infty} \Phi(z)\mathrm{d}z = \infty. \tag{30.44}$$

If $\beta \geqslant 3$, then condition (30.40) may be replaced by

$$\lim_{u\to\infty} u^{\frac{1}{\beta-1}}\Phi(u) = \infty, \tag{30.45}$$

which in many situations is less restrictive than condition (30.40) with κ given by (30.39) or (30.42).

The next comment refers to Theorem 30.7. Specifically, suppose that $\gamma_H \neq R(0,W) = W(0)$ and that $\pi(M,L)$ is odd and different from 1. Then the number $\kappa = [2\pi(M,L) - 2]^{-1}$ in condition (30.40) can be replaced by $\kappa = [2\pi(M,L) - 3]^{-1}$.

It should be mentioned that in the general case none of the inequalities i)–iv) listed above can be improved. Indeed, let us suppose that $\pi(M,L) > m$. If $\pi(M,L)$ is even, then for the polynomials

$$L(p) = \left(p^2 + \omega_1^2\right) \cdots \cdot \left(p^2 + \omega_{\pi(M,L)}^2\right) + p + 1, \quad M(p) \equiv 1, \tag{30.46}$$

we can easily check that equality (30.43) is true for $2\pi(M,L)$ distinct values of ω, namely,

$$\pm\omega_1\mathrm{i}, \pm\omega_2\mathrm{i}, \ldots, \pm\omega_{\pi(M,L)}\mathrm{i},$$

hence $\beta = 2\pi(M,L)$. If $\pi(M,L) > 1$ is odd and we consider the polynomials

$$L(p) = p\left(p^2 + \omega_1^2\right) \cdots \cdot \left(p^2 + \omega_{\pi(M,L)-1}^2\right) + p + 1, \quad M(p) = p + 1, \tag{30.47}$$

then equality (30.43) is true for $2\pi(M,L) - 1$ distinct values of ω, namely,

$$0, \pm\omega_1\mathrm{i}, \pm\omega_2\mathrm{i}, \ldots, \pm\omega_{\pi(M,L)-1}\mathrm{i},$$

hence $\beta = 2\pi(M, L) - 1$. The polynomials (30.46) and (30.47) also provide examples when condition (30.40) in Theorem 30.7 cannot be improved. Analogous examples can be constructed in case $\pi(M, L) \leqslant m$.

30.7. PROOFS OF THEOREMS 30.7 AND 30.8. The proofs given below deal with the case when the function $f(t, x)$ in equation (30.31) is subject to the constraint (30.37). The proofs are completely analogous if instead of (30.37) we use the constraint (30.38).

We start by choosing an arbitrary number $\gamma < \gamma_H$. Instead of equation (30.31) we may consider an equivalent one:

$$L_\gamma\left(\frac{\mathrm{d}}{\mathrm{d}t}\right) x = M\left(\frac{\mathrm{d}}{\mathrm{d}t}\right) f_\gamma(t, x), \qquad (30.48)$$

where $L_\gamma(p)$ stands for the polynomial $L(p) - \gamma M(p)$ and $f_\gamma(t, x)$ denotes the function $f(t, x) - \gamma x$. According to Lemma 26.4 it follows that the polynomial $L_\gamma(p)$ has no roots of the form (30.3), hence equation (30.48) can be replaced by the operator equation

$$x = A_\gamma f_\gamma(t, x), \qquad (30.49)$$

where the operator A_γ of the periodic problem for the linear link with the transfer function

$$W_\gamma(p) = \frac{M(p)}{L(p) - \gamma M(p)} \qquad (30.50)$$

is positive.

LEMMA 30.1. *Under the assumptions in either Theorem 30.7, or Theorem 30.8, the operator A_γ is strictly positive.*

Proof. We first suppose that condition (30.22) in Theorem 30.7 is fulfilled. In this case we have

$$2\pi(M, L) = \deg\{\Pi(\omega) - \gamma[M_1^2(\omega) + M_2^2(\omega)]\} = 2\pi(M, L) > 2m.$$

Therefore $\bar{\mu}(A_\gamma) = 0$, and Lemma 30.1 follows from Theorem 30.2.

Let us next suppose that $\pi(M, L) \leqslant m$ and $\bar\gamma \neq \gamma_H$ (i.e., $\bar\gamma > \gamma_H$). Then

$$\bar{\mu}(A_\gamma) = \lim_{\omega \to \infty} [R(\omega_k, W_\gamma)]^{-1} = \lim_{\omega \to \infty} \frac{M_1^2(\omega) + M_2^2(\omega)}{\Pi(\omega) - \gamma\,[M_1^2(\omega) + M_2^2(\omega)]} = \frac{1}{\bar\gamma - \gamma},$$

and

$$\mu(A_\gamma; L_2) = \sup_{k=0,1,\ldots;\ M(\omega_k i \neq 0)} [R(\omega_k, W_\gamma)]^{-1} =$$

$$= \sup_{k=0,1,\ldots;\ M(\omega_k i \neq 0)} \frac{M_1^2(\omega) + M_2^2(\omega)}{\Pi(\omega) - \gamma\,[M_1^2(\omega) + M_2^2(\omega)]} = \frac{1}{\bar\gamma_H - \gamma} \neq \frac{1}{\bar\gamma - \gamma}.$$

It remains to observe that once more Lemma 30.1 follows from Theorem 30.2.

The proof is complete.

In order to state the next result we let $E_0(A_\gamma)$ denote the subspace (28.3) which in our particular case is described by

$$E_0(A_\gamma) = \left\{ u(t) : u(t) = \sum (\xi_k \cos \omega_k t + \eta_k \sin \omega_k t) \right\}. \tag{30.51}$$

The sum in (30.51) above is extended over the set of all those values of k for which $R(\omega_k, W_\gamma) = [\mu(A_\gamma; L_2)]^{-1}$, or, equivalently, $R(\omega_k, W) = \gamma_H$.

LEMMA 30.2. *If $\pi(M, L) > m$, then*

$$\dim E_0(A_\gamma) \leqslant \begin{cases} 2\pi(M, L), & \textit{if } \pi(M, L) \textit{ is even,} \\ 2\pi(M, L) - 1, & \textit{if } \pi(M, L) \textit{ is odd;} \end{cases} \tag{30.52}$$

if $\pi(M, L) \leqslant m$ ($m > 0$), then

$$\dim E_0(A_\gamma) \leqslant \begin{cases} 2m, & \textit{if } m \textit{ is even,} \\ 2m - 1, & \textit{if } m \textit{ is odd.} \end{cases} \tag{30.53}$$

Proof. The dimension of the subspace $E_0(A_\gamma)$ coincides with the number β of all the distinct roots of the form (30.3) of equation (30.43). Every root ω of equation (30.43) is a root of the polynomial $\Pi(\omega) - \gamma_H[M_1^2(\omega) + M_2^2(\omega)]$. Therefore, the estimates $\dim E_0(A_\gamma) \leqslant 2\pi(M, L)$ if $\pi(M, L) > m$ and $\dim E_0(A_\gamma) \leqslant 2m$ if $\pi(M, L) \leqslant m$ are obvious. The estimates (30.52) and (30.53), in the case when $\pi(M, L)$ and m are both odd, follow from Lemma 11.3.

Lemma 30.2 is proved.

From Lemma 30.2 and based on equality (30.40) it follows that under the conditions in either Theorem 30.7, or Theorem 30.8, we have

$$\lim_{u \to \infty} u^{\frac{1}{\dim E_0(A_\gamma) - 1}} \Phi(u) = \infty, \quad \dim E_0(A_\gamma) > 1. \tag{30.54}$$

LEMMA 30.3. *The equality*

$$\lim_{\delta \to 0} \sup_{e(t) \in E_0(A_\gamma),\, \|e\|=1} \frac{\chi(\delta; e)}{\displaystyle\int_0^T \Phi\left[t, u_* + R\delta^{-1}|e(t)|\right] dt} = 0 \tag{30.55}$$

holds true for every $R > 0$ and each $u_* \geqslant u_0$.

Proof. If $\dim E_0(A_\gamma) = 1$, then equality (30.55) is obvious. Indeed, the subspace $E_0(A_\gamma)$ consists but of constant functions and $\chi(\delta; e) = 0$ for all $e \in E_0(A_\gamma)$ and every sufficiently small $\delta > 0$.

Suppose now that $\dim E_0(A_\gamma) > 1$. Since there exists $c > 0$ such that $|e(t)| \leqslant c$ for all normalized functions $e(t) \in E_0(A_\gamma)$, in order to prove equality (30.55) it will be enough to show that

$$\lim_{\delta \to 0} \sup_{e(t) \in E_0(A_\gamma),\ \|e\|=1} \frac{\chi(\delta; e)}{\int\limits_0^T \Phi\left[t, u_* + R\delta^{-1}c\right] dt} = 0,$$

an equality that can be rewritten as

$$\lim_{\delta \to 0} \left\{ \frac{1}{\int\limits_0^T \Phi\left[t, u_* + R\delta^{-1}c\right] dt} \sup_{e(t) \in E_0(A_\gamma),\ \|e\|=1} \chi(\delta; e) \right\} = 0. \tag{30.56}$$

But

$$\int\limits_0^T \Phi\left[t, u_* + R\delta^{-1}c\right] dt \geqslant \int\limits_{\Omega_0} \Phi\left[t, u_* + R\delta^{-1}c\right] dt = \mu\Omega \cdot \Phi\left[t, u_* + R\delta^{-1}c\right],$$

and from Lemma 11.2 we get the inequality

$$\sup_{e(t) \in E_0(A_\gamma),\ \|e\|=1} \chi(\delta; e) \leqslant c_1 \delta^{\frac{1}{\dim E_0(A_\gamma)-1}}.$$

Consequently, (30.56) is true as soon as

$$\lim_{\delta \to 0} \frac{\delta^{\frac{1}{\dim E_0(A_\gamma)-1}}}{\Phi\left[t, u_* + R\delta^{-1}c\right]} = 0.$$

The last equality follows straightforwardly from (30.54).

Lemma 30.3 is proved.

We are now in a position to conclude both the proofs of Theorems 30.7 and 30.8.

Let us start with Theorem 30.7. According to Theorem 30.4 the auxiliary operator $A_\gamma^\square$ acts continuously from L_2 into L_∞. It follows that equation (30.49) satisfies all the assumptions in Theorem 28.2. Therefore, there exists $\varepsilon_1 > 0$ such that the inequality

$$x f_\gamma(t, x) \leqslant k_1 x^2 - \Phi[t, |x|],$$

where $k_1(\gamma_H - \gamma)^{-1} < 1 + \varepsilon_1$, implies the solvability of equation (30.49). The solvability of equation (30.31) follows from condition (30.37) if $k < \gamma_H + \varepsilon$, where $\varepsilon = \varepsilon_1(\gamma_H - \gamma)$. Thus Theorem 30.7 is proved.

The proof of Theorem 30.8 can be concluded analogously. All we have to do is to invoke Theorem 28.2 instead of Theorem 28.1.

30.8. SECOND ORDER EQUATIONS. Our subsequent goal is to study equation (30.31) in the particular case when

$$l = \deg L(p) = 2, \quad m = \deg M(p) = 1.$$

Since in this case clearly $\pi(M, L) \leqslant m$ we can handle equation (30.31) using Theorem 30.5 (or Theorem 30.8 if the corresponding operator is strictly positive), i.e., assuming that the nonlinearity $f(t, x)$ satisfies the quadratric estimate (30.35). The interesting feature of the method developed below is that it yields existence theorems without relying upon the estimate (30.35).

Throughout this subsection we assume that the nonlinearity $f(t, x)$ is of the form

$$f(t, x) \equiv f(x) + u(t), \quad t, x \in \mathbb{R}^1,$$

where both the functions $f(x)$ and $u(t)$ are continuous and $u(t)$ is T-periodic.

THEOREM 30.9. *Suppose that $f(t)$ is subject to the constraint*

$$x f(x) \leqslant k x^2 + c, \quad x \in \mathbb{R}^1, \tag{30.57}$$

where $k < \gamma_H$, and that the coefficient b_1 in the polynomial $M(p)$ is nonpositive, i.e., $M(0) \leqslant 0$. Then the equation

$$L\left(\frac{\mathrm{d}}{\mathrm{d}t}\right) x = M\left(\frac{\mathrm{d}}{\mathrm{d}t}\right) [f(x) + u(t)] \tag{30.58}$$

has at least one T-periodic solution.

Theorem 30.9 has a natural analog for equations with nonlinearities $f(x)$ satisfying estimates of the form

$$x f(x) \geqslant k x^2 - c, \quad x \in \mathbb{R}^1.$$

We will prove Theorem 30.9 under the additional assumption that both the functions $f(x)$ and $u(t)$ are sufficiently smooth. For the case of arbitrary continuous functions $f(x)$ and $u(t)$ it will be enough to approximate first these functions by smooth ones and then to take the limits in the operator equations corresponding to that smooth approximations.

Proof. Let $\gamma \in (k, \gamma_H)$ be fixed and replace equation (30.58) by the equivalent operator equation

$$x = A_\gamma \mathfrak{f}_\gamma x. \tag{30.59}$$

The operator A_γ in (30.59) is defined according to (26.9) and $\mathfrak{f}_\gamma x = f[x(t)] - \gamma x(t) + u(t)$. Clearly A_γ is a positive operator on L_2 (since $\gamma < \gamma_H$). On the other hand, there exists $c_1 > 0$ such that the function $f_\gamma(t, x) = f(x) - \gamma x + u(t)$ satisfies the estimate

$$x f_\gamma(t, x) \leqslant -\kappa x^2 + c_1, \quad 0 \leqslant t \leqslant T, \ x \in \mathbb{R}^1, \tag{30.60}$$

where $\kappa = \frac{1}{2}[\gamma - k] > 0$. Estimate (30.60) follows from the relations

$$x f_\gamma(t, x) = x f(x) - \gamma x^2 + x u(t) \leqslant k x^2 + c - \gamma x^2 + |x| \max u(t) \leqslant$$
$$\leqslant -\kappa x^2 + [-\kappa x^2 + c + |x| \max u(t)] \leqslant -\kappa x^2 + c_1.$$

We will consider equation (30.59) in the space $C = C[0, T]$. Since the function $f[x(t)]$ — and, consequently, the function $\mathfrak{f}_\gamma x(t)$ — is continuous for every function $x(t) \in C$, we get that the operator $\mathfrak{f}_\gamma$ acts continuously on the space C. Therefore, the operator $A_\gamma \mathfrak{f}_\gamma$ is completely continuous on C. In view of the Leray-Schauder Principle, the proof of Theorem 30.9 will be complete as soon as we establish a general a priori norm estimate

$$\|x\|_C \leqslant \text{const} < \infty \tag{30.61}$$

in the space C of all the solutions of the equations

$$x = \xi A_\gamma \mathfrak{f}_\gamma x, \quad 0 \leqslant \xi \leqslant 1. \tag{30.62}$$

As an intermediate step towards the proof of (30.61) we will first establish an estimate of the form

$$\|x\|_{L_2} \leqslant c_2, \tag{30.63}$$

for the norms in L_2 of all the solutions $x \in C$ of the equations (30.62). To this end, we consider the quantity

$$\mathcal{J} = \int_0^T x(t) \mathfrak{f}_\gamma x(t) \mathrm{d}t. \tag{30.64}$$

Based on estimate (30.60) we clearly get

$$\mathcal{J} \leqslant -\kappa\|x\|_{L_2}^2 + c_1 T.$$

On the other hand, the positivity of the operator A_γ leads to

$$\mathcal{J} = \xi[A_\gamma f_\gamma x, f_\gamma x] \geqslant \xi\mu(A_\gamma; L_2)\|A_\gamma f_\gamma x\|_{L_2}^2 \geqslant 0.$$

The estimate (30.63) follows from the last two relations by taking

$$c_2 = \sqrt{c_1 T/\kappa}.$$

To prove estimate (30.61) we proceed as follows.

Start with the simple remark that each continuous solution $x(t)$ of every equation (30.62) satisfies the equation

$$L\left(\frac{d}{dt}\right)x = M\left(\frac{d}{dt}\right)[\xi f(x) + \xi u(t) + \gamma(1-\xi)x],$$

i.e.,

$$x'' + a_1 x' + a_2 x = \left[b_0 \frac{d}{dt} + b_1\right][\xi f(x) + \xi u(t) + \gamma(1-\xi)x].$$

Suppose next that $x(t)$ is fixed and consider the quantity

$$\mathcal{J}_1 = \int_0^T [x'(t)]^2 dt. \qquad (30.65)$$

Since

$$\mathcal{J}_1 = -\int_0^T x''(t)x(t)dt =$$

$$= -\int_0^T x(t)\left(-a_1 x' - a_2 x + b_0 \frac{d}{dt}\xi f(t) + b_1 f\xi(x) + \left[b_0\frac{d}{dt} + b_1\right][\xi u(t) + \gamma(1-\xi)x]\right) dt,$$

the equalities

$$\int_0^T x'(t)x(t)dt = \int_0^T x(t)\frac{d}{dt}f[x(t)]dt = 0,$$

which are true for arbitrary T-periodic functions $x(t)$, together with estimate (30.63) and the inequality

$$-b_1 \int_0^T x(t)f[x(t)]dt \leqslant |b_1 k|c_2^2 + |b_1|cT,$$

imply the existence of a constant $c_3 = c_3(a_2, a_2, b_0, b_1, c, c_1, c_2, T)$ such that

$$\mathcal{J}_1 \leqslant c_3.$$

Let now $t_0 \in [0, T]$ be a point where the continuous function $|x(t)|$ attains its minimum. We clearly have

$$|x(t_0)| \leqslant c_1 T^{-\frac{1}{2}}.$$

The estimate (30.61) follows from the next chain of relations:

$$|x(t)| = \left| \int_0^T x'(t)dt + x(t_0) \right| \leqslant \int_0^T |x'(t)|dt + |x(t_0)| \leqslant$$

$$\leqslant c_1 T^{-\frac{1}{2}} + \left\{ T \int_0^T [x'(t)]^2 dt \right\}^{\frac{1}{2}} \leqslant c_1 T^{-\frac{1}{2}} + (c_3 T)^{\frac{1}{2}}.$$

Theorem 30.9 is proved.

30.9. LIENARD EQUATIONS. As a last example to which the methods developed along this section can be applied, we consider the equation

$$x'' + g(x)x' - x = h(t). \tag{30.66}$$

Equation (30.66) coincides with equation (30.58), where $L(p) = p^2 - 1$, $M(p) = p$, and

$$f(x) = \int_0^x g(z)dz, \quad [u(t)]' = h(t).$$

We assume that $g(x)$ is a continuous function and $u(t)$ is periodic with period T.

Based on the results proved in the previous subsection we conclude that each of the estimates

$$x \int_0^x g(z)dz \leqslant -\varepsilon x^2 + c \quad \text{or} \quad x \int_0^x g(z)dz \geqslant \varepsilon x^2 - c$$

provides a sufficient condition for the existence of at least one T-periodic solution of equation (30.66). From Theorem 7.1 we get another sufficient condition that yields the same conclusion. More precisely, it is enough to have the estimate

$$\left| \int_0^x g(z) \mathrm{d}z \right| \leqslant kx^2 + c,$$

where k satisfies the constraint

$$k < \frac{2\pi T}{4\pi^2 + T^2}. \tag{30.67}$$

Condition (30.67) can be weakened using some results from Section 15.

If the T-periodic function $h(t)$ is given as the derivative of a T-periodic function, then obviously its average over the interval $[0, T]$ equals zero. The proofs of all the above mentioned facts about equation (30.66) with an arbitrary function $h(t)$ follow by carrying out appropriate arguments for the operator equation $x = A\mathfrak{f}x + h(t)$.

§31. Multiply-connected control systems

31.1. EQUATIONS DESCRIBING THE DYNAMICS OF COMPLEX NONLINEAR SYSTEMS. In this section we study systems whose dynamics are described by the equations

$$L_j\left(\frac{\mathrm{d}}{\mathrm{d}t}\right) x_j = M_j\left(\frac{\mathrm{d}}{\mathrm{d}t}\right) f_j(t, x_1, \ldots, x_n), \quad j = 1, \ldots, n. \tag{31.1}$$

We assume that

$$L_j(p) = p^{l(j)} + a_{j,1} p^{l(j)-1} + \cdots + a_{j,l(j)}, \quad j = 1, \ldots, n, \tag{31.2}$$

and

$$M_j(p) = b_{j,0} p^{m(j)} + b_{j,1} p^{m(j)-1} + \cdots + b_{j,m(j)}, \quad j = 1, \ldots, n, \tag{31.3}$$

are polynomials with constant real coefficients, and that $L_j(p)$ and $M_j(p)$ are coprime for every j. In addition, we suppose that all the coefficients $b_{j,0}$ are different from zero and

$$l(j) > m(j), \quad j = 1, \ldots, n. \tag{31.4}$$

The functions $f_j(t, x_1, \ldots, x_n)$ are jointly continuous in their real variables $t, x_1, \ldots, x_n$, and peridic with period T in the variable t, i.e.,

$$f_j(t, x_1, \ldots, x_n) \equiv f_j(t + T, x_1, \ldots, x_n), \quad -\infty < t, x_1, \ldots, x_n < \infty; \ j = 1, \ldots, n.$$

Equations of the form (31.1) occur in the study of the dynamics of multi-connected control systems (see, for instance, [Voronov, 1985], [Meerov, 1965 and 1986], [Wonham, 1985], [Rosenbrock, 1970], as well as the references indicated there).

The results and constructions developed so far in the previous chapters provide powerful tools to investigate various problems on T-periodic solutions of systems of equations like (31.1) above. We confine ourselves in the sequel to indicate some results that rely on sectional estimates of nonlinearities.

31.2. TWO-SIDED ESTIMATES. In this subsection we deal with equations (31.1) in case the nonlinearities satisfy either the conditions

$$|f_j(t, x_1, \ldots, x_n)| \leqslant \sum_{s=1}^{n} d_{js}|x_s| + b_j(t),$$

(31.5)

$$-\infty < t, x_1, \ldots, x_n < \infty; \ j = 1, \ldots, n,$$

where $b_j(t) \in L_2$, or the Lipschitz conditions

$$|f_j(t, x_1, \ldots, x_n) - f_j(t, y_1, \ldots, y_n)| \leqslant \sum_{s=1}^{n} d_{js}|x_s - y_s|,$$

(31.6)

$$-\infty < t, x_1, y_1 \ldots, x_n, y_n < \infty; \ j = 1, \ldots, n.$$

We assume that none of the polynomials $L_j(p)$ has roots of the form (30.3), i.e., the numbers

$$w_j = \max_{k=0,1,\ldots} |W_j(\omega_k i)| \quad (j = 1, \ldots, n),$$

(31.7)

are well-defined and finite, where by

$$W_j(p) = \frac{M_j(p)}{L_j(p)}$$

(31.8)

we denote the transfer functions of the corresponding linear links.

We next introduce the matrix

$$\mathcal{D} = \begin{bmatrix} d_{11}w_1 & \ldots & d_{1n}w_n \\ \ldots & \ldots & \ldots \\ d_{n1}w_1 & \ldots & d_{nn}w_n \end{bmatrix}$$

(31.9)

with the nonnegative entries $d_{js}w_j$, where d_{js} $(j, s = 1, \ldots, n)$ are the coefficients in conditions (31.5) or (31.6) and w_j $(j = 1, \ldots, n)$ are given by (31.7). Based on a well-known theorem of Frobenius (see, for instance, [Gantmaher, 1967], [Collatz,

1966]), the spectral radius $\rho(\mathcal{D})$ of the matrix $\mathcal{D}$ equals the largest real root λ of the characteristic equation

$$\det[\mathcal{D} - \lambda I] = 0.$$

There are many effective ways to get upper and lower estimates for the spectral radius of nonnegative matrices (see, for instance, [Gantmaher, 1967], [Krasnoselskii, M. A. et. al., 1969]).

THEOREM, 31.1. *Suppose conditions (31.5) hold true and $\rho(\mathcal{D}) < 1$. Then the system (31.1) has at least one T-periodic solution*

$$\boldsymbol{x}(t) = \{x_1(t), \ldots, x_n(t)\}. \tag{31.10}$$

If, in addition, condition (31.6) is fulfilled, then the system (31.1) has a unique T-periodic solution.

The proof of Theorem 31.1 will be given in Subsection 31.3.

Our next goal is to describe briefly two methods that yield approximations of the solution (31.10) of system (31.1). Both of them — the harmonic balance method, in short HBM, and the method of successive approximations — have already been discussed in Section 11 for a single equation. Their extension to the case of a system of equations does not require essentially new ideas. In the following description of HBM we will use the notations introduced in Subsection 11.2. To start with, we choose a vector

$$\boldsymbol{N} = \{N_1, \ldots, N_n\}$$

with natural components N_j and search for an approximate solution

$$\boldsymbol{x}^{\boldsymbol{N}}(t) = \left\{x_1^{\boldsymbol{N}}(t), \ldots, x_n^{\boldsymbol{N}}(t)\right\} \tag{31.11}$$

of the form

$$x_j^{\boldsymbol{N}}(t) = \eta_{j0}^{\boldsymbol{N}} + \sum_{s=1}^{N_j} \left(\xi_{js}^{\boldsymbol{N}} \sin \omega_s t + \eta_{js}^{\boldsymbol{N}} \cos \omega_s t\right). \tag{31.12}$$

The unknown coefficients ξ and η in the trigonometric polynomials (31.12) are defined using the equations

$$L_j \left(\frac{\mathrm{d}}{\mathrm{d}t}\right) x_j^{\boldsymbol{N}} = M_j \left(\frac{\mathrm{d}}{\mathrm{d}t}\right) P(N_j) f_j \left(t, x_1^{\boldsymbol{N}}, \ldots, x_n^{\boldsymbol{N}}\right), \quad j = 1, \ldots, n. \tag{31.13}$$

The system of n equations (30.31) is equivalent to a system of $\sum(1 + 2N_j)$ scalar equations in $\sum(1 + 2N_j)$ unknowns. HBM is called realizable if the system (31.13) has at least one solution for every $\boldsymbol{N}$. HBM is said to be (uniformly) convergent in case the Hausdorff distance (with respect to the uniform metric) from the set of all approximate solutions (31.11) provided by HBM to the set of all exact solutions (31.10) of equations (31.1) approaches zero as $\inf N_j \to \infty$.

It can be proved that HBM is realizable and convergent under the conditions in Theorem 31.1. Moreover, if the additional uniqueness conditions (31.6) are fulfilled, then the approximate solution (31.11) is unique for every $\boldsymbol{N}$. A more detailed presentation of HBM for the system (31.1) can be found in [Krasnoselskii, A. M., 1983], [Dementieva *et. al*, 1987].

In case $\rho(\mathcal{D}) < 1$ and conditions (31.6) hold true, we may construct approximations of the T-periodic solutions of system (31.1) using the successive approximations method. This method starts with an arbitrary initial approximation $\boldsymbol{x}^0(t) = \{x_1^0(t), \ldots, x_n^0(t)\}$ and yields a sequence of approximations $\boldsymbol{x}^N(t) = \{x_1^N(t), \ldots, x_n^N(t)\}$ ($N \in \mathbb{N}$), where $\boldsymbol{x}^N(t)$ is the unique T-periodic solution of the linear equations

$$L_j\left(\frac{\mathrm{d}}{\mathrm{d}t}\right) x_j^N = M_j\left(\frac{\mathrm{d}}{\mathrm{d}t}\right) f_j\left(t, x_1^{N-1}, \ldots, x_n^{N-1}\right), \quad j = 1, \ldots, n.$$

31.3. PROOF OF THEOREM 31.1. In view of the basic condition $\rho(\mathcal{D}) < 1$ in Theorem 31.1 we may introduce a new norm $\| \cdot \|_*$ on the space $\mathbb{R}^n$ (see [Functional Analysis, 1972]) such that

$$\|\mathcal{D}\xi\|_* \leqslant \rho_0 \|\xi\|_*, \quad \xi \in \mathbb{R}^n, \tag{31.14}$$

where $\rho(\mathcal{D}) < \rho_0 < 1$. We let $\boldsymbol{L_2}$ denote the direct sum of m copies of $L_2 = L_2([0, T], \mathbb{R}^1)$ and define the norm on L_2 by the equalities

$$\|\boldsymbol{x}(t)\|_{\boldsymbol{L_2}} = \|\{x_1(t), \ldots, x_n(t)\}\|_{\boldsymbol{L_2}} = \|\{\xi_1, \cdots, \xi_n\}\|_*,$$

$$\xi_j = \|x_j\|_{L_2}, \quad j = 1, \ldots, n. \tag{31.15}$$

We next introduce the operators $\boldsymbol{f}_j \boldsymbol{x}(t) : \boldsymbol{L_2} \to L_2$ given by

$$\boldsymbol{f}_j \boldsymbol{x}(t) = f_j[t, x_1(t), \ldots, x_n(t)], \quad j = 1, \ldots, n,$$

and let A_j denote the periodic problem operators for the transfer functions (31.8). In addition, we set

$$\boldsymbol{A}\boldsymbol{x}(t) = \{A_1 \boldsymbol{f}_1 \boldsymbol{x}(t), \ldots, A_n \boldsymbol{f}_n \boldsymbol{x}(t)\}. \tag{31.16}$$

Under the conditions in Theorem 31.1 the operator (31.16) acts on the space $\boldsymbol{L_2}$ and is completely continuous. Every fixed point $\boldsymbol{x}(t)$ of this operator defines a T-periodic solution $\{x_1(t), \ldots, x_n(t)\}$ of the system (31.1). Since the numbers (31.7) coincide with the norms of the operators A_j on the space $\boldsymbol{L}_2$, from (31.5) it follows that

$$\|A_j \boldsymbol{f}_j \boldsymbol{x}(t)\|_{L_2} \leqslant w_j \cdot \left[\sum_{s=1}^{n} d_{js} \|x_s\|_{L_2} + \|B_j(t)\|_{L_2}\right], \tag{31.17}$$

for any vector-function $\boldsymbol{x}(t) \in \boldsymbol{L}_2$. Let ξ denote the vector $\{\|x_1\|_{L_2}, \ldots, \|x_n\|_{L_2}\}$. The estimates (31.17) lead to the inequality

$$\|\boldsymbol{A}\boldsymbol{x}(t)\|_{\boldsymbol{L}_2} \leqslant \|\mathcal{D}\xi\|_* + \bar{b}, \quad \bar{b} = \max \|b_j(t)\|.$$

From it and from (31.14) we get the estimate

$$\|\boldsymbol{A}\boldsymbol{x}(t)\|_{\boldsymbol{L}_2} \leqslant \rho_0 \|\boldsymbol{x}(t)\|_{\boldsymbol{L}_2} + \bar{b}.$$

The last estimate shows that the operator $\boldsymbol{A}$ sends the ball

$$\left\{\boldsymbol{x}(t) : \boldsymbol{x}(t) \in \boldsymbol{L}_2, \ \|\boldsymbol{x}(t)\|_{\boldsymbol{L}_2} \leqslant \bar{b}(1 - \rho_0)^{-1}\right\}$$

into itself and that all its fixed points lie in this ball. Since $\boldsymbol{A}$ is a completely continuous operator on the space $\boldsymbol{L}_2$, the Schauder Principle implies the existence of at least one fixed point of the operator (31.16). Based on a previous remark, each fixed point (31.10) is a T-periodic solution of equations (31.1). The first assertion in Theorem 31.1 is proved.

In order to prove the second assertion we claim that under the additional condition (31.6) the operator $\boldsymbol{A}$ is a contraction on $\boldsymbol{L}_2$. This fact follows from the relations

$$\|\boldsymbol{A}\boldsymbol{x} - \boldsymbol{A}\boldsymbol{y}\|_{\boldsymbol{L}_2} = \|\{\|A_1\boldsymbol{f}_1\boldsymbol{x} - A_1\boldsymbol{f}_1\boldsymbol{y}\|_{L_2}, \ldots, \|A_n\boldsymbol{f}_n\boldsymbol{x} - A_n\boldsymbol{f}_n\boldsymbol{y}\|_{L_2}\}\|_* \leqslant$$
$$\leqslant \|\{w_1\|\boldsymbol{f}_1\boldsymbol{x} - \boldsymbol{f}_1\boldsymbol{y}\|_{L_2}, \ldots, w_n\|\boldsymbol{f}_n\boldsymbol{x} - \boldsymbol{f}_n\boldsymbol{y}\|_{L_2}\}\|_* \leqslant$$
$$\leqslant \|\mathcal{D}\{\|x_1(t) - y_1(t)\|_{L_2}, \ldots, \|x_n(t) - y_n(t)\|_{L_2}\}\|_* \leqslant \rho_0 \|\boldsymbol{x} - \boldsymbol{y}\|_{\boldsymbol{L}_2},$$

and it clearly proves the second assertion in Theorem 31.1.

The theorem is completely proved.

31.4. ONE-SIDED ESTIMATES. To achieve our next goal we need a few new notations and definitions. First, similarly to (30.6) we introduce the polynomials

$$\Pi_j(\omega) \stackrel{\text{def}}{=} \text{Re}[M_j(\omega i)\overline{L_j(\omega i)}], \quad j = 1, \ldots, n, \tag{31.18}$$

and let $2\pi(M_j, L_j)$ denote their degrees. Second, similarly to (30.21) we define the quantities

$$\gamma_H(M_j, L_j) = \inf_{k=0,1,\ldots;\ M_j(\omega_k i)\neq 0} \Big| \frac{\Pi_j(\omega_k)}{|M_j(\omega_k i)|^2}\Big|, \tag{31.19}$$

and

$$\gamma_B(M_j, L_j) = \sup_{k=0,1,\ldots;\ M_j(\omega_k i)\neq 0} \frac{\Pi_j(\omega_k)}{|M_j(\omega_k i)|^2}. \tag{31.20}$$

Suppose the polynomials (31.18) satisfy the conditions

$$\pi(M_j, L_j) \leqslant m(j) \quad \text{for } j = 1, \ldots, r;$$

$$\pi(M_j, L_j) > m(j) \quad \text{for } j = r+1, \ldots, n. \tag{31.21}$$

We will say that the functions $f_j(t, x_1, \ldots, x_n)$ have an admissible growth in the phase variables if for any given $\rho > 0$ there exists $R(\rho) < \infty$ such that

$$|f_j(t, x_1, \ldots, x_n)| \leqslant R(\rho)\left(1 + x_1^2 + \cdots + x_r^2\right), \quad |x_{r+1}|, \ldots, |x_n| \leqslant \rho, \tag{31.22}$$

for every $j = 1, \ldots, n$. The estimates (31.22) are always true if $r = 0$ (for instance, in case all the differences $l(j) - m(j)$ are even).

In the same setting, we say that the functions $f_j(t, x_1, \ldots, x_n)$ are properly compatible with the linear links W_j, if there exist a collection $\mathfrak{M}$ of indices j and some positive constants $\rho_1, \ldots, \rho_n$, such that $\gamma_H(M_j, L_j)$ is finite for $j \in \mathfrak{M}$, $\gamma_B(M_j, L_j)$ is finite for $j \notin \mathfrak{M}$, and

$$\sum_{j\in\mathfrak{M}} \rho_j x_j f_j(t, x_1, \ldots, x_n) - \sum_{j\notin\mathfrak{M}} \rho_j x_j f_j(t, x_1, \ldots, x_n) \leqslant \delta_1 x_1^2 + \cdots + \delta_n x_n^2 + \beta,$$

$$-\infty < t, x_1, \ldots, x_n < \infty, \tag{31.23}$$

for some coefficients $\delta_1, \ldots, \delta_n$ subject to the conditions

$$\delta_j < \rho_j \gamma_H(M_j, L_j) \quad \text{for } j \in \mathfrak{M}, \tag{31.24}$$

$$\delta_j < -\rho_j \gamma_B(M_j, L_j) \quad \text{for } j \notin \mathfrak{M}. \tag{31.25}$$

THEOREM 31.2. *Suppose the functions $f_j(t, x_1, \ldots, x_n)$ have admissible growth in the phase variables and are properly compatible with the linear links W_j. Then the system (31.1) has at least one continuous T-periodic solution.*

31.5. PROOF OF THEOREM 31.2. For the sake of convenience we confine ourselves to the case when $\mathfrak{M}$ consists of all the numbers $1, 2, \ldots, n$. In fact, this is not a real restriction because all the other cases can be easily reduced to this particular case. Accordingly, the estimate (31.23) becomes

$$\sum_{j=1}^{n} \rho_j x_j f_j(t, x_1, \ldots, x_n) \leqslant \sum_{j=1}^{n} \delta_j x_j^2 + \beta, \tag{31.26}$$

where the coefficients δ_j satisfy (31.25) for every $j = 1, \ldots, n$.

Based on (31.24) we can find some numbers

$$\gamma(1), \gamma(2), \ldots, \gamma(n),$$

satisfying the inequalities

$$\delta_j < \rho_j \gamma(j) < \rho_j \gamma_H(M_j, L_j), \quad j = 1, \ldots, n \tag{31.27}$$

(recall that the numbers $\gamma_H(M_j, L_j)$ may have arbitrary signs).

Since $\gamma_H(M_j, L_j)$ are finite for every j, by Theorem 30.3 we get that the operators A_j of the T-periodic problems for the linear links (31.8) are potentially positive. Consequently we may define the operators

$$A_{j,\gamma(j)} \stackrel{\text{def}}{=} [I - \gamma(j)A_j]^{-1} A_j, \quad j = 1, \ldots, n. \tag{31.28}$$

The operators $A_{j,\gamma(j)}$ are the periodic problem operators for the links with the transfer functions

$$W_{j,\gamma(j)}(p) = \frac{M_j(p)}{L_j(p) - \gamma(j)M_j(p)}, \quad j = 1, \ldots, n. \tag{31.29}$$

As we already know, the operators (31.28) are positive, normal, and completely continuous on L_2. For each of them we consider its self-adjoint part

$$B_j = \frac{1}{2}\left[A_{j,\gamma(j)} + A_{j,\gamma(j)}^*\right], \quad j = 1, \ldots, n, \tag{31.30}$$

and the corresponding auxiliary operator (see (26.21))

$$C_j = A_{j,\gamma(j)}^{\square} = B_j^{-\frac{1}{2}} A_{j,\gamma(j)}, \quad j = 1, \ldots, n. \tag{31.31}$$

By Lemma 26.5, the operators (31.31) are continuous on the space L_2 for any $j = 1, \ldots, n$. Moreover, Theorem 30.4 implies that the operators (31.31) act continuously from L_2 into C for every $j = r + 1, \ldots, n$.

We next replace the problem on T-periodic solutions of the system (31.1) by the equivalent problem on the solvability of the system of integral equations

$$x_j(t) = A_{j,\gamma(j)}\{f_j[t, x_1(t), \ldots, x_n(t)] - \gamma(j)x_j(t)\}, \quad j = 1, \ldots, n. \tag{31.32}$$

We pass from system (31.32) to the new system

$$y_j(t) = f_j[t, A_{1,\gamma(1)}y_1(t), \ldots, A_{n,\gamma(n)}y_n(t)] - \gamma(j)A_{j,\gamma(j)}y_j(t), \quad j = 1, \ldots, n. \tag{31.33}$$

If $\boldsymbol{y}(t) = \{y_1(t), \ldots, y_n(t)\}$ is a solution of the system (31.33) with all the components in L_2, then these components are continuous functions, since the operators (31.28) act from L_2 into C. Consequently, the functions

$$x_1(t) = A_{1,\gamma(1)}y_1(t), \quad \ldots, \quad x_n(t) = A_{n,\gamma(n)}y_n(t) \tag{31.34}$$

provide a solution of the system (31.32), i.e., they define a continuous T-periodic solution of the system (31.1).

Let $\boldsymbol{L}_2$ denote the usual Hilbert space (with the usual inner product and norm) of all vector-functions $\boldsymbol{y}(t) = \{y_1(t), \ldots, y_n(t)\}$ with components in L_2. Since every operator (31.28) is completely continuous as an operator from L_1 into C and the functions are jointly continuous in their variables, it turns out that each operator

$$\boldsymbol{f}_j\boldsymbol{y}(t) = f_j[t, A_{1,\gamma(1)}y_1(t), \ldots, A_{n,\gamma(n)}y_n(t)] - \gamma(j)A_{j,\gamma(j)}y_j(t),$$
$$j = 1, \ldots, n, \tag{31.35}$$

acts from $\boldsymbol{L}_2$ into C and is completely continuous. Consequently, any operator (31.35) is completely continuous as an operator from $\boldsymbol{L}_2$ into L_2. Therefore, according to the Leray-Schauder Principle, it follows that in order to prove the solvability in $\boldsymbol{L}_2$ of the system (31.32) (i.e., in order to conclude the proof of Theorem 31.2) it is enough to establish a general a priori norm estimate in $\boldsymbol{L}_2$ of all the solutions $\boldsymbol{y}(t) = \{y_1(t), \ldots, y_n(t)\}$ of the systems $\boldsymbol{L}_2$

$$y_j(t) = \lambda\{f_j[t, A_{1,\gamma(1)}y_1(t), \ldots, A_{n,\gamma(n)}y_n(t)] - \gamma(j)A_{j,\gamma(j)}y_j(t)\},$$
$$j = 1, \ldots, n, \tag{31.36}$$

with $\lambda \in [0, 1]$.

Let

$$y_j^*(t) = \lambda^*\{f_j[t, A_{1,\gamma(1)}y_1^*(t), \ldots, A_{n,\gamma(n)}y_n^*(t)] - \gamma(j)A_{j,\gamma(j)}y_j^*(t)\},$$
$$j = 1, \ldots, n, \tag{31.37}$$

be the components of such a solution, where $\lambda^* \in [0, 1]$ is fixed. Since $y_j^*(t) \in L_2$ and the operators (31.28) act continuously from L_2 into C, both the left and the right sides in equality (31.37) are continuous (and T-periodic) functions. We next take the inner products in L_2 of $y_j^*(t)$ and $\rho_j A_{j,\gamma(j)} y_j^*(t)$, for every j. Using (31.37) and (31.26), after a few easy computations we get the inequality

$$\sum_{s=1}^{n} \rho_j [y_j^*, A_{j,\gamma(j)} y_j^*] \leqslant \lambda^* \sum_{s=1}^{n} (\delta_j - \gamma(j)\rho_j)[A_{j,\gamma(j)} y_j^*, A_{j,\gamma(j)} y_j^*] + \lambda^* \beta T.$$

Since $\delta_j < \gamma(j)\rho_j$ (see (31.27)) and

$$[y_j^*, A_{j,\gamma(j)} y_j^*] = \left\| B_j^{-\frac{1}{2}} y_j^* \right\|_{L_2}^2, \quad j = 1, \dots, n,$$

the previous inequality leads to

$$\sum_{s=1}^{n} \rho_j \left\| B_j^{-\frac{1}{2}} y_j^* \right\|_{L_2}^2 \leqslant \beta T.$$

In its turn this relation implies

$$\left\| B_j^{-\frac{1}{2}} y_j^* \right\|_{L_2} \leqslant \sqrt{\frac{\beta T}{\rho_0}}, \quad j = 1, \dots, n, \tag{31.38}$$

where $\rho_0 = \min\{\rho_1, \dots, \rho_n\}$.

We are now ready to take advantage of some continuity properties of the auxiliary operators (31.31). Specifically, from (31.38) we get the relations

$$\|A_{j,\gamma(j)} y_j^*(t)\|_{L_2} = \left\| C_j B_j^{-\frac{1}{2}} y_j^* \right\|_{L_2} \leqslant \|C_j\|_{L_2 \to L_2} \sqrt{\frac{\beta T}{\rho_0}} \stackrel{\text{def}}{=} c_1, \quad j = 1, \dots, n, \tag{31.39}$$

and

$$\|A_{j,\gamma(j)} y_j^*(t)\|_C = \left\| C_j B_j^{-\frac{1}{2}} y_j^* \right\|_C \leqslant \|C_j\|_{L_2 \to C} \sqrt{\frac{\beta T}{\rho_0}} \stackrel{\text{def}}{=} c_2, \quad j = r+1, \dots, n. \tag{31.40}$$

Further, in view of condition (31.22) and based on (31.39) and (31.4) we find that

$$\|f_j[t, A_{1,\gamma(1)} y_1^*(t), \dots, A_{n,\gamma(n)} y_n^*(t)]\|_{L_1} \leqslant R(c_2)(1 + r \cdot c_1^2) \stackrel{\text{def}}{=} c_3,$$

whence

$$\|f_j[t, A_{1,\gamma(1)}y_1^*(t), \ldots, A_{n,\gamma(n)}y_n^*(t)] - \gamma(j)A_{j,\gamma(j)}y_j^*(t)\|_{L_1} \leqslant$$
$$\leqslant c_3 + |\gamma(j)|T^{1/2}c_1 \leqslant c_3 + T^{1/2}c_1 \max_j |\gamma(j)| \overset{\text{def}}{=} c_4.$$

Finally, from (31.37) we get the estimates

$$\|y_j^*(t)\|_{L_1} \leqslant c_4, \quad j = 1, \ldots, n. \tag{31.41}$$

The rest of the proof is now at hand. Since every operator (31.28) acts continuously from L_1 into C, by the estimates (31.41) we clearly obtain

$$\|A_{j,\gamma(j)}y_j^*\|_C \leqslant \|A_{j,\gamma(j)}\|_{L_1 \to C} \cdot c_4 \leqslant c_5, \quad j = 1, \ldots, n.$$

It remains to observe, using once again (31.22), that for each $j = 1, \ldots, n$ we have the estimate

$$|y_j^*(t)| \leqslant |f_j[t, A_{1,\gamma(1)}y_1^*(t), \ldots, A_{n,\gamma(n)}y_n^*(t)] - \gamma(j)A_{j,\gamma(j)}y_j^*(t)| \leqslant$$
$$\leqslant R(c_5)[1 + c_5^2] + |\gamma(j)|c_5 \leqslant c_6 < \infty.$$

The conclusion is obvious and even stronger than we want it to be. The components of all the solutions of any system (31.36) satisfy not only an a priori norm estimate in L_2, but also an a priori norm estimate in the space C of all continuous functions on $[0, T]$.

The theorem is proved.

31.6. UNIQUENESS CONDITIONS. Sticking around the same setting as above, it is quite easy to complete Theorem 31.2 with some criteria for the uniqueness of the solutions of system (31.1). We state, without proof, but one result of this kind. It is based on the following definition.

We will say that the functions $f_j(t, x_1, \ldots, x_n)$ satisfy the one-sided Lipschitz conditions compatible with the linear links W_j if there exist a collection $\mathfrak{M}$ of indices and some positive constants $\rho_1, \ldots, \rho_n$, such that $\gamma_H(M_j, L_j)$ is finite for $j \in \mathfrak{M}$, $\gamma_B(M_j, L_j)$ is finite for $j \notin \mathfrak{M}$, and

$$\sum_{j \in \mathfrak{M}} \rho_j(x_j - y_j)[f_j(t, x_1, \ldots, x_n) - f_j(y_1, \ldots, y_n)] -$$
$$- \sum_{j \notin \mathfrak{M}} \rho_j(x_j - y_j)[f_j(t, x_1, \ldots, x_n) - f_j(y_1, \ldots, y_n)] \leqslant$$
$$\leqslant \sum_{s=1}^n \delta_j(x_s - y_s)^2, \quad -\infty < t, x_1, y_1, \ldots, x_n, y_n < \infty,$$

where the coefficients $\delta_1, \ldots, \delta_n$ are subject to conditions (31.24) and (31.25).

THEOREM 31.3. *Suppose the functions $f_j(t, x_1, \ldots, x_n)$ satisfy the one-sided Lipschitz conditions compatible with the linear links W_j. Then the system (31.1) has no more than one continuous T-periodic solution.*

It should be mentioned that under the conditions in Theorem 31.2 the harmonic balance method is realizable and converges, thus it may be used to find approximate T-periodic solutions of the system (31.1). Moreover, in case the assumptions in both Theorems 31.2 and 31.3 are fulfilled, the HBM equations (31.13) are solvable and have a unique solution of the form (31.12) for any given $\boldsymbol{N} = \{N_1, \ldots, N_n\}$.

§32. One-sided estimates in nonquasilinear problems

32.1. SETTING THE PROBLEM. In this section we consider operator equations of the form

$$x = K\mathfrak{f}(Kx, Vx) \tag{32.1}$$

in a real Hilbert space H. We let $\mathfrak{f} : H \times H \to H$ denote a nonlinear operator, and suppose that both K and V are continuous linear operators on H, the former being self-adjoint. There are plenty of boundary value problems related to ordinary or partial differential equations that lead to equations like (32.1) above. A possible example is given in the next subsection. Quasilinear equations quite close to (32.1) have been already discussed in Subsection 27.4.

We assume that the space H is equipped with a moduls M (see Section 7) and that the nonlinearity $\mathfrak{f}(x, y)$ $(x, y \in H)$ satisfies the estimate

$$[x, \mathfrak{f}(x, y)] \leqslant q_1 \|x\|^2 + 2q_2[Mx, My] + q_3\|y\|^2 + b, \quad x, y \in H. \tag{32.2}$$

Let $\mathfrak{J}$ denote the quantity (13.7), i.e.,

$$\mathfrak{J} = \mathfrak{J}(q_1, q_2, q_3; K, V) = \sup_{\|x\|=1} \left\{ q_1\|Kx\|^2 + 2q_2[MKx, MVx] + q_3\|Vx\|^2 \right\}. \tag{32.3}$$

THEOREM 32.1. *Suppose the estimate (32.2) holds true with some coefficients q_1, q_2, and q_3 such that quantity (32.3) satisfies the condition*

$$\mathfrak{J}(q_1, q_2, q_3; K, V) < 1. \tag{32.4}$$

Then any solution x of every equation

$$x = \xi K\mathfrak{f}(Kx, Vx), \quad 0 \leqslant \xi \leqslant 1, \tag{32.5}$$

satisfies the general a priori estimate

$$\|x\| \leqslant \sqrt{b(1-\mathfrak{J})^{-1}}.$$ (32.6)

Proof. Let x be a solution of equation (32.5) for a fixed $\xi \in [0, 1]$. Then

$$\|x\|^2 = \xi[x, K\mathfrak{f}(Kx, Vx)]$$

and, since the operator K is self-adjoint on H,

$$\|x\|^2 = \xi[Kx, \mathfrak{f}(Kx, Vx)].$$

The last equality clearly implies the estimate

$$\|x\|^2 \leqslant [Kx, \mathfrak{f}(Kx, Vx)],$$

which, according to (32.3), leads to the inequality

$$\|x\|^2 = q_1\|Kx\|^2 + 2q_2[MKx, MVx] + q_3\|Vx\|^2 + b.$$

Hence we get the relation

$$\|x\|^2 \leqslant \mathfrak{J}(q_1, q_2, q_3; K, V)\|x\|^2 + b,$$

that clearly proves the estimate (32.6).

Theorem 32.1 is proved.

The estimate (32.2) holds true for any superposition operator

$$\mathfrak{f}(x, y) = f[t, x(t), y(t)],$$ (32.7)

corresponding to a function $f(t, x, y) : \Omega \times \mathbb{R}^n \times \mathbb{R}^n \to \mathbb{R}^n$ subject to the condition

$$(x, \mathfrak{f}(t, x, y)) \leqslant q_1|x|^2 + 2q_2(|x|, |y|) + q_3|y|^2 + b(t),$$
(32.8)
$$t \in \Omega, \ x, y \in \mathbb{R}^n; \ b(t) \in L_1; \ |\{x_1, \ldots, x_n\}| = \{|x_1|, \ldots, |x_n|\}.$$

Theorem 32.1 can be used to find solvability criteria for equation (32.1). Some of these criteria may require additional assumptions on the growth at infinity of the nonlinearity.

32.2. A TWO-POINT BOUNDARY VALUE PROBLEM. The arguments developed in this section rest upon the article [Krasnoselskii, M. A., 1956a]. Specifically, we consider the scalar two-point boundary value problem

$$x'' = f(t, x, x'), \quad x(0) = x(\pi) = 0. \tag{32.9}$$

The nonlinearity $f(t, x, y) : [0, \pi] \times \mathbb{R}^1 \times \mathbb{R}^1 \to \mathbb{R}^1$ is supposed to be jointly continuous in its variables. The solvability conditions for problem (32.9) we are interested in are provided by inequalities of the form

$$f(t, x, y) \leqslant k_1|x| + k_2|y| + a(t), \quad 0 \leqslant t \leqslant \pi, \ x \geqslant 0, \ -\infty < y < \infty,$$
$$\tag{32.10}$$

$$f(t, x, y) \geqslant -k_1|x| - k_2|y| - a(t), \quad 0 \leqslant t \leqslant \pi, \ x \leqslant 0, \ -\infty < y < \infty.$$

We introduce the notation

$$\kappa(k_1, k_2) = \begin{cases} k_1 + k_2 & \text{if } \dfrac{k_2}{k_1} \geqslant \dfrac{12\pi - 24}{16 + \pi}, \\[3mm] K & \text{if } \dfrac{k_2}{k_1} < \dfrac{12\pi - 24}{16 + \pi}, \end{cases} \tag{32.11}$$

where

$$K = \frac{1}{2}\sqrt{\left[\frac{3}{4}k_1 + \left(\frac{2}{\pi} - \frac{1}{2}\right)k_2\right]^2 + \frac{9}{4}k_2^2} + \frac{5}{8}k_1 + \left(\frac{1}{\pi} + \frac{1}{4}\right)k_2.$$

Under the previous conditions we always have $\kappa(k_1, k_2) \leqslant k_1 + k_2$; if $\dfrac{k_2}{k_1} < \dfrac{12\pi - 24}{16 + \pi}$, then we get the strict inequality $\kappa(k_1, k_2) < k_1 + k_2$.

THEOREM 32.2. *Suppose the constraints (32.10) hold true, and*

$$\kappa(k_1, k_2) < 1. \tag{32.12}$$

Then the boundary value problem (32.9) has at least one twice continuously differentiable solution.

A first version of this theorem was proved in [Krasnoselskii, M. A., 1956a]. There, the solvability conditions for problem (32.9) were given by the constraints (32.10) under the assumption that the coefficients k_1 and k_2 satisfy a more restrictive than (32.12) condition, namely $k_1 + k_2 < 1$.

Proof. We replace (32.9) by the equivalent operator equation (32.1) in the space $H = L_2 = L_2([0, 1], \mathbb{R}^1)$, where $\mathfrak{f}$ stands for the superposition operator (32.7), and the linear operators K and V are given by

$$Kx(t) = \frac{2}{\pi} \sum_{n=1}^{\infty} \frac{1}{n}(x(t), \sin nt)\sin nt,$$

$$V x(t) = \frac{2}{\pi} \sum_{n=1}^{\infty} (x(t), \sin nt) \cos nt.$$

We clearly have the equality $\dfrac{d}{dt} K x(t) = V x(t)$, as well as the equality $K^2 = -A$, where A is the operator that takes any integrable function $u(t)$ into the solution $x(t)$ of equation $x'' = u(t)$ satisfying the boundary conditions $x(0) = x(\pi) = 0$. The operator V is an isometry on the space L_2. The operator K is completely continuous, self-adjoint, and positive definite on L_2, and, moreover, it acts as a completely continuous operator from L_2 into C. Every solution $x \in L_2$ of equation (32.1) (each such solution is obviously continuous) corresponds to a solution $K x \in C^2$ of problem (32.9). As usual, the modulus on L_2 is given by the equality $M x(t) = |x(t)|$.

The proof of Theorem 32.2 goes along the same lines as the proof of Theorem 1 in [Krasnoselskii, M. A., 1956a] and is based on the existence of an a priori norm estimate in L_2 of the solutions of equation (32.1). More precisely, such an estimate has the form

$$\|x\| \leqslant \|a(t)\| \left[1 - \Im \left(k_1, \frac{1}{2} k_2, 0; K, V \right) \right]^{-1}$$

and follows from Theorem 32.1 (it is enough to observe that (32.10) implies (32.2) with $q_1 = k_1$, $q_2 = \frac{1}{2} k_2$ and $q_3 = 0$). Therefore, all we need in order to conclude the proof of our theorem is to show that

$$\Im \left(k_1, \frac{1}{2} k_2, 0; K, V \right) \leqslant \kappa(k_1, k_2). \tag{32.13}$$

For the proof of estimate (32.13) we use Theorem 13.3. Let E_0 and E_1 be the subspaces defined by

$$E_0 = \{ e(t) : e(t) = \xi \sin t, \ \xi \in \mathbb{R} \},$$

$$E_1 = \{ e(t) : e(t) \in L_2, \ (e(t), \sin t) = 0 \}.$$

Since

$$\|H x_0\| \leqslant \|x_0\|, \quad \|V x_0\| = \|x_0\|, \qquad x_0 \in E_0,$$

$$\|H x_1\| \leqslant \frac{1}{2} \|x_1\|, \quad \|V x_1\| = \|x_1\|, \qquad x_1 \in E_1,$$

and

$$[|H \sin t|, |V \sin t|] = \frac{2}{\pi} \|\sin t\|^2,$$

by Theorem 13.3 we get

$$2\Im \leqslant \sqrt{ \left[\frac{3}{4} k_1 + \left(\frac{2}{\pi} - \frac{1}{2} \right) k_2 \right]^2 + \frac{9}{4} k_2^2 } + \frac{5}{4} k_1 + \left(\frac{2}{\pi} + \frac{1}{2} \right) k_2.$$

Estimate (32.13) follows from this last relation and from the obvious inequality $\mathfrak{J} \leqslant k_1 + k_2$.

Theorem 32.2 is proved.

§33. First order equations with variable coefficients

33.1. PROPERTIES OF THE ASSOCIATED LINEAR OPERATOR. The problem we are next interested in consists of the scalar differential equation

$$\frac{\mathrm{d}x}{\mathrm{d}t} + q(t)x = u(t) \tag{33.1}$$

and the periodic boundary condition

$$x(0) = x(T). \tag{33.2}$$

We will assume that the coefficient $q(t)$ $(0 \leqslant t \leqslant T)$ is continuous. If the condition

$$q_0 \overset{\text{def}}{=} \frac{1}{T} \int_0^T q(t)\mathrm{d}t \neq 0, \tag{33.3}$$

is fulfilled, then problem (33.1), (33.2) has a unique solution $x(t)$ for every function $u(t)$ integrable on $[0, T]$. Specifically, the solution is given by

$$x(t) = Au(t), \tag{33.4}$$

where the linear operator A is defined by the formula

$$Au(t) = \frac{\rho(0, t)}{1 - \rho(0, T)} \int_0^T \rho(s, T)u(s)\mathrm{d}s + \int_0^t \rho(s, t)u(s)\mathrm{d}s, \tag{33.5}$$

with

$$\rho(s, t) = \exp\left\{ -\int_s^t q(\tau)\mathrm{d}\tau \right\}, \quad 0 \leqslant s, t \leqslant T. \tag{33.6}$$

The function (33.6) is jointly continuous in its variables.

Our main goal in this subsection is to investigate the basic properties of the operator (33.5). To this end, we first introduce a new measure on the interval $[0, T]$ defined by the equality

$$\mu\Omega = \int_\Omega \exp\left\{ -2\int_0^t [q_0 - q(\tau)]\mathrm{d}\tau \right\} \mathrm{d}t. \tag{33.7}$$

A set $\Omega \subset [0, T]$ is measurable with respect to the measure μ if and only if it is Lebesgue measurable. Next, using the measure (33.7) we define the corresponding space $L_2(\mu)$ of all square-integrable functions, with the inner product

$$[x, y]_\mu = \int\limits_0^T x(t)y(t)\mathrm{d}\mu(t), \quad x, y \in L_2. \tag{33.8}$$

Let $e_0(t)$, $e_k(t)$, and $g_k(t)$ be the functions given by formulas (7.24); then the functions

$$\tilde{e}_0(t) = \rho_0(t)e_0(t), \quad \tilde{e}_k(t) = \rho_0(t)e_k(t), \quad \tilde{g}_k(t) = \rho_0(t)g_k(t), \qquad k \in \mathbb{N}, \tag{33.9}$$

where

$$\rho_0(t) = \exp\left\{ \int\limits_0^t [q_0 - q(\tau)]\mathrm{d}\tau \right\}, \quad 0 \leqslant t \leqslant T, \tag{33.10}$$

provide a complete orthogonal system in $L_2(\mu)$. Direct computations yield simple formulas for the values of the operator (33.5) corresponding to each basic function (33.9). In their turn, these formulas lead to the spectral decomposition of the operator (33.5), explicitly given by

$$Au(t) = \sum_{k=0}^{\infty} \frac{1}{\sqrt{\omega_k^2 + q_0^2}} \tilde{U}_k \tilde{P}_k u(t), \quad u(t) \in L_2(\mu). \tag{33.11}$$

The precise meaning of every symbol involved in (33.11) is the following:

i) $\omega_k = 2k\pi T^{-1}$, $k = 0, 1, 2, \ldots$;

ii) the operator $\tilde{P}_0$ is the orthogonal projection from the space $L_2(\mu)$ onto the one-dimensional subspace $\tilde{\Pi}_0$ spanned by the function $\tilde{e}_0(t)$;

iii) the operators $\tilde{P}_k$ ($k = 1, 2, \ldots$) are the orthogonal projections from $L_p(\mu)$ onto the two-dimensional subspaces

$$\tilde{\Pi}_k = \{x(t) : x(t) = \xi\tilde{e}_k(t) + \eta\tilde{g}_k(t), \ \xi, \eta \in \mathbb{R}^1\}; \tag{33.12}$$

iv) the unitary operators $\tilde{U}_k : \tilde{\Pi}_k \to \tilde{\Pi}_k$ ($k = 1, 2, \ldots$) are defined by the equalities

$$\tilde{U}_0\tilde{e}_0 = (\operatorname{sign} q_0)\tilde{e}_0,$$

$$\tilde{U}_k(\xi\tilde{e}_k + \eta\tilde{g}_k) = \frac{1}{\sqrt{\omega_k^2 + q_0^2}}[(q_0\xi - \omega_k\eta)\tilde{e}_k + (q_0\eta + \omega_k\xi)\tilde{g}_k]. \tag{33.13}$$

The representations (33.5) and (33.11) may now be used to derive all the properties of the operator A stated below. We begin by observing that A is a normal and completely continuous operator on the space $L_2(\mu)$. It also acts continuously from L_1 to C and is completely continuous as an operator from L_2 or $L_2(\mu)$ into C. The spectrum $\sigma(A)$ of A consists of all the numbers $q_0 + \omega_k\mathrm{i}$ $(k = 0, \pm 1, \pm 2, \ldots)$.

Moreover, the operator A is potentially positive on $L_2(\mu)$ both from below and from above; actually, $\gamma_H(A) = \gamma_B(A) = q_0$. However, A is not potentially strictly positive.

As a last remark, we mention that A is an integral operator, and its kernel is positive if $q_0 > 0$, and negative if $q_0 < 0$.

33.2. ONE-SIDED ESTIMATES. In this subsection we state and prove a criterion for the existence of periodic solutions of the equation

$$\frac{\mathrm{d}x}{\mathrm{d}t} + q(t)x = f(t, x). \tag{33.14}$$

We assume that the nonlinearity $f(t, x) : [0, T] \times \mathbb{R}^1 \to \mathbb{R}^1$ is jointly continuous in its variables.

THEOREM 33.1. *Suppose there exists $\varepsilon > 0$ such that either the estimate*

$$xf(t, x) \leqslant (q_0 - \varepsilon)x^2 + c, \quad 0 \leqslant t \leqslant T, \ -\infty < x < \infty, \tag{33.15}$$

or the estimate

$$xf(t, x) \geqslant (q_0 + \varepsilon)x^2 - c, \quad 0 \leqslant t \leqslant T, \ -\infty < x < \infty, \tag{33.16}$$

holds true. Then the problem (33.14), (33.2) has at least one continuously differentiable solution.

Theorem 33.1 (as well as Theorem 30.9) is a consequence of some theorems proved in Section 27. However, it should be noticed that the quadratic constraint (27.13) — which eventually made it possible to reduce Theorem 31.1 to some already proved results — is not included as a hypothesis in Theorem 33.1.

Proof. We will but deal with the case when the nonlinearity is subject to estimate (33.15). Set $\gamma = q_0 - \varepsilon$ and consider the operator A_γ (defined by (26.9)). This operator is positive on $L_2(\mu)$ and its positivity coefficient is given by $\mu(A_\gamma; L_2(\mu)) = \varepsilon^{-1}$.

Since the superposition operator $\mathfrak{f}x = f[t, x(t)]$ is continuous on the space C, and A_γ acts as a completely continuous operator on C, we get that the operator

$$Bx(t) = A_\gamma[\mathfrak{f}x(t) - \gamma x(t)] \tag{33.17}$$

is completely continuous on C, too. Every fixed point $x(t) \in C$ of the operator (33.17) is a solution of problem (33.14), (33.2). Consequently, based on the Leray-Schauder Principle, all we need in order to prove our theorem is to establish an a priori norm estimate

$$\|x(t)\|_C \leqslant \text{const} < \infty \tag{33.18}$$

of all the solutions $x(t)$ of any equation

$$x = \xi B x, \tag{33.19}$$

with $0 \leqslant \xi \leqslant 1$. The proof of estimate (33.18) has two steps.

Step 1: An a priori estimate in L_2. Let $x(t)$ be a solution of equation (33.19) for a fixed $\xi \in [0, 1]$. We first observe that

$$\|x\|^2_{L_2(\mu)} \leqslant \xi^2 \|A_\gamma[\mathfrak{f}x - \gamma x]\|^2_{L_2(\mu)} \leqslant \xi \|\|A_\gamma[\mathfrak{f}x - \gamma x]\|^2_{L_2(\mu)}.$$

Since the operator A_γ is positive and its positivity coefficient equals ε^{-1}, we next get

$$\|x\|^2_{L_2(\mu)} \leqslant \varepsilon^{-1} \xi [A_\gamma(\mathfrak{f}x - \gamma x), \mathfrak{f}x - \gamma x]_\mu,$$

i.e., we have the estimate

$$\|x\|^2_{L_2(\mu)} \leqslant \varepsilon^{-1} [x, \mathfrak{f}x - \gamma x]_\mu. \tag{33.20}$$

On the other hand, condition (33.15) implies that

$$[x, \mathfrak{f}x - \gamma x]_\mu \leqslant b_1,$$

for any $x \in L_2(\mu)$, where

$$b_1 = b \int_0^T \exp\left\{ -2 \int_0^t [q_0 - q(\tau)] d\tau \right\} dt.$$

Therefore, from (33.20) it follows that

$$\|x\|^2_{L_2(\mu)} \leqslant b_1 \varepsilon^{-1}. \tag{33.21}$$

Since the norm on $L_2(\mu)$ is equivalent to the norm inherited from the usual space L_2 (as it easily follows from the estimate

$$0 < m \leqslant \exp\left\{ -2 \int_0^t [q_0 - q(\tau)] d\tau \right\} \leqslant M < \infty, \quad 0 \leqslant t \leqslant T),$$

relation (33.21) implies the existence of a constant $b_2 > 0$ such that

$$\|x\|_{L_2} \stackrel{\text{def}}{=} \sqrt{\int_0^T [x(t)]^2 dt} \leqslant b_2. \tag{33.22}$$

Step 2: An a priori estimate in C. Every solution $x(t) \in C$ of equation (33.19) is continuously differentiable and satisfies the equality

$$\frac{dx}{dt} + q(t)x = \xi f(t, x) + \gamma(1 - \xi)x.$$

Therefore, given two arbitrary points t_1 and t_2 in $[0, T]$ we may write

$$[x(t_1)]^2 - [x(t_2)]^2 = \int_{t_2}^{t_1} \frac{d}{dt}([x(t)]^2) dt = 2 \int_{t_2}^{t_1} x(t)x'(t) dt =$$

$$= 2 \int_{t_2}^{t_1} x(t)(-q(t)x(t) + \xi f[t, x(t)] + \gamma(1 - \xi)x(t)) dt,$$

whence — based on (33.15) — we get the the estimate

$$[x(t_1)]^2 - [x(t_2)]^2 \leqslant -2 \int_{t_2}^{t_1} [q(t) - \gamma(1 - \xi)]^2 dt + 2\xi \int_{t_2}^{t_1} \{\gamma[x(t)]^2 + b\} dt =$$

$$= \int_{t_2}^{t_1} 2[x(t)]^2 [q(t) + \gamma] dt + 2\xi b(t_1 - t_2).$$

The last inequality clearly leads to

$$[x(t_1)]^2 - [x(t_2)]^2 \leqslant \int_0^T 2[x(t)]^2 [|q(t)| + |\gamma|] dt + 2bT,$$

a relation that together with (33.22) proves the estimate

$$[x(t_1)]^2 - [x(t_2)]^2 \leqslant b_3 \stackrel{\text{def}}{=} 2b_2 \left(\max_{0 \leqslant t \leqslant T} |q(t)| + |\gamma| \right) + 2bT. \tag{33.23}$$

Assume next that t_2 is a point where the continuous function $[x(t)]^2$ attains its minimum. Then

$$[x(t_2)]^2 \leqslant \frac{1}{T^2} \int_0^T [x(t)]^2 dt \leqslant b_2^2 T^{-2}.$$

Thus, from (33.23) it follows that

$$|x(t_1)| \leqslant \sqrt{b_3 + b_2^2 T^{-2}}, \quad t_1 \in [0, T],$$

an inequality that concludes the proof of the a priori estimate (33.18).

Theorem 33.1 is proved.

The second step in the previous proof follows an idea of M. G. Yumagulov.

33.3. TWO-SIDED ESTIMATES. The present subsection continues the study of problem (33.14), (33.2). The main difference is that besides two one-sided estimates analogous to (33.15) and (33.16) above, we impose an additional two-sided linear estimate of the form

$$|f(t, x)| \leqslant c(|x| + 1), \quad 0 \leqslant t \leqslant T, \quad -\infty < x < \infty. \tag{33.24}$$

Throughout this subsection we let $\Phi(t, u)$ denote a given function in one of the classes $\mathfrak{N}(u_o)$.

THEOREM 33.2. *Suppose that the estimate (33.24) holds true. Then there exists $\varepsilon > 0$ such that each of the estimates*

$$xf(t, x) \leqslant (q_0 + \varepsilon)x^2 - \Phi(t, |x|), \quad 0 \leqslant t \leqslant T, \quad -\infty < x < \infty, \tag{33.25}$$

or

$$xf(t, x) \geqslant (q_0 - \varepsilon)x^2 + \Phi(t, |x|), \quad 0 \leqslant t \leqslant T, \quad -\infty < x < \infty, \tag{33.26}$$

implies the existence of at least one continuously differentiable solution of problem (33.14), (33.2).

It should be observed that among the conditions in Theorem 33.2 there are no constraints of type (8.7) or (28.5) related to the asymptotic behavior of the function $\Phi(t, u)$. The reason is that in the given setting, the subspace E_0 is one-dimensional, and the function $\tilde{e}_0(t)$ that spans this subspace is such that $\chi(\delta; \tilde{e}_0) \equiv 0$ for small values of $\delta > 0$.

Theorem 33.2 is a corollary to Theorem 8.1. All we need is merely to transform the initial equation into an equation of the form $x = Bx$, where B is the operator (33.17) corresponding to a suitable choice of γ.

33.4. FIRST-ORDER EQUATIONS WITH DELAY. In this subsection we consider the problem of T-periodic solutions for the equation

$$\frac{dx(t)}{dt} + q(t)x = f[t, x(t), x(t - h)]. \tag{33.27}$$

The function $f(t, x, y)$ is jointly continuous in its variables and periodic with period T in t. The function $g(t)$ is also continuous and T-periodic. For the delay h we assume that $0 < h < T$. We let S_h denote the operator that takes any function $x(t) : [0, T] \to \mathbb{R}$ into the function

$$S_h x(t) = \begin{cases} x(t - h), & h \leqslant t \leqslant T, \\ x(t - h + T), & 0 \leqslant t < h. \end{cases} \tag{33.28}$$

Suppose next that condition (33.3) is fulfilled. Then the problem of T-periodic solutions of equation (33.27) is equivalent to the operator equation

$$y = f[t, Ay(t), S_h Ay(t)] \tag{33.29}$$

in the sense that every continuous solution $y(t)$ of equation (33.29) defines the T-periodic continuously differentiable solution $x(t) = Ay(t)$ of equation (33.27). In (33.29) above, as well as in the remaining part of this section, A stands for the operator (33.5). We notice that the operators (33.28) commute with A if and only if $q(t) \equiv \text{const}$.

In addition, we suppose that the function $f(t, x, y)$ satisfies the constraint

$$|f(t, x, y)| \leqslant k_1 |x| + k_2 |y| + b(t), \quad 0 \leqslant t \leqslant T, \ -\infty < x, y < \infty; \ b(t) \in L_1. \tag{33.30}$$

Let us now choose and fix a norm $\| \cdot \|_*$ on L_2 equivalent to the usual one, and subject to the following two natural conditions:

a) if $x(t) \geqslant y(t) \geqslant 0$ $(0 \leqslant t \leqslant T)$, then $\|x\|_* \geqslant \|y_*\|$;

b) $\| |x(t)| \|_* \geqslant \|x(t)\|_*$.

If the joint norm $N = N(k_1 A, k_2 S_h A) = N(k_1 A, k_2 S_h A; |\cdot|, \|\cdot\|_*)$ is less than 1, then equation (33.29) has at least one continuous solution $y(t)$. As the next lemma shows, the joint norm N can be computed exactly.

LEMMA 33.1. *The joint norm N equals the norm*

$$\|B\|_* = \|B(k_1, k_2; h)\|_* = \sup_{\|x\|_* = 1} \|Bx\|_* \tag{33.31}$$

of the linear operator

$$B = B(k_1, k_2; h) = k_1 A + k_2 S_h A. \tag{33.32}$$

For the proof of Lemma 33.1 it is enough to observe that both A and $S_h A$ are integral operators, and their kernels are either simultaneously positive for $0 \leqslant t, s \leqslant T$

(if $q_0 > 0$), or simultaneously negative for $0 \leqslant t, s \leqslant T$ (if $q_0 < 0$). Consequently, for any $x(t)$ and all $0 \leqslant t \leqslant T$, we have the inequality

$$k_1|Ax(t)| + k_2|S_h Ax(t)| \leqslant |k_1 A|x(t)| + k_2 S_h A|x(t)||=|B|x(t)||, \qquad (33.33)$$

that becomes an equality for every $x(t) \geqslant 0$. The rest of the proof is obvious.

THEOREM 33.3. *Let the numbers k_1, k_2, and h be such that the spectral radius $r(B)$ of the linear operator (33.32) is less than 1. Then the constraint (33.30) implies the existence of at least one T-periodic continuously differentiable solution $x(t)$ of equation (33.27).*

Proof. We limit ourselves to the case $q_0 > 0$. Accordingly, the integral operator (33.32) has a positive kernel. Therefore (see [M. A. Krasnoselskii, 1962]), for any given $\varepsilon > 0$ there exists a norm $\| \cdot \|_*$ on the space L_2 subject to both the conditions a) and b) above, and such that the norm (33.31) satisfies the condition

$$\|B\|_* \leqslant r(B) + \varepsilon. \qquad (33.34)$$

If $r(B) < 1$, then by choosing $\varepsilon = \frac{1}{2}[1 - r(B)]$ we clearly get

$$\|B\|_* < 1.$$

According to Lemma 33.1, the last inequality concludes the proof of Theorem 33.3.

Results on the solvability of problem (33.14), (33.2) analogous to Theorem 33.3 can be also proved in case the nonlinearity satisfies one-sided constraints of the form

$$xf(t, x, y) \leqslant k_1 x^2 + k_2|xy| + k_3 y^2 + b(t),$$

or

$$xf(t, x, y) \geqslant k_1 x^2 + k_2|xy| + k_3 y^2 - b(t).$$

As before, the main ingredient in their proofs is provided by the joint norm of some concrete linear operators.

§34. Variational methods

34.1. A GENERAL SCHEME. Let A be a self-adjoint positive definite operator on the space $L_2 = L_2(\Omega, \mathbb{R}^1)$, that acts as a completely continuous operator from E into E^*, where $E^* \subset L_2 \subset E$. Let $\mathfrak{f}$ be the superposition operator (28.2), and assume that $\mathfrak{f}$ acts from E^* into E. Then equation (27.1) can be replaced (see Subsection 27.4) by the equation

$$x = K^* \mathfrak{f} K x, \tag{34.1}$$

where K stands for the positive definite self-adjoint square root of the operator A. We let κ denote the norm of the operator K on L_2, i.e., its largest eigenvalue; clearly

$$\|A\|_{L_2 \to L_2} = \kappa^2.$$

Let us next consider the nonlinear functional

$$\Gamma[x(t)] = \int_\Omega F[t, Kx(t)] \mathrm{d}\mu, \tag{34.2}$$

where

$$F(t, x) = \int_0^x f(t, z) \mathrm{d}z. \tag{34.3}$$

This functional is defined and weakly continuous on L_2. If the function (34.3) satisfies the estimate

$$F(t, x) \leqslant \frac{1}{2} k x^2 + b(t), \quad t \in \Omega, \ x \in \mathbb{R}, \ b(t) \in L_1, \tag{34.4}$$

where

$$k\kappa^2 < 1, \tag{34.5}$$

then

$$V[x(t)] \overset{\text{def}}{=} \frac{1}{2}[x, x] - \Gamma[x(t)] \geqslant \frac{1}{2}[x, x] - \int_\Omega \left(\frac{1}{2} k \|Kx(t)\|^2 + b(t) \right) \mathrm{d}\mu =$$

$$= \frac{1}{2}(1 - k\kappa^2)\|x(t)\|^2 = \|b\|_{L_2}.$$

Therefore, if the estimate (34.4) and condition (34.5) hold true, then the functional $V[x(t)]$ is bounded from below and

$$\lim_{\|x\| \to \infty} V[x(t)] = \infty.$$

Consequently, there exists at least one point $x_* \in L_2$ where the functional $V[x(t)]$ attains its minimum. At every such point x_* the gradient

$$\operatorname{grad} V[x(t)] = x - K^* \mathfrak{f} K x$$

of the functional $V[x(t)]$ equals zero, i.e., x_* is a solution of equation (32.1). This scheme for proving the existence of solutions of equation $x = A\mathfrak{f}x$ has been used in many papers and books (see, for instance, [Vainberg, 1956], [Fučik, Kufner, 1980], and the references included there).

In case of equations with scalar-valued functions, condition (34.4) — considered as a restriction imposed to the nonlinearity $f(t, x)$ — is "qualitatively better" than, although quite close to, one-sided estimates of type (27.3), or of other particular types. For instance, if $g(t, x)$ is a function satisfying a one-sided constraint of type (27.3), then every function $f(t, x) = cx^a \sin(x^{a+1}) + g(t, x)$ satisfies condition (34.4).

The general scheme sketched above may be improved in various directions. As an illustration, we notice that by passing from the equation $x = A\mathfrak{f}x$ to a suitably chosen equation $x = A_\gamma f_\gamma x$, we reduce the former equation to the case when the self-adjoint operator A has but a finite number of negative eigenvalues. In the same respect, it should be mentioned that the variational methods can be successfully used for operators A with a "bad" behavior, e.g., in case A is not completely continuous as an operator from E into E^*. Cases like these may be handled by following the procedure developed in the proof of Theorem 27.5.

The main result proved in the present section is based on previous theorems on integral-functional inequalities.

34.2. AN APPLICATION OF INTEGRAL-FUNCTIONAL INEQUALITIES. Throughout the remaining part of this section instead of estimate (34.4) we will use the constraint

$$F(t, x) \leqslant \frac{1}{2}kx^2 - \Phi(t, |x|), \quad t \in \Omega, \ x \in \mathbb{R}^1, \tag{34.6}$$

where $\Phi(t, u)$ $(t \in \Omega, u \geqslant 0)$ is a given function in one of the classes $\mathfrak{N}(u_0)$. As in other chapters of this book, our main concern is to show that a suitable asymptotic behavior at infinity of the function $\Phi(t, u)$ assures the solvability of equation (34.1).

Before stating the main result, let us denote by E_0 the finite dimensional subspace

$$E_0 = \{x(t) : Kx(t) = \kappa x(t)\} \subset L_2. \tag{34.7}$$

THEOREM 34.1. *Suppose the function* $\Phi(t, u) \in \mathfrak{N}(u_0)$ *satisfies the equality*

$$\lim_{\delta \to 0} \sup_{e(t) \in E_0; \ \|e\|=1} \frac{\chi(\delta; e)}{\displaystyle\int_\Omega \Phi\left[t, u_* + R\delta^{-1}|e(t)|\right] \mathrm{d}\mu} = 0, \tag{34.8}$$

for any $R > 0$ and all $u_* \geqslant u_0$. Then there exists $\varepsilon > 0$ such that the constraint (34.6) with a coefficient k subject to the condition

$$k\kappa^2 < 1 + \varepsilon, \tag{34.9}$$

implies the existence of at least one solution $x(t) \in L_2$ of equation (34.1)

Condition (34.8) in Theorem 34.1 has been already used several times in different chapters of this book.

From Theorem 34.1 we get the next result.

COROLLARY 34.1. *Suppose the function $\Phi(t, u) \in \mathfrak{N}(u_0)$ satisfies equality (34.8) for any $R > 0$ and all $u_* \geqslant u_0$. Then the constraint (34.6) with a coefficient k such that $k\kappa^2 = 1$ implies the existence of at least one solution $x(t) \in L_2$ of equation (34.1).*

There are many examples of specific boundary value problems for which Theorem 34.1 may be applied. For instance, all the theorems on boundary value problems stated in Section 29 can be accordingly improved. We mention but one example.

Let $\Phi(t, u)$ be a function satisfying condition (29.13). Then there exists $\varepsilon_0 > 0$ such that the estimate (34.6) with a coefficient $k \leqslant \lambda_0^4 + \varepsilon_0$ implies the existence of at least one classical solution of problem (29.9), (29.10).

However, by passing to equations with vector-valued functions the situation changes significantly. The one-sided estimates of type (26.3) can be used without other additional conditions. For a variational approach we have to assume that the vector-function $f(t, x)$ is a gradient in the variable x of a scalar function.

34.3. PROOF OF THEOREM 34.1. For the proof we will introduce the orthogonal projections P and Q onto the subspace (34.7) and its orthogonal complement E_1, respectively ($E_0 \oplus E_1 = L_2$). In addition, we set $q = \dfrac{1}{k}\|K\|_{E_1 \to E_1}$, $q \in [0, 1)$. The following relations are obvious

$$\|KPx\| = \kappa\|Px\|, \quad \|KQx\| \leqslant q\kappa\|Qx\|, \qquad x \in L_2. \tag{34.10}$$

We next consider the family of functions

$$\mathcal{F} = \{e(t) : e(t) \in E_0, \ \|e\| = 1\}. \tag{34.11}$$

According to Theorem 2.2, the family (34.11) is compatible with the function $\Phi(t, u)$. Therefore, for every $\beta > 0$ there exist a positive nondecreasing function $\alpha(u)$ ($u \geqslant 0$)

and a number $c = c(\beta) > 0$, such that all the solutions $y(t) = \xi e(t) + h(t)$ $(e(t) \in \mathcal{F})$ of the inequality

$$\|h(t)\|^2 \leqslant -\beta \int_{\Omega} \Phi[t, |y(t)|] \mathrm{d}\mu + \beta \alpha(\|y(t)\|) \tag{34.12}$$

satisfy the estimate

$$\|y(t)\| \leqslant c(\beta). \tag{34.13}$$

In the sequel we will need the number

$$\beta_1 = \kappa^2 \varepsilon_1^{-1}, \tag{34.14}$$

where $\varepsilon_1 = \frac{1}{2}(1 - q^2)$; let $c_1 = c(\beta_1)$. We also set

$$M = \frac{2}{\varepsilon_1} \mu \Omega \sup_{t \in \Omega,\, u \geqslant 0} |\Phi(t, u)|, \quad \rho = \sqrt{2c_1^2 + M\kappa^2} + 1.$$

LEMMA 34.1. *Suppose that the coefficient k in constraint (34.6) satisfies condition (34.9), where*

$$\varepsilon = \min\left\{\varepsilon_1, 2\kappa^2 \cdot \frac{\alpha(\rho)}{\rho^2}\right\}. \tag{34.15}$$

Then the functional $V[x(t)]$ is positive at every x such that

$$\rho - 1 < \kappa\|x\| < \rho. \tag{34.16}$$

Theorem 34.1 follows from Lemma 34.1: the functional $V[x(t)]$ equals zero at the point $x(t) \equiv 0$, whereas on the annulus (34.16) it is positive. Hence inside that annulus $V[x(t)]$ attains its minimum at a point $x_* \in L_2$ which is a solution of equation (34.1).

Thus, all we have to do in order to conclude the proof of Theorem 34.1 is to prove Lemma 34.1.

Proof. The next chain of relations

$$V[x(t)] \geqslant \frac{1}{2}\|x\|^2 - \int_{\Omega} \left\{\frac{1}{2}k|Kx(t)|^2 - \Phi[t, |Kx(t)|]\right\} \mathrm{d}\mu \geqslant$$

$$\geqslant \frac{1}{2}\|x\|^2 - \frac{1}{2}k\|Kx\|^2 + \int \Phi[t, |Kx(t)|] \mathrm{d}\mu \geqslant$$

$$\geq \frac{1}{2}\|Qx\|^2 + \frac{1}{2}\|Px\|^2 - \frac{1}{2}\left(\frac{1}{\kappa^2} + \frac{\varepsilon}{\kappa^2}\right)\|Kx\|^2 + \int_{\Omega} \Phi[t, |Kx(t)|]d\mu \geq$$

$$\geq \frac{1}{2}\|Qx\|^2 + \frac{1}{2}\|Px\|^2 - \frac{1}{2}\|Px\|^2 - \frac{1}{2}q^2\|Qx\|^2 - \frac{\varepsilon}{2\kappa^2}\|Kx\|^2 + \int_{\Omega} \Phi[t, |Kx(t)|]d\mu \geq$$

$$\geq \frac{1}{2}(1 - q^2)\|Qx\|^2 - \frac{\varepsilon}{2\kappa^2}\|Kx\|^2 + \int_{\Omega} \Phi[t, |Kx(t)|]d\mu \geq$$

$$\geq \varepsilon_1\|Qx\|^2 - \frac{\varepsilon}{2\kappa^2}\|Kx\|^2 + \int_{\Omega} \Phi[t, |Kx(t)|]d\mu$$

leads to the inequality

$$V[x(t)] \geq \varepsilon_1\|Qx\|^2 - \frac{\varepsilon}{2\kappa^2}\|Kx\|^2 + \int_{\Omega} \Phi[t, |Kx(t)|]d\mu. \tag{34.17}$$

Along these computations we used the definition of ε_1, the inequality $k \leq \kappa^{-2}(1 + \varepsilon)$ satisfied by the coefficient k in (34.6), and the estimates (34.10).

Suppose now that $V[x(t)] \leq 0$. Then from (34.17) we get the inequality

$$\varepsilon_1\|Qx\|^2 \leq \frac{\varepsilon}{2\kappa^2}\|Kx\|^2 + \mu\Omega \sup_{t\in\Omega,\ u\geq 0} |\Phi(t, u)|,$$

therefore,

$$\varepsilon_1\|Qx\|^2 \leq \frac{1}{2}\varepsilon\|x\|^2 + \frac{1}{2}\varepsilon_1 M$$

and

$$\varepsilon_1\|Qx\|^2 \leq \frac{1}{2}\varepsilon_1\|x\|^2 + \frac{1}{2}\varepsilon_1 M.$$

Thus, we have

$$2\|Qx\|^2 \leq \|x\|^2 + M,$$

i.e., every $x(t)$ for which $V[x(t)] \leq 0$ satisfies the estimate

$$\|Qx\|^2 \leq \|Px\|^2 + M. \tag{34.18}$$

Assume next that $V[x(t)] \leq 0$ and $\|x\| \leq \kappa^{-1}\rho$. Using once again (34.17) we get

$$\varepsilon_1\|Qx\|^2 \leq \frac{\varepsilon}{2\kappa^2}\|Kx\|^2 - \int_{\Omega} \Phi[t, |Kx(t)|]d\mu,$$

an inequality that implies the relation

$$\beta^{-1}\|KQx\|^2 \leqslant \frac{\varepsilon}{2\kappa^2}\|Kx\|^2 - \int_\Omega \Phi[t,|Kx(t)|]\mathrm{d}\mu, \qquad (34.19)$$

where β is the number (34.14). On the other hand, from $\|x\| \leqslant \kappa^{-1}\rho$ we obtain the estimate $\|Kx\| \leqslant \rho$. Therefore, in view of (34.19) and (34.15) it follows that

$$\beta^{-1}\|KQx\|^2 \leqslant \alpha(\rho) - \int_\Omega \Phi[t,|Kx(t)|]\mathrm{d}\mu,$$

hence

$$\|KQx\|^2 \leqslant -\beta \int_\Omega \Phi[t,|Kx(t)|]\mathrm{d}\mu + \beta\alpha(\|Kx\|). \qquad (34.20)$$

Further we introduce the notations $y(t) = Kx(t)$, $h(t) = Qy(t)$, $\xi = \|Px\|$, and let $e(t)$ denote either $\kappa\dfrac{Px}{\|Px\|}$ if $\|Px\| \neq 0$, or any fixed element in $\mathcal{F}$ if $\|Px\| = 0$. In terms of these notations, we easily check that inequalities (34.12) and (34.20) coincide, i.e., for all x satisfying $V[x(t)] \leqslant 0$ and $\|x\| \leqslant \kappa^{-1}\rho$ we have the estimate $\|y(t)\| \leqslant c_1$, that is, $\|Kx\| \leqslant c_1$. But $\|Kx\| \leqslant c_1$ clearly implies $\|Px\|^2 \leqslant c_1^2\kappa^{-2}$ and therefore, based on (34.18), we conclude with $\|x\|^2 \leqslant 2c_1^2\kappa^{-2} + M$. The last inequality shows that the inequalities $V[x(t)] \leqslant 0$ and $\|x\| \leqslant \kappa^{-1}\rho$ imply the relation $\|x\| \leqslant \kappa^{-1}(\rho - 1)$.

The proof of Lemma 34.1 and, consequently, the proof of Theorem 34.1, are now complete.

Theorem 34.1 was formulated in [Krasnoselskii, A. M., Krasnoselskii, M. A., 1991].

References

APPELL, YU.; KRASNOSELSKII, A. M.,
New theorems on asymptotic bifurcation points, *Nonlinear Anal. Theory, Meth. Appl.*, **18**:1(1992), 269–276.

BABITSKII, V. I.; KRUPENIN, V. M.,
Oscillations in strongly nonlinear systems (Russian), Nauka, Moscow, 1985.

CARATHEODORY, C.,
Vorlesungen über reele Funktionen, Leipzig—Berlin, 1918.

COLLATZ, L.,
Functional analysis and numerical mathematics (German), Springer-Verlag, Berlin—Göttingen—Heidelberg, 1964; English translation: Academic Press, New York, 1966.

DANCER, E. N.,
— On the Dirichlet problem for weakly nonlinear elliptic partial differential equations, *Proc. Roy. Soc., Edinburgh, Sect. A*, **76**(1977), 283–300.
— On the use of asymptotics in nonlinear boundary value problems, *Ann. Math. Pura Appl.*, **131**:4(1982), 167–185.

DEMENTEVA, A. M.; GRACHEV, N. I.; KRASNOSELSKII, A. M.; PETROV, D. I.,
Remarks on the harmonic balance method in problems on forced oscillations in multiply-connected systems (Russian), *Tr. VNIISI Acad. Nauk SSSR*, **6**(1987), 68–71.

DEMENTEVA, A. M.; KRASNOSELSKII, A. M.; YUMAGULOV, M. G.,
The operator of a periodic problem for a first order equation (Russian), *Dokl. Akad. Nauk Tadzhik SSR*, **32**:2(1988), 802–805.

DRABEK, P.,

Nonlinear operator equations, Dissertation, Prague, 1977.

DUNFORD, N.; SCHWARTZ, J.,

Linear operators.I: General Theory, Interscience Publishers, New York, 1958.

FELLER, W.,

An introduction to probability theory and its applications, John Wiley & Sons, Inc., New York, 1950.

FOURNIER, G.; MAWHIN, J.,

On periodic solutions of forced pendulum-like equations, *J. Diff. Equations*, **59**(1985), 214–237.

FUČIK, S.,

— Nonlinear equations with non-invertible linear part, *Czech. Math. J.*, **24**(1974a), 467–495.

— Further remarks on a theorem by E. M. Landesman and A. C. Laser, *Comment. Math. Univ. Carolinae*, **15**(1974b), 259–271.

— *Solvability of nonlinear equations and boundary value problems*, Prague, 1980.

FUČIK, S.; HESS, S.,

Nonlinear perturbations of linear operators having null-space with strong unique continuation property, *Nonlinear Anal. Theory, Meth. Appl.*, **3**(1979), 271–277.

FUČIK, S.; KRBEC, M,

Boundary value problem with bounded nonlinearity and general null-space of the linear part, *Math. Z.*, **155**(1977), 129–138.

FUČIK, S.; KUFNER, A.,

Nonlinear differential equations, Elsevier, Amsterdam—Oxford—New York, 1980.

FUČIK, S.; NEČAS, J.; KUČERA, M.,

Ranges of nonlinear asymptotically linear operators, *J. Diff. Equations*, **17**(1975), 375–394.

GANTMAHER, F. R.,

Theory of matrices (Russian), Nauka, Moscow, 1967; English translation: Chelsea Publ. Co., New York, 1959.

GOLOMB, M.,

Theorie der nichtlinearen Integralgleichungen, Integralgleichungssysteme und allgemeinen Funktionalgleichungen, *Math. Z.*, **39**(1935), 45–75.

HAMMERSTEIN, A.,

Nichtlineare Integralgleichungen nebst Anwendungen, *Acta Math.*, **54**(1929), 117–176.

HARDY, G. H.; LITTLEWOOD, J. E.; POLYA, G.,

Inequalities, Cambridge University Press, London, 1934.

HESS, P.,

On a theorem by Landesman and Lazer, Indiana Univ. Math. J., 23(1974), 827–829.

KATO, T.,

Perturbation theory for linear operators, Springer-Verlag, Berlin—Heidelberg—New York, 1966.

KAZDAN, J. L.; WARNER, F. W.,

Remarks on some quasilinear elliptic equations, Commun. Pure Appl. Math., 28 (1975), 567–597.

KOLMOGOROV, A. N.; FOMIN, S. V.,

Introductory real analysis (Russian), Nauka, Moscow, 1972; English translation: Prentice-Hall, Englewood Cliffs, 1970.

KRASNOSELSKII, A.M.,

— A priori estimates of the norms of solutions of scalar inequalities (Russian), Dokl. Akad. Nauk SSSR, 243:5(1978), 1119–1122; translated in Soviet Math. Dokl., 19:6(1978), 1481–1485.

— New a priori estimates of the solutions of a two-point boundary problem and their applications (Russian), Sib. Mat. J., 21:1(1980a), 115–124.

— Frequency criteria in the problem of forced oscillations in controled systems (Russian), Avtomat. i Telemekh., 9(1980b), 23–30; translated in Automat. & Remote Control, 41:9(1981), 1203–1209.

— Forced periodic oscillations in complex nonlinear systems (Russian), Avtomat. i Telemekh., 10(1983), 76–82; translated in Automat. & Remote Control, 44:10 (1984), 1301–1306.

— Forced periodic oscillations in nonlinear systems (Russian), Dokl. Akad. Nauk SSSR, 276:6(1984a), 1356–1359; translated in Soviet Phys. Dokl., 29:6(1984), 438–440.

— On the problem of forced oscillations in nonlinear control systems (Russian), Izv. Akad. Nauk SSSR Ser. Tekhn. Kibernet., 4(1984b), 176–182; translated in Engrg.-Cybernetics, 22:5(1984), 92–99.

— One-sided estimates in a problem on forced oscillations in nonlinear control systems (Russian), Dokl. Akad. Nauk SSSR, 283:2(1985), 284–286; translated in Soviet Math. Dokl., 32:1(1985), 74–77.

— New theorems on forced periodic oscillations in nonlinear systems (Russian), Prikl. Mat. Mekh., 50:2(1986a), 224–230; translated in Appl. Math. Mech., 50:2(1987), 165–170.

— Nonlocal theorems on forced oscillations in nonlinear systems (Russian), in *Mathematical systems theory*, by Bobylev, N. A., et al., Nauka, Moscow, 1986b.

— Forced oscillations in systems with hysteresis nonlinearities (Russian), *Dokl. Akad. Nauk SSSR*, **292**:5(1987), 1078–1082.

— Conditions for the applicability of Fredholm theorems in the analysis of forced oscillations in nonlinear control systems (Russian), *Avtomat. i Telemekh.*, **3**(1988), 162–164.

— Joint norm of operators and the study of nonlinear problems (Russian), *Ukrain. Mat. J.*, **42**:12(1990), 1624–1635.

— On a method of analysis of resonance problems, *Nonlinear Anal. Theory, Meth. Appl.*, **15**:4(1991), 321–345.

— On bifurcation points of equations with Landesman-Lazer type nonlinearities, *Nonlinear Anal. Theory, Meth. Appl.*, **18**:12(1992), 1187–1199.

KRASNOSELSKII, A. M.; KRASNOSELSKII, M. A.,

On some class of equations with potential operators, *Note-Mat.* , **11**(1991), 237–245.

KRASNOSELSKII, A. M.; MAWHIN, J.,

Remark on some type of bifurcation at infinity, *Bulletin de la Classe des Sciences, Ac. Roy. de Belgique, 6e série*, **III**:12(1992), 293–297.

KRASNOSELSKII, M. A.,

— Eigenfunctions of nonlinear operators, asymptotically close to linear (Russian), *Dokl. Akad. Nauk SSSR*, **74**:2(1950), 177-179.

— Continuity of the operator $fu(x) = f[x, u(x)]$ (Russian), *Dokl. Akad. Nauk SSSR*, **77**:2(1951), 185-188.

— On a boundary value problem (Russian), *Izv. Akad. Nauk SSSR, Ser. Mat.*, **20**(1956a), 241–252.

— *Topological methods in the theory of nonlinear integral equations* (Russian), Gostehizdat, Moscow, 1956b; English translation: Pergamon Press, Oxford, 1964.

— *The translation operator along trajectories of differential equations* (Russian), Nauka, Moscow, 1960; English translation: *Transl. Math. Monographs*, **19**, AMS, Providence, 1968.

— *Positive solutions of operator equations* (Russian), Fizmatgiz, Moscow, 1962; English translation: Noordhoff Publ., Groningen, 1972.

KRASNOSELSKII, M. A.; VAINIKKO, G. M.; ZABREIKO, P. P.; RUTITSKII, YA. B.; STETSENKO, V. YA.,

Approximate solutions of operator equations (Russian), Nauka, Moscow, 1969; English translation: Noordhoff Publ., Groningen, 1972.

KRASNOSELSKII, M. A.; ZABREIKO, P. P.,

    Geometric methods in nonlinear analysis (Russian), Nauka, Moscow, 1975; English translation: Springer Verlag, Berlin—Heidelberg—New York—Tokyo, 1984.

KRASNOSELSKII, M. A.; ZABREIKO, P. P.; PUSTYLNIK, P. P.; SOBOLEVSKII, P. E.,

    Integral operators in spaces of summable functions (Russian), Nauka, Moscow, 1966; English translation: Noordhoff Publ., Leyden, 1976.

KRASNOSELSKII, M. A.; LIFSHITS, E. A.; SOBOLEV, A. B.,

    Positive linear systems, Nauka, Moscow, 1985; English translation: Hilderman, Berlin, 1990.

LANDESMAN, E. N.; LAZER, A. C.,

— Nonlinear perturbations of linear elliptic boundary value problems at resonance, *J. Math. Mech.*, **19**(1970a), 609–623.

— Linear eigenvalues and a nonlinear boundary value problem, *Pacific J. Math.*, **33**(1970b), 311–328.

LAZER, A. C.,

    On Schauder's fixed point theorem and forced second order nonlinear oscillations, *J. Math. Anal. Appl.*, **21**(1968), 421–425.

MARTELLI, M.,

    On forced nonlinear oscillations, *J. Math. Anal. Appl.*, **69**(1979), 456–504.

MAWHIN, J.,

— Nonlinear perturbations of Fredholm mappings to differential equations, *Trab. Mat. Univ. Brasil.*, **61**(1974a).

— Periodic solutions of some vector retarded functional differential equations, *J. Math. Anal. Appl.*, **45**(1974b), 588–603.

MAWHIN, J.; WARD, J. R.,

    Periodic solutions of some forced Lienard differential equations at resonance, *Arch. Math.*, **41**:4(1983), 337–351.

MEEROV, M. V.,

— *Multivariable control systems* (Russian), Nauka, Moscow, 1965.

— *Integration and optimization of multiply-connected control systems* (Russian), Nauka, Moscow, 1986.

NAKAO, M.,

    An existence theorem for a nonlinear Dirichlet problem, *Mem. Fac. Sci. Kyushu Univ.*, **26**(1972), 201–217.

NEČAS, J.,

    On the range of nonlinear operators with linear asymptotes which are not invertible, *Comment. Math. Univ. Carolinae*, **14**(1973), 63–72.

NEMYCKII, V. V.,

Théorèmes d'existence et d'unicité des solutions de quelques équations intégrales non-linéaires, *Mat. Sb.*, **41**:3(1934), 438–452.

ROSENBROCK, H. H.,

State-space and multivariable theory, New York, 1970.

ROSENVASSER, E. N.,

Oscillation in nolinear systems. Integral equations method (Russian), Nauka, Moscow, 1969.

SCHAEFER, H.,

Neue Existenzsätze in der Theorie nichtlinearer Integralgleichungen, Abh. Sächs. Acad. Wiss. Leipzig Math. Natur. K1., **101**(1955).

VAINBERG, M. M.,

Variational methods for the study of nonlinear operators (Russian), Gostehizdat, Moscow, 1956; English translation: Holden-Day, Inc., San Francisco—London— Amsterdam, 1964.

VAINIKKO, E. G.,

Approximate construction of forced periodic oscillations by the collocation method (Russian), *Avtomat. i Telemekh.*, **11**(1989), 161–164.

VORONOV, A. A.,

Introduction to the dynamics of complex controllable systems (Russian), Nauka, Moscow, 1985.

WONHAM, W. M.,

Linear multivariable control, Springer-Verlag, New York—Berlin—Heidelberg— Tokyo, 1985.

WILLIAMS, S. A.,

A sharp sufficient condition for solution of a nonlinear elliptic boundary value problem, *J. Diff. Equations*, **8**(1970), 580–586.

YOSIDA, K.,

Functional analysis, Springer-Verlag, Berlin—Göttingen—Heidelberg, 1965.

ZYGMUND, A.,

Trigonometric series, Warsaw, 1935.

* * *

Functional analysis — a reference text (Russian), Nauka, Moscow, 1972; English translation: Noordhoff Publ., Groningen, 1972.

List of Symbols

A. GENERAL SYMBOLS

$\{t \in \Omega : \mathrm{P}(t)\}$	the set of all elements t of Ω satisfying a property $\mathrm{P}(t)$
$\mathbb{R}^n$	the n-dimensional Euclidean space
$\mathbb{R}$, $\mathbb{R}^1$	the set of real numbers
Ω	subset of $\mathbb{R}^n$
t	point in Ω
$\partial\Omega$	the boundary of Ω
$[a, b]$	the closed interval $\{t \in \mathbb{R} : a \leqslant t \leqslant b\}$
(a, b)	the open interval $\{t \in \mathbb{R} : a < t < b\}$
$[a, b)$, $(a, b]$	the half-open intervals $\{t \in \mathbb{R} : a \leqslant t < b\}$, $\{t \in \mathbb{R} : a < t \leqslant b\}$
$f(t) : X \to Y$	function defined for $t \in X$ with values $f(t) \in Y$.
A, Ax, $Ax(t)$	linear continuous operator
$\|A\|_{X \to Y}$	the norm of a linear operator $A : X \to Y$
A^*	the adjoint of a linear operator $A : X \to Y$
E^*	the topological dual of a space E
μ	measure on Ω
$\displaystyle\int_\Omega f(t)\mathrm{d}\mu$	the integral of a function $f(t)$ on Ω with respect to a measure μ
$[\cdot, \cdot]$	the inner product on L_2
$\deg P$	the degree of a polynomial P

B. SPECIAL SYMBOLS. Each entry is followed by two numbers in parantheses indicating the subsection in which the symbol first occurs.

$(\cdot, \cdot)$	the inner product on $\mathbb{R}^n$ **(1.1)**
$|\cdot|$	the norm on $\mathbb{R}^n$ **(1.1)**
$\chi(\delta)$, $\chi(\delta; e)$	the distribution of a function $e(t)$ **(1.1)**
$e^*(t)$	the arrangement of a function $e(t)$ **(1.1)**
$\mathcal{F}$	compact family of functions **(1.3)**
$\chi_{\mathrm{L}}(\delta; \mathcal{F})$, $\chi_{\mathrm{U}}(\delta; \mathcal{F})$	uniform lower and upper estimates of distributions **(1.3)**
$\Phi(t, u)$	scalar-valued functions of variables $t \in \Omega$ and $u \geqslant 0$ **(2.1)**
$\mathfrak{N}(u_0)$	special class of functions $\Phi(t, u)$ **(2.1)**
L_p	the space of p-integrable functions **(2.3)**
$\|\cdot\|_{L_p}$, $\|\cdot\|_p$	the norm on the space L_p **(2.3)**
$\alpha(u)$	nonincreasing positive function, $u \geqslant 0$ **(2.3)**
$\mathfrak{W}(\Phi_0)$	special class of functions $\Phi(t, u)$ **(2.4)**
Mes	measure on Ω, distinct from the basic measure μ **(4.1)**
$\chi(\delta; e, g)$	the distribution of a function $e(t)$ with respect to the measure Mes **(4.1)**
$\Psi(u)$	nondecreasing positive function, $u \geqslant 0$ **(4.2)**
$\mathfrak{V}(\Phi, \Psi)$	class of functions $\varphi(t, x) : \Omega \times \mathbb{R}^1 \to \mathbb{R}^1$ **(4.2)**
$H(u)$	monotonous positive function, $u \geqslant 0$ **(4.2)**
p'	the number $p(p-1)^{-1}$ **(4.4)**
$\theta(\delta)$	special function associated with the distribution $\chi(\delta; e, g)$ **(4.4)**
Ker A	the kernel of an operator A **(7.1)**
T	period **(7.3)**
Π_n	invariant subspaces of a normal operator **(7.3)**
P_n	the orthogonal projection onto Π_n **(7.3)**
$\sigma(A)$	the spectrum of an operator A **(7.3)**
$w(T)$	the norm of the periodic problem operator on L_2 **(7.3)**
$L(p)$, $M(p)$	differential polynomials **(7.4)**
$W(p)$	transfer function **(7.4)**
ω_k	frequencies $2k\pi T^{-1}$ **(7.4)**
$\mathfrak{f}$	superposition operator **(7.4)**
$\mathfrak{B}$, $\mathfrak{B}(\rho)$	balls in spaces of functions **(7.5)**

E_0	finite dimensional invariant subspace **(8.1)**
E_1	infinite dimensional invariant subspace **(8.1)**
P, Q	the orthogonal projections onto E_0 and E_1 **(8.1)**
I	the identity operator $Ix(t) \equiv x(t)$ **(8.1)**
$\Psi(t, u)$	scalar-valued function, $t \in \Omega$, $u \geqslant 0$ **(8.2)**
S_ρ	sphere of radius ρ centered at the origin **(8.3)**
$\mathfrak{L}(\Psi_0, E_0)$	special class of nonlinear equations **(8.4)**
Λ	spectral radius **(10.1)**
Δ	the Laplace operator **(10.3)**
$\theta(G_1, G_2; X)$	Hausdorff deviation of a set G_1 from a set G_2 in a space X **(11.3)**
$\mu(q_1, q_2)$	the root of equation (12.1) **(12.1)**
$\mathcal{B}$	special class of nonlinear equations **(12.3)**
H	Hilbert space **(13.1)**
Mx	modulus on a Hilbert space **(13.1)**
$N(A, B)$	the joint norm of two operators A and B **(13.1)**
$\mathfrak{R}(q)$	set of ordered quadruples of numbers **(13.4)**
ζ	the largest root of the quadratic equation (13.12) **(13.4)**
$L(x, \lambda)$	Lagrangean **(13.5)**
Lx	differential operator **(14.1)**
$Q(\tau)$	the function (15.11) **(15.1)**
$S(h)$	the shift operator (16.3) **(16.1)**
A_γ	the operator $(I - \gamma A)^{-1}A$ **(20.2)**
$\Psi_p(u)$	the function $[\Psi(u^{1/p})]^p$ **(21.2)**
$G(t, s)$	kernel of a linear integral operator **(23.1)**
$\Xi(\lambda, x)$	homotopy of vector fields **(24.2)**
$\mathrm{ind}(\lambda)$	the index at infinity of a vector field **(25.2)**
$\mu(A; H)$	the positivity coefficient of an operator A on H **(26.1)**
K_μ	disk in the complex plane **(26.3)**
$\gamma_H(A)$, $\gamma_B(A)$	the numbers (26.17) and (26.18) **(26.4)**
$A^\square$	auxiliary operator **(26.5)**
$\Pi(\omega)$	the polynomial (30.6) **(30.1)**
$\pi(M, L)$	the degree of the polynomial $\Pi(\omega)$ **(30.1)**

Subject Index

61. **A. Gheondea, D. Timotin, F.-H. Vasilescu** (Eds.): Operator Extensions, Interpolation of Functions and Related Topics, 1993, (3-7643-2902-5)
62. **T. Furuta, I. Gohberg, T. Nakazi** (Eds.): Contributions to Operator Theory and its Applications. The Tsuyoshi Ando Anniversary Volume, 1993, (3-7643-2928-9)
63. **I. Gohberg, S. Goldberg, M.A. Kaashoek:** Classes of Linear Operators, Volume 2, 1993, (3-7643-2944-0)
64. **I. Gohberg** (Ed.): New Aspects in Interpolation and Completion Theories, 1993, (3-7643-2948-3)
65. **M.M. Djrbashian:** Harmonic Analysis and Boundary Value Problems in the Complex Domain, 1993, (3-7643-2855-X)
66. **V. Khatskevich, D. Shoiykhet:** Differentiable Operators and Nonlinear Equations, 1993, (3-7643-2929-7)
67. **N.V. Govorov †:** Riemann's Boundary Problem with Infinite Index, 1994, (3-7643-2999-8)
68. **A. Halanay, V. Ionescu:** Time-Varying Discrete Linear Systems Input-Output Operators. Riccati Equations. Disturbance Attenuation, 1994, (3-7643-5012-1)
69. **A. Ashyralyev, P.E. Sobolevskii:** Well-Posedness of Parabolic Difference Equations, 1994, (3-7643-5024-5)
70. **M. Demuth, P. Exner, G. Neidhardt, V. Zagrebnov** (Eds): Mathematical Results in Quantum Mechanics. International Conference in Blossin (Germany), May 17-21, 1993, 1994, (3-7643-5025-3)
71. **E.L. Basor, I. Gohberg** (Eds): Toeplitz Operators and Related Topics. The Harold Widom Anniversary Volume. Workshop on Toeplitz and Wiener-Hopf Operators, Santa Cruz, California, September 20–22, 1992, 1994 (3-7643-5068-7)
72. **I. Gohberg, L.A. Sakhnovich** (Eds): Matrix and Operator Valued Functions. The Vladimir Petrovich Potapov Memorial Volume, (3-7643-5091-1)
73. **A. Feintuch, I. Gohberg** (Eds): Nonselfadjoint Operators and Related Topics. Workshop on Operator Theory and Its Applications, Beersheva, February 24–28, 1994, (3-7643-5097-0)
74. **R. Hagen, S. Roch, B. Silbermann:** Spectral Theory of Approximation Methods for Convolution Equations, 1994, (3-7643-5112-8)
75. **C.B. Huijsmans, M.A. Kaashoek, B. de Pagter**: Operator Theory in Function Spaces and Banach Lattices. The A.C. Zaanen Anniversary Volume, 1994 (ISBN 3-7643-5146-2)
76. **A.M. Krasnoselskii**: Asymptotics of Nonlinearities and Operator Equations, 1995 (ISBN 3-7643-5175-6)
77. **J. Lindenstrauss, V.D. Milman** (Eds): Geometric Aspects of Functional Analysis Israel Seminar GAFA 1992-94, 1995 (ISBN 3-7643-5207-8)
78. **M. Demuth, B.-W. Schulze** (Eds): Partial Differential Operators and Mathematical Physics: International Conference in Holzhau (Germany), July 3-9, 1994, 1995 (ISBN 3-7643-5208-6)